THE
TRUE STATE
OF THE
PLANET

RONALD BAILEY, EDITOR

A Project of the Competitive Enterprise Institute

THE FREE PRESS

New York   London   Toronto   Sydney   Tokyo   Singapore

The Free Press
A Division of Simon & Schuster Inc.
1230 Avenue of the Americas, New York, N.Y. 10020

Printed in the United States of America

printing number

1 2 3 4 5 6 7 8 9 10

Library of Congress Cataloging-in-Publication Data

Bailey, Ronald.
    The true state of the planet / Ronald Bailey.
       p. cm.
    "A project of the Competitive Enterprise Institute."
    Includes index.
    ISBN 0-02-874010-6
    1. Environmentalism.   2. Environmental policy.   3. Environmental protection.
    I. Competitive Enterprise Institute.   II. Title.
    GE195.B35   1995
    363.7–dc20   95-937
                                                               CIP
                                                               AC

Credits for the photographs reproduced in this book are listed after the Index.

CONTENTS

CONTRIBUTORS

Bruce Ames is a Professor of Biochemistry and Molecular Biology and Director of the National Institute of Environmental Health Sciences Center at the University of California, Berkeley. He is a member of the National Academy of Sciences, and he is on its Commission on Life Sciences. He was formerly a member of the board of directors of the National Cancer Institute (National Cancer Advisory Board). He was the recipient of the most prestigious award for cancer research, the General Motors Cancer Research Foundation Prize (1985), as well as the Gold Medal Award of the American Institute of Chemists (1991) and the Glenn Foundation Award of the Gerontological Society of America (1992). He has been elected to the Royal Swedish Academy of Sciences, the Japan Cancer Association, and the Academy of Toxicological Sciences. Because of his 300 scientific publications, he is the twenty-third most-cited scientist in all fields for the period 1973–1984.

Terry L. Anderson is Professor of Economics at Montana State University and Senior Associate of PERC, a think tank specializing in environmental and natural resource issues headquartered in Bozeman, Montana. He is author of *Water Crisis: Ending the Policy Drought* (1983) and editor of *Water Rights* (1983). Dr. Anderson is co-author (with Donald Leal) of *Free Market Environmentalism* (1991) and editor of PERC's Political Economy Forum series, which includes *Property Rights and Indian Economies* (1992), *The Political Economy of Customs and Culture* (1993), *Taking the Environment Seriously* (1993), and *The Political Economy of the American West* (1994). He is currently writing a new book (with Donald Leal) entitled *Enviro-Capitalists*.

Dennis T. Avery is the Director for Global Food Issues at the Hudson Institute. Mr. Avery is the principal author of *Global Food Process* (1991) and *Saving the Planet with Pesticides and Plastics* (forthcoming). For nearly a decade, he was the senior agricultural analyst in the U.S. Department of State, and he has done policy analysis for the U.S. Department of Agriculture and President Johnson's National Advisory Commission on Food and Fiber.

Ronald Bailey is producer of the national weekly public television series *Think Tank with Ben Wattenberg*. He is the author of *Eco-Scam: The False Prophets of Ecological Apocalypse* (1993) and was a science and technology writer at *Forbes* for three years. Mr. Bailey was the 1993 Warren T. Brookes Fellow in Environmental Journalism at the Competitive Enterprise Institute and was the producer of the national PBS series *Technopolitics*. He is a contributing editor to *Reason* magazine.

Robert Balling, Jr., is Director of the Office of Climatology and Associate Professor of geography at Arizona State University. He is author of *The Heated Debate: Greenhouse Prediction Versus Climate Reality* (1992) and co-author with Martin Williams of *Interactions of Desertification and Climate* (1994). An active participant in the greenhouse debate, Dr. Balling has published over seventy articles in the professional scientific literature. Dr. Balling is currently serving as climate consultant to the United Nations, the World Meteorological Organization, and the Intergovernmental Panel on Climate Change. He received an M.A. from Bowling Green State University and a Ph.D. from the University of Oklahoma.

Nicholas Eberstadt is a Visiting Fellow at Harvard University's Center for Population Studies and a Visiting Scholar at the American Enterprise Institute. He has been a consultant to the World Bank, the State Department, and the Agency for International Development. His books include *The Tyranny of Numbers* (1995), *Korea Approaches Unification* (1994), *The Population of North Korea* (in Korean 1992, forthcoming in English and Japanese), *The Poverty of Communism* (1988), and *Foreign Aid and American Purpose* (1989). He has written numerous articles on population and poverty. Mr. Eberstadt received an A.B. from Harvard College, an M.Sc. from the London School of Economics, and an M.P.A. from the Kennedy School of Government at Harvard University.

Stephen R. Edwards earned his Ph.D in ecology and systematics from the University of Kansas in 1974. He is currently Global Support Team Coordinator of the Sustainable Use Initiative at the IUCN–The World Conservation Union. During the prior six years Dr. Edwards was Coordinator of IUCN's Species Conservation Programme at the world headquarters in Gland, Switzerland. With over twenty years of experience in international wildlife conservation, Dr. Edwards is committed to working with others to find rational approaches to wildlife conservation that enhances sustainable use of wildlife resources. He pursues this goal through several means: working with local institutions to develop projects for sustainable use, advising government and private agencies on rational policies, gaining better understanding of economic implications of conservation, providing a bridge between conservation and business communities, and applying pertinent laws to conservation needs. Dr. Edwards is a noted author of many articles on sustainable use.

Indur M. Goklany is Manager of Science and Engineering in the Office of Policy Analysis at the U.S. Department of the Interior. He has worked on environmental and energy issues for over twenty years, and he has several publications on global climate change, sustainable development, adaptation to global change, risk analysis, and air pollution. He helped develop the Environmental Protection Agency's emissions trading policy and Interior's global change research program.

Lois Swirsky Gold is Director of the Carcinogenic Potency Project at Lawrence Berkeley Laboratory and the University of California, Berkeley. She has published more than fifty articles on the results of animal cancer tests, interspecies extrapolation in carcinogenesis, and issues related to the use of bioassay data to assess cancer risks to humans. These analyses are based on the Carcinogenic Potency Database, which she has directed for fifteen years; the database includes analyses of 5,000 chronic, long-term carcinogenesis bioassays on 1,200 chemicals. Dr. Gold has served on the panel of expert reviewers for the bioassay program of the National Toxicology Program. She is currently on the board of the Harvard Center for Risk Analysis and is a member of the Harvard Risk Management Reform Group.

Kent Jeffreys is a Senior Fellow at the National Center for Policy Analysis focusing on environmental issues. He is the former Director of Environmental Studies at the Competitive Enterprise Institute, where he contributed (with Fred Smith) the essay "A Free-Market Environmental Vision" to the book *Market Liberalism, A Paradigm for the 21st Century* (1993) and authored "Who Should Own the Ocean?" in addition to other policy studies. Mr. Jeffreys previously served as an environmental and energy policy analyst with the Heritage Foundation and the Republican Study Committee.

Stephen Moore is an economist at the Cato Institute in Washington, D.C. He is the author of *Still an Open Door: U.S. Immigration Policy and the American*

Economy (1994) and was a fellow on the Joint Economic Committee in 1993. He has appeared on "MacNeil/Lehrer," "CNN Moneyline," "NBC Nightly News," and "Fox Morning News," and his writings are frequently in such newspapers as the *Wall Street Journal, National Review, Human Events,* and the *Christian Science Monitor.* He has conducted several studies on the economics of natural resources.

Roger A. Sedjo is a Senior Fellow in the Energy and Natural Resources Division at Resources for the Future. Since 1977 he has been director of RFF's Forest Economics and Policy Program, and was an economist with the U.S. Agency for International Development. He is co-author (with Kenneth Lyon) of *The Long-Term Adequacy of World Timber Supply, America's Renewable Resources: Past Trends and Current Challenges* (with Kenneth Frederick), and *Forests for a World of Six Billion People* (with Jan Laarman).

Fred L. Smith, Jr., is the President and founder of the Competitive Enterprise Institute, a nonprofit, nonpartisan public interest group active in a wide range of economic and environmental public policy issues. His writings have appeared in leading newspapers such as the *Wall Street Journal, Washington Post, New York Times,* and the *Washington Times,* as well as numerous public policy journals. He is editor (with Michael Greve) of the book *Environmental Politics: Public Costs, Private Rewards* (1992). He has contributed chapters to numerous books, including *Steering the Elephant: How Washington Works* (1987), *Assessing the Reagan Years* (1988), and (with Kent Jeffreys) *Market Liberalism: A Paradigm for the 21st Century* (1993), among others. Before founding CEI, Mr. Smith served as Director of Government Relations for the Council for a Competitive Economy and as a senior research economist for the Association of American Railroads. For five years Mr. Smith was a senior policy analyst at the Environmental Protection Agency.

PROLOGUE
Environmentalism for the Twenty-first Century

Ronald Bailey

In 1970, the first Earth Day brought together more than 20 million Americans to launch the first wave of the modern environmental movement. Since then, public concern about the state of the planet has steadily grown. The membership rolls and budgets of leading environmental activist organizations have swollen by millions. The federal government has adopted thousands of pages of environmental regulations. Cities and industries are spending billions every year to clean up pollution. Activity on the international level has also increased. In 1992, leaders from more than 170 countries gathered at Rio de Janeiro for the first Earth Summit at which sweeping treaties dealing with global climate and biodiversity were signed. And in September 1994, the International Conference on Population and Development, sponsored by the United Nations Population Fund, was held in Cairo. Again, more than 170 national governments adopted a sweeping twenty-year Program of Action, which, among other things, aims to restrain global population to a target of 7.27 billion by the year 2015.

The first wave has scored some major successes in its twenty-five-year history: in the Western developed world, air and water are much cleaner; automobiles are far cleaner to operate; belching smokestacks are far fewer and generally more efficient than ever before. Clearly developed societies can come together to clean up much of the pollution produced by industries and cities.

But the first wave has also turned out to be spectacularly wrong about certain things. The good news is that many of the looming threats predicted in the early days of the environmental movement turned out to be exaggerated. For example, the global famines expected to occur in the 1970s never happened. Fears that the United States and Europe would cut down all of their forests have been belied by increases in forest area. Global warming, despite so many continuing reports, does not appear to be a major problem. And it turns out that the damage to human health and the natural world caused by pesticides is far less than Rachel Carson feared it would be when she wrote *Silent Spring* in 1962.

It is inevitable, perhaps, that the first wave would begin to run its course and give way to a new strategy. The problem with the past twenty-five years of environmentalism has been a simple one: a failure of theory. From Rachel Carson to the Club of Rome (*The Limits to Growth*) to Paul Ehrlich (*The Population Bomb*), the leaders of the first wave all operated under a false assumption, Malthusianism.

In 1798, the Reverend Thomas Malthus in *An Essay on Population* argued that humanity had a propensity to reproduce far faster than the food supply could increase. Thus, some portion of humanity would always be condemned to starvation. First-wave environmentalists embraced this notion, arguing that humanity and technological civilization are heedlessly and profligately using up the earth's resources. They liken our situation to that of deer uncontrolled by disease or predators whose burgeoning numbers will quickly overgraze their pasturage, bringing starvation to all. But they fail to take into account the fact that people, unlike deer, can react and adjust their behavior using their intelligence to increase resources and modifying their activities to avert Malthusian disasters. In short, if the earth's population was once 1 billion and the earth's forests had been cut, say, 10 percent, it is a mistake to assume that 2 billion people would cut 20 percent or that 6 billion would cut 60 percent. Instead, people have figured out how to use less wood.

Malthus's theory was wrong. This failure of theory has been compounded by failures of information gathering. Modern environmentalists have learned to question the efficacy of computer models to help set priorities and guide policies. The famous limits-to-growth computer model of the 1970s, for example, has been drastically wrong in its predictions. Atmospheric models designed to estimate the effects of the refrigerants called chlorofluorocarbons on stratospheric ozone completely failed to predict the development of the Antarctic "ozone hole." More recently, the original global climate models that predicted significant global warming as a result of higher levels of human-generated carbon dioxide in the atmosphere appear to have overestimated potential increases in global temperatures by as much as an order of magnitude. The earth's atmosphere has actually *cooled* by 0.10 degrees Celsius since 1979, according to highly accurate satellite-based atmospheric temperature measurements. Taking into account the effects of volcanoes and El Niño, scientists calculate that global temperatures are rising at only 0.09 degrees centigrade per decade—or less than 1 degree centigrade per century. This increase is far less than the earlier predictions that sparked so many apocalyptic pronouncements.

But even with accurate information and a recognition that behavior in the past cannot be extrapolated to the future, a major leap is required to exit the problems of the first wave, for hidden deep within Malthus's assumption about behavior is an even worse assumption about how to change it. Malthus assumed that past behavior would continue in the future. And if behavior does not change on its own, it can be changed only by force—by direct orders from above, as, for example, with gasoline rationing. Americans were ordered to use less oil in the 1970s, and with disastrous results. People hoarded gas; they formed longer gas lines out of fear, and the energy "crisis" was thereby made worse. And consider the case of a dioxin-contaminated waste site in Arkansas. The Environmental Protection Agency (EPA) required the owner to burn the dioxin-contaminated wastes in an incinerator, but in a classic catch-22 situation, the incinerator cannot meet the EPA's standards for destroying dioxin. Consequently, the incinerator produces two barrels of dioxin-laced ash for every barrel that is burned. Under federal law, this ash cannot be put in a landfill, so it is stored in a $300,000 building, apparently forever.

The greatest problem with the first wave has been its solutions, which involve the top-down imposition of laws and regulations, some of which, in turn, impair the capacity of people to change their behavior on their own. This seems paradoxical, but it is all too often true.

A final problem with the first wave has been its priorities. For many pollutants, the industrialized countries have reached the point that Supreme Court justice Stephen Breyer calls the "problem of the last 10%." We have taken care of the first 90 percent of the pollutants, but cleaning up the last 10 percent is exceedingly difficult and expensive. It is at this point of diminishing returns that we must consider whether devoting resources to cleaning up the last 10 percent is better for the natural environment than directing those resources to other problems.

Resources are being poured into areas that pose little harm to either the natural world or to human beings, while other more critical problems receive relatively little attention. Unfortunately, many decisions made at the Earth Summit at Rio de Janeiro and the Cairo population conference were based on outdated scientific, demographic, and economic information. The Global Climate Change Treaty's call for a massive and costly shift in the world's economy from fossil fuels is based in part on a belief that the world's average temperature has increased significantly in recent years. It has not. Also, total fertility rates are dropping much faster than the negotiators at the United Nations's Cairo population conference assumed they are. And closer to home, the so-called Superfund program is, at the cost of billions of dollars, forcing the cleanup of waste sites that scientists say pose no real risks to people or the natural environment.

Meanwhile, our most serious instances of environmental degradation have proved hard to fix by law. The deplorable state of global fisheries is a case in point. Overfishing results from the all-too-familiar problem known as the "tragedy of the commons." The analogy to overfishing is overgrazing of lands held in common. When land is open to anyone who wants to use it, the tragedy of the commons is an almost inevitable result. In the case of commonly held grazing land, herdsmen have no incentives to restrain the number of cows grazing on the commons. In fact, the reverse is true. If a herdsman does not put a cow on the land, his neighbor will, and thus reap the benefits of raising an additional cow. This "logic" leads inexorably to overgrazing and the eventual destruction of the common pastureland. This is what has happened to many of the world's fisheries. First-wave environmentalists fail to realize that the problem lies in the commons, not in the herdsmen. They typically want to regulate the herdsmen instead of abolishing the commons. History shows that the better way to avoid the tragedy of the commons is through privatizing resource ownership. If individual herdsmen (or fishermen) can fence in portions of the commons and secure ownership rights and responsibilities, their incentives to protect the land (or sea) from overgrazing dramatically increase.

This is precisely why second-wave environmentalists propose that private owners, individual or group, commercial or noncommercial, offer the best defense against environmental degradation. Simply by protecting their property—trees, animals, fish, grazing areas, rivers—they incidentally protect the earth for the rest of us.

While governments and interest groups spend a lot of time fighting about every last drop of "toxic" waste, the ecological systems of ocean fisheries have deteriorated far more than any damage that landfills have caused. Creative thinking must be devoted to figuring out what arrangements and institutions can successfully restore these imperiled oceanic ecosystems. As another example, indoor air pollution in the form of smoke and carbon monoxide—the result of burning biofuels like wood and dung in houses in the developing world—is one of the chief global threats to human health. It is also one of the least discussed.

It is time for environmentalism's second wave. Second-wave environmentalists must recognize that people modify their activities to avert environmental crises and disasters. Humanity's growing store of knowledge about resources and nature can help us identify more clearly the problems and the opportunities for environmental improvement. First-wave environmentalists' concerns about running out of nonrenewable resources or poisoning the environment by overusing pesticides may not have been unreasonable two and half decades ago. But twenty-five years of research has dispelled many of the first wave's earlier fears. There has been a growing gap between the mounting scientific evidence about the actual status of various environmental problems and the often bleaker views promoted by environmental activists. It is time to close that gap.

Many conventional environmentalists advocate the "precautionary principle," which says that humanity must not interfere with nature until all the consequences of an action can be taken into account. But it is impossible to know all the consequences of even the most trivial action. The second wave of environmentalism must recognize that following the precautionary principle can lead to greater environmental degradation. It is better to move forward using intelligent trial and error to uncover new knowledge. Moving forward can increase resources and wealth. Greater knowledge and wealth give human communities resilience, enabling them to respond flexibly and effectively to the unexpected. In other words, if something does go wrong, our increased knowledge and greater economic resources can be mobilized to solve the problem. This is why impoverished people in Bangladesh die by the thousands when cyclones strike their villages, while only fourteen Americans died when Hurricane Andrew hit Florida and Louisiana in August 1992. Better roads, housing, medical facilities, and emergency response measures made possible by American wealth make it far easier to weather storms in Florida than in Bangladesh.

Modern smart environmentalism must avoid the "crisis of the month" media mentality, and it must beware the dangers of interest group politics. It must focus on setting priorities and dealing with the biggest problems first. It must recognize that all problems are not equally bad and that not all can be solved at the same time. The new, smarter environmentalism must also understand that there is no perfect solution to any problem; trade-offs have to be made. The good cannot be held hostage to the perfect.

In the twenty-first century it will be clear that the preservation of natural resources and the expansion of human ones are tightly linked. This concept may be very hard for traditionalists to accept, but history has shown that environmental improvement depends directly on rapid economic progress. If poor countries do not adopt modern high-yield agriculture, for example, then their impoverished farmers will be forced by hunger to level millions of square miles of wildlands. Agricultural intensification is essential to forestall famine and the plowdown of massive amounts of wildlife habitat. Currently, more than 75 percent of the land on every continent except Europe is still available for wildlife. It is this "undeveloped" land in developing countries that is of greatest importance to conserving biodiversity over the long term.

Modern forestry also helps preserve wildlife habitat. Although nearly 75 percent of total industrial wood production comes from industrialized countries in the Northern Hemisphere, the temperate forestlands of this region are expanding. With modern technology, the world's current industrial wood consumption requirements could be produced on only 5 percent of the world's total current forestland. Technology and progress are not the problem; they are the solutions.

Technology has vastly increased the ability of couples around the world to choose their family sizes. Already total fertility rates are plummeting in most of the world. A recent report by the World Bank and Winrock International concluded that global zero population growth will be achieved by 2035 and population will actually decrease throughout the rest of the twenty-first century. This forecast stands in stark contrast to those of world population growth relied on by conventional environmentalists at the Cairo population conference.

In the developed countries, air and water pollution have been significantly cut in the last twenty-five years. For example, in the United States, sulfur dioxide emissions per capita are now down 60 percent from their peak in 1920. Both particulates and carbon monoxide emissions are down 79 percent and 53 percent, respectively, since the end of World War II. This steep reduction in air pollution makes it clear that as societies become wealthier, they use some of their wealth for environmental improvement. Impoverished societies have far fewer resources to devote to protecting and cleaning up their natural environments.

Only if policymakers and citizens have access to sound scientific information and careful analyses of past policy successes and failures can they make the critical decisions about how best to preserve the world's natural heritage for future generations. *The True State of the Planet* is dedicated to providing that information. This book seeks to close the widening gap between environmental activists and environmental science. In these pages, eleven leading researchers present the latest facts about resource use and availability, environmental cleanliness, and other trends. They will puncture several first-wave illusions, but they will also refocus our priorities on the very real problems still to be faced. The key difference for the second wave is how to solve them: not by fiat but by freely available and accurate information; not by doomsaying but by developing new structures of responsibility that allow the vast human resources that we already enjoy to be employed for ensuring the safety and abundance of the natural resources we all desire.

Chapter 1

POPULATION, FOOD, AND INCOME
Global Trends in the Twentieth Century

Nicholas Eberstadt

HIGHLIGHTS

- *World population has more than tripled since 1900, rising from 1.6 billion to 5.3 billion today.*

- *Global life expectancy more than doubled this century from 30 to 64 years, while global infant mortality fell from 170 infant deaths per 1,000 births in 1950 to just 60 in 1990. Rapid population growth has occurred not because human beings suddenly started breeding like rabbits but because they finally stopped dropping like flies.*

- *Despite a tripling of the world's population in this century, global health and productivity have exploded. Today human beings eat better, produce more, and consume more than ever before in the past.*

- *The current rate of annual worldwide population growth is down to 1.6 percent and continues to fall from its 1960s peak of 2.0 percent. The average annual growth rate is 0.4 percent for developed countries and 1.9 percent for developing regions.*

- *"Overpopulation" is a problem that has been misidentified and misdefined. The term has no scientific definition or clear meaning. The problems typically associated with overpopulation (hungry families, squalid and overcrowded living conditions) are more properly understood as issues of poverty.*

- *The total fertility rate for the world as a whole dropped by nearly two-fifths between 1950/55 and 1990/95—from about 5 children per woman down to about 3.1 children per woman. Average fertility in the more developed regions fell from 2.8 to 1.7 children per woman, well below biological replacement. Meanwhile total fertility rates in less developed regions fell by 40 percent, falling from 6.2 to 3.5 children per woman.*

- *Global per capita calorie availability rose by nearly a third between the 1930s and late 1980s, with per capita food supplies rising by 40 percent in Africa, Asia, and Latin America.*

- *Worldwide per capita productivity has surged alongside the population boom. Between 1950 and 1989, per capita output tripled in Western Europe, North America, and Australia, doubled in India and Pakistan, and increased eightfold in South Korea and tenfold in Taiwan. Even parts of Africa may have enjoyed a near tripling of per capita output between 1913 and 1989.*

- *Although some blame dwindling natural resources for the reversals and catastrophes that have recently befallen heavily populated low-income countries, such episodes are directly traceable to the policies or practices of presiding governments.*

8

Will humanity's burgeoning numbers outstrip the world's supplies of food and other resources, plunging the planet into disaster? This question can be traced back to T. R. Malthus's original *Essay on the Principle of Population* in 1798[1] and to less famous inquiries from even earlier times.[2] The specific concerns enunciated have varied from one generation to the next. Shortly after World War II, for example, the catchword was "overpopulation"; in the mid-1960s, the issue was framed as "the world food problem." More recently, the world population question has been discussed in terms of "carrying capacity" and "sustainable development." Whatever the particulars of these formulations, however, the underlying themes in much of this evolving commentary on the world population would be immediately familiar to Malthus himself: that the globe cannot support the ongoing and enormous increase in human numbers and that unless we somehow check this uncontrolled demographic growth, we will face rising poverty, mass starvation, perhaps even worldwide catastrophe.

"The battle to feed all of humanity is over. In the 1970s the world will undergo famines—hundreds of millions of people are going to starve to death in spite of any crash programs embarked upon now," is the arresting way that biologist Paul Ehrlich began his 1968 best-seller, *The Population Bomb*.[3] Recently, Physicians for Social Responsibility updated this warning: "Without population control environmental degradation can only increase. The Malthusian prediction that population growth will finally be limited by war, pestilence, and famine seems likely to be realized."[4] Lester Brown of the Worldwatch Institute agrees: "Seldom has the world faced an unfolding emergency whose dimensions are as clear as the growing imbalance between food and people."[5]

Such sentiments and concerns have come to inform state actions on a global scale. The degree to which such views are now translated into policy was suggested by the results of the International Conference on Population and Development, held in Cairo in September 1994, and organized by the United Nations Population Fund (UNFPA). At this conference, over 170 national governments agreed to a twenty-year "Program of Action," which proposed, among other things, to attempt to slow the growth of world population. One of the key objectives specified in the document was a global population target of 7.27 billion for the year 2015—an objective, the conference had been warned, that would not be obtained without dramatically expanding population programs in low-income, high-fertility countries. The Program of Action endorsed a spending target for population and family planning programs in developing countries of $17 billion, more than a threefold increase from the estimated $5 billion level for 1994. Two years before the Cairo conference, the UNFPA's executive director, Nafis Sadik, spelled out in preparatory discussions the rationale of the proposals she hoped the upcoming conference would embrace: "a sustained and concerted program starting immediately" to curb worldwide population growth, since the continuation of current trends would trigger a "crisis [that] heightens the risk of future economic and ecological catastrophes."[6]

To an unsettling extent such pronouncements and assessments sound like the utterances of a faith. Matters of faith are not meant to be held up to scientific scrutiny or to be treated as testable and potentially falsifiable hypotheses. If we wish to treat the population issue as an empirical question, however, we have considerable evidence available for analysis: estimates of global trends in population growth, fertility, and mortality; estimates of patterns and trends in the production and consumption of food in the world's diverse regions; and estimates of long-term trends in income and productivity within and between countries. Most of this evidence appears to weigh very heavily against the more pessimistic interpretation so often given favorable exposure in the mass media, the universities, and even the corridors of government.

The facts, briefly, are these: World population *has* increased tremendously in our century—more than tripling, it appears, between 1900 and 1990—and it continues to grow (by any historical benchmark) with extraordinary speed. (See the Appendix at the end of this book for warnings about the reliability of these and other data.) This demographic explosion, however, has not plunged humanity into penury

The global population boom has coincided with an explosion of health, and of productivity, around the world. On average, the human population today lives longer, eats better, produces more, and consumes more than at any other time in the past.

and deprivation. Quite the contrary, the global population boom has coincided with an explosion of health, and of productivity, around the world. On average, the human population today lives longer, eats better, produces more, and consumes more than at any other time in the past. And although dramatic, and sometimes appalling, disparities in living standards can be identified between and within countries, considerable evidence points to long-term improvements in the material condition of the most vulnerable elements in the world population.

To take note of these basic facts is not to argue that rapid population growth of itself has accelerated the pace of material advance in the modern world. The relationship between population change and economic change is complex and varied; despite intensive research into the topic over the past several generations, our understanding of the linkages in this relationship remains distinctly limited.[7] At the very least, however, the coincidence of rapid population growth and long-term improvements in material conditions should conclusively demonstrate that population increase and economic development are not inherently incompatible.

To observe that the world has been moving in the general direction of affluence over the course of our century, moreover, is by no means to ignore the setbacks and tragedies that have befallen large populations over the past several generations. To the contrary, in living memory entire regions of the globe have suffered severe economic reversals or sustained economic declines; harvest failures and other disasters have threatened the nutritional security of hundreds of millions of persons; and millions of men, women, and children have perished outright from famine.

At first glance, these grim realities may seem to affirm the neo-Malthusian claim that poor, rapidly breeding populations will tend to exhaust the resources at their disposal and will thereafter suffer the terrible consequences of an "overshoot" of "carrying capacity." A closer look leaves a rather different impression.

For the most part, the reversals and catastrophes that have afflicted large populations in low-income countries in recent decades can be traced directly to the policies or practices of presiding governments. Ill-advised, injurious, or positively destructive programs have been enthusiastically embraced by more than a few regimes in Asia, Africa, and Latin America over the past two generations. Such programs have direct and predictable consequences. Most of the troubling episodes that proponents of "population crisis" invoke to make their case can be more easily explained in terms of specific, albeit perverse, governmental actions. A "population crisis" that can be explained without any reference to demographic forces whatsoever is a crisis misdefined.

World Population Trends, "Limits to Growth," and "Overpopulation"

No one knows what the exact world population total was at the start of the twentieth century, but a conventional figure is around 1.6 billion in 1900.[8] In 1990, by the estimate of the United Nations, world population totaled about 5.3 billion persons.[9] By these numbers, world population would have increased by a factor of about 3.3 over the intervening decades—this despite two world wars, a number of major famines, and innumerable brutal but less famous conflicts. Between 1900 and 1990, by these numbers, the average pace of global population growth would have been about 1.3 percent a year. However such a rate may look to the reader, it would be extremely rapid by any historical perspective. It would, for example, be well over twice the tempo of growth demographers guess the nineteenth century to have experienced[10] and vastly faster than the rate of increase that could conceivably have characterized earlier periods.

It should be no surprise that the pace of world population growth has not been steady over the course of our century or even among its diverse regions. Figure 1-1 outlines the trends from 1950/55 to 1990/95, as estimated and projected by the United Nations. For the world as a whole, the rate of natural increase is believed to have peaked in the 1960s at about 2.0 percent a year; at present it is said to be around 1.6 percent a year. Today the pace of population growth in the "more developed regions," defined as the member countries of the Organization for Economic Cooperation and Development (OECD) plus the former Warsaw Pact region, is said to be about 0.4 percent a year. By contrast, the rate in the "less developed regions"—the rest of the world—is put at about 1.9 percent a year, or over four times as fast. Within these less developed regions, in turn, distinct differences are also evident. Africa's rate of natural increase is placed at about 2.8 percent a year, while average rates for Asia and Latin America are put, respectively, at roughly 1.6 and 1.8 percent per year—significantly slower tempos.

With these differentials in population growth, the distribution of the world's population has been shifting quite rapidly in recent decades. In 1950, by the United Nations's estimates, North America and "Europe" (defined here to include "European" republics from the former Soviet Union) together accounted for an estimated 28 percent of the world's population; by 1990, their share was down to an estimated 19 percent. Conversely, Africa is thought to have accounted for less than 9 percent of the world total in 1950 but for nearly 12 percent by 1990.[11]

Such patterns, it should be noted, mark a sharp reversal from the trends of the previous several centuries. During the Age of Exploration and the Age of Colonization, the world's European population grew much more rapidly than its non-European population.

A rough indication of those differences is provided in figures used by Nobel economics laureate Simon Kuznets. By those numbers, the rate of population increase for what he termed the "area of European settlement" (Europe, the Americas, Oceania, and the Russian Empire/Soviet Union) was well over twice as fast as for the rest of the world over the years 1800 to 1930. Over that same period, the share of world population living in this "area of European settlement" is said to have increased dramatically: from 24 percent to nearly 40 percent.[12]

It was not until after World War II that demographers generally recognized that the pace of growth for the world's non-European populations had finally come to exceed that of its populations of European descent. It is also true that the international movement for worldwide population control is essentially a postwar phenomenon that has drawn upon the revitalization of Malthusian con-

FIGURE 1-1 Estimated and Projected Rates of Population Growth, 1950/55–1990/95

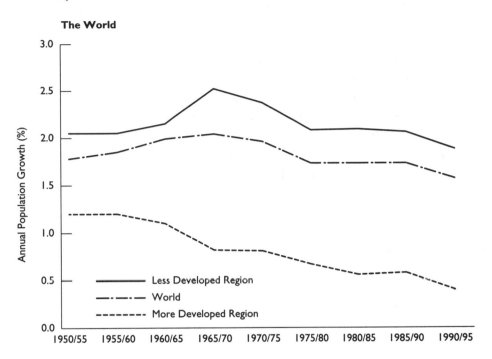

The World

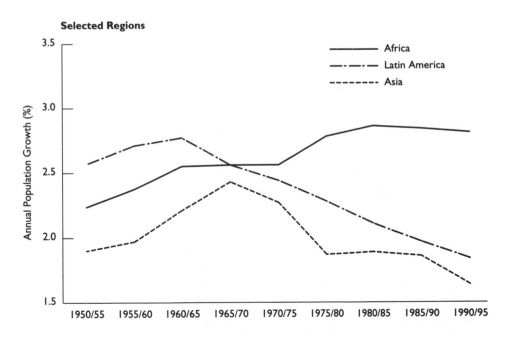

Selected Regions

Source: United Nations, *World Population Prospects: The 1994 Revision* (New York: UN Department for Economic and Social Information and Policy Analysis, 1994) pp. 56–58, 62, 64. (*Note:* "medium variant" projections used for 1990/95).

cerns since the end of World War II. The significance of this coincidence is worth pondering.[13]

In reviewing global population trends and prospects, two points, both usually neglected in conventional presentations of the topic, deserve emphasis. The first is that there is no accurate method for forecasting the future rate of population growth for a country or a region, much less the world. Demographers have no procedure for reliably predicting birthrates or death rates into the future. Consequently, long-term projections coincide with actual trends only by complete chance. (This shortcoming as a predictive science is widely acknowledged by demographers,[14] but it has not been—and cannot be—be corrected.)

> **There is no accurate method for forecasting the future rate of population growth for a country or a region, much less the world. Demographers have no procedure for reliably predicting birthrates or death rates into the future.**

Demographers have no way of knowing how fast a population will be growing a generation or two hence—or whether it will be growing at all. Long-term projections often assume that the rate of growth will settle at zero, but this is little more than convention adopted for simplicity's sake. No theoretical grounds exist for preferring that hypothesis to an alternative—say, that negative rates of growth are the prospect for postindustrial populations. We may note that a number of countries—Austria, Denmark, Germany, and Switzerland among them—have reported bouts of negative population growth over the past generation.[15] More recently, the collapse of the Soviet empire has apparently thrown the former Warsaw Pact area into negative population growth.[16] That this event was completely unforeseen, and that demographers have no idea how long this new regional pattern will continue, should further caution us against the presumption that there is a "natural" trajectory for the growth of modern populations, explicable through theory and identifiable in advance. In short, when we look at the "demand side" of the population-resources equation, we really have very little idea of the "limits to growth."

The second point is that population trends and demographic data provide no basis for defining "overpopulation." Familiar as the term may be, the fact remains that the concept cannot be described consistently and unambiguously by demographic indicators.

What are the criteria by which to judge a country "overpopulated"? Population density is one possibility that comes to mind. By this measure, Bangladesh would be one of the contemporary world's most "overpopulated" countries—but it would not be as "overpopulated" as Bermuda. By the same token, the United States would be more "overpopulated" than the continent of Africa, West Germany would be every bit as "overpopulated" as India, Italy would be more "overpopulated" than Pakistan, and virtually the most "overpopulated" spot on the globe would be the kingdom of Monaco.[17]

Rates of population growth offer scarcely more reliable guidance for the concept of "overpopulation." In the contemporary world, Africa's rates of increase are the very highest, yet rates of population growth were even higher in North America in the second half of the eighteenth century.[18] Would anyone seriously suggest that frontier America suffered from "overpopulation"?

What holds for density and rates of growth obtains for other demographic variables as well: birthrates, "dependency ratios" (the proportion of children and elderly in relation to working age groups), and the like. If "overpopulation" is a demographic problem, why can't it be described unambiguously in terms of population characteristics? The reason is that "overpopulation" is a problem that has been misidentified and misdefined.

The images evoked by the term *overpopulation*—hungry families; squalid, overcrowded living conditions; early death—are real enough in the modern world, but these are properly described as problems of poverty. As a human characteristic, poverty, like all other possible human attributes, is represented in individual members of a population. It is an elementary lapse in logic to conclude that poverty is a "population problem" simply because it exists.

Fertility and Contraception

At the end of the eighteenth century—ironically, just around the time of Malthus's first famous pamphlet on population—a completely new population pattern was introduced to the world. This was sustained fertility decline: long-term shifts toward smaller family size, driven and enforced by changing parental attitudes.

The phenomenon of sustained fertility decline was first evident in France, where it seems to have gotten underway by the time of the Jacobin revolution. Sustained fertility decline subsequently spread to the rest of Europe, and the areas of overseas European settlement, during the course of the nineteenth century. By the early decades of the twentieth century, fertility levels in several European populations had fallen temporarily below the level of net replacement: to levels so low, in other words, that their continuation would eventually result in national population decline, barring offsetting improvements in mortality or inflows of immigrants. By the early postwar era, it was clear that sustained fertility decline was not exclusively a European phenomenon; populations in East Asia (Japan), Western Asia (Cyprus), the Caribbean (Puerto Rico), and Latin America (Costa Rica) were also embarked upon it. By the mid-1990s, sustained fertility decline has come to be characteristic of the great majority of the world's populations.

Perhaps the clearest and most intuitively meaningful measure of fertility is the total fertility rate (TFR): the average number of births per woman over the course of the childbearing ages.[19] Figure 1-2 presents the United Nations's estimates and projections of global and regional TFR trends over the postwar era.

According to these numbers, the TFR for the world as a whole dropped by nearly two-fifths between 1950/55 and 1990/95, from about 5 children per woman down to about 3.1 children per woman. (By way of perspective, the current projected global TFR would be somewhat lower than that of the United States in the early to mid-1960s.) Over these four decades, fertility in the more developed regions is believed to have dropped by nearly two-fifths, from 2.8 to 1.7 children per woman. Proportionately and absolutely, the fertility drop for the less developed regions as a group is thought to have been even greater; there the TFR is estimated to have fallen from 6.2 to 3.5, or by well over 40 percent.

Fertility patterns, of course, differ markedly among low-income countries and even among entire regions. The TFR is believed to have been almost halved between the early 1950s and the early 1990s for Latin America and the Caribbean (5.9 to 3.1) and for Asia (5.9 to 3.0). The situation in Africa to date has been completely different. The TFR for the continent as a whole in the early 1990s is projected at 5.8—nearly twice the current projected Asian or Latin American levels and well over three times the level in the world's more developed regions. Moreover, the continent as a whole seems to have witnessed very little in the way of sustained fertility decline over the past four decades. Although sustained drops in fertility have been noted in some of the Arab countries in Northern Africa, only one sub-Saharan country on the continent proper (South Africa) is thought to have experienced a decline in fertility of as much as one-third between the early 1950s and the early 1990s. By contrast, in a number of sub-Saharan countries, fertility levels are thought to have been higher in the early 1990s than they were in the early 1950s.[20]

Sub-Saharan Africa thus constitutes an enormous exception to the fertility trends evident in nearly all the rest of the contemporary world. Three points should be emphasized here. First, there is no scientific method for accurately predicting the onset of sustained fertility decline for any population. A revolution in family size may not begin in most African countries for some time; alternatively, it may begin tomorrow or, given the limits of our statistical knowledge, it may even be underway.[21] Second, once initiated, many sustained fertility declines have been

FIGURE 1-2 Estimated and Projected Total Fertility Rates, 1950/53–1990/95

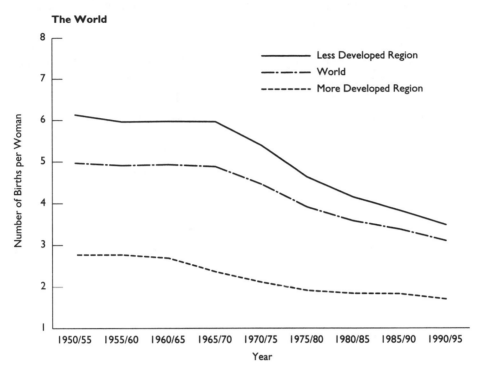

The World

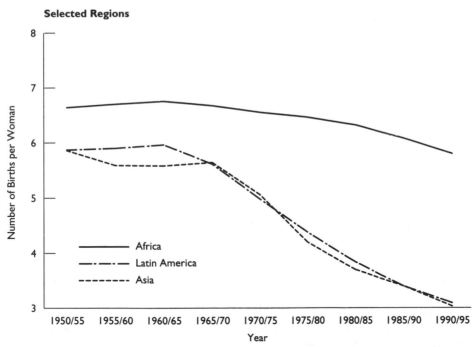

Selected Regions

Source: Ibid., pp. 122, 124, 126, 130, 136.

extraordinarily rapid in the postwar era. Costa Rica and Thailand, for example, cut their TFRs by half or more in only twenty years; over the course of twenty-five years, Colombia's TFR fell by nearly 60 percent and Hong Kong's by over 75 percent.[22] Third, there is no evidence of any "natural" tendency for sustained fertility declines to halt at the level of net reproduction; to the contrary, many modern populations have entered into extended periods of subreplacement fertility. Nearly all of Europe has been below net replacement fertility for the past two decades. Nor are European populations unique in this regard. Japan has been below the net replacement level for nearly four decades; Taiwan, South Korea, Hong Kong, and Singapore all report subreplacement fertility today; and subreplacement fertility levels are currently believed to be characteristic of such otherwise different Caribbean societies as Cuba and Martinique.[23]

Rapid fertility declines in low-income countries in the postwar era have been facilitated by the postwar period's revolution in contraceptive technologies. Medical breakthroughs since the end of World War II have brought forth such

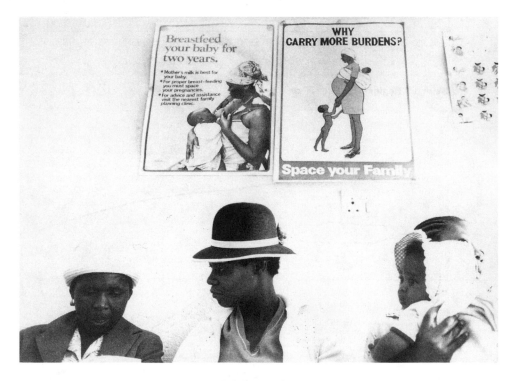

Total fertility rates are dropping in most developing countries as couples take advantage of modern contraceptive techniques. Breast-feeding reduces fertility and helps women space their children.

products and services as the birth control pill, the intrauterine device (IUD), injectable contraceptives, and routine sterilization procedures; technological advances have significantly improved such preexisting products as the condom and the diaphragm. With the advent of these new opportunities for contracepting, motivated populations in low-income settings have effected dramatic shifts in family planning behavior in only a few years.

These shifts in attitude and behavior are indicated by the fertility survey data presented in figure 1-3. Like other forms of polling, these survey data are not without their limitations and defects; nevertheless, they may be illustrative. Between the mid-1960s and the late 1980s, the proportion of couples of reproductive age deliberately practicing contraception is reported to have risen from less than 10 percent to nearly 80 percent in South Korea. Of course, South Korea may be viewed as an exception in the light of its recent exceptional economic performance, yet rapid shifts in contraceptive behavior are also reported in Colombia, Ecuador, Paraguay, Turkey, and Thailand. Acceptance and utilization of modern contraceptive methods is common to populations of dis-

FIGURE 1-3 Reported Prevalence of Contraceptive Practice Among Couples of Reproductive Ages, Selected Countries, 1963–1990

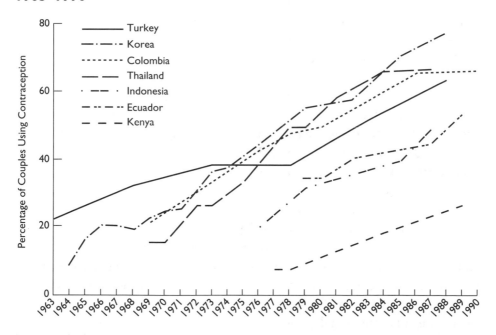

Source: Family Planning and Child Survival Programs (New York: Population Council, 1992), pp. 56–62.

tinctly different income level, educational attainment, religion, and cultural or historical background. The extent to which modern contraceptive practices have already been embraced in various low-income locales may sometimes seem surprising; as of 1988/89, for example, survey results reported a greater prevalence of modern method use for India than in Italy a decade earlier.

It is currently conventional for population specialists and family planning authorities to ascribe a decisive role in contemporary fertility reductions to the new methods of contraception. Such mechanistic faith is misplaced. The availability of modern methods of contraception is neither a necessary nor a sufficient condition for sustained fertility declines. Fertility in most of Europe, after all, dropped substantially during the nineteenth century, long before modern contraceptives were even imagined. Moreover, fertility in some Western countries had fallen below the replacement level before the modern postwar contraceptive revolution. Conversely, the fact that modern contraceptives are available does not ensure their widespread use. After a quarter of a century of government-sponsored family planning efforts in Pakistan, for example, less than 10 percent of the couples surveyed in 1990/91 said they were using modern contraceptive methods.

If modern contraceptive technology were the decisive factor in determining national fertility levels, we would expect TFRs to track closely with modern contraceptive utilization rates. They do not. In Pakistan, the TFR around 1990 is believed to have been over 6 births per woman. The Bulgarian utilization rate for modern contraceptives in 1977 was reported at just about the same as Pakistan's today, yet in the late 1970s the Bulgarian TFR was around 2.2.

The reason fertility levels can differ by a factor of 3 when the reported prevalence of modern contraceptive usage appears virtually identical is that parental preferences rather than medical technology are the decisive factor in determining a society's average family size. If this seems an obvious point, it is nevertheless an idea that supporters of the international birth control movement have typically ignored.[24]

Parental preferences rather than medical technology are the decisive factor in determining a society's average family size.

One recent World Bank study has estimated that at least 90 percent of the variations in fertility levels in developing countries in the postwar era can be explained by differences in the desired levels of fertility reported by local women.[25] If this is correct, the prospects for voluntary reductions of current fertility through expansion of voluntary family planning programs in low-income regions are at best marginal.

What explains the shift to lower desired levels of fertility on the part of parents? We do not know. "Modernization" is often said to be the engine powering sustained fertility decline. But if that were truly so, why would France—poor, rural, agrarian, and largely illiterate at the time—have been the first country in the world to begin the process? Some sociologists and economists have argued that the transition from high to low levels of fertility is explained by the advent of new familial arrangements whereby wealth begins to flow from parents to their children or by the rising value of (and opportunity costs attendant upon) human time.[26] Other seemingly reasonable theories have been offered as well. "The problem," as Charles Tilly once observed about fertility decline as a historical phenomenon, "is that we have too many explanations which are individually plausible in general terms, which contradict each other to some degree, and which fail to fit some significant part of the facts."[27]

Mortality and Health

From the standpoint of the individual, there is no preordained "optimum" for fertility. The situation is quite different for health and mortality; apart from pathological cases, the desire for long and healthy lives is all but universal. In this respect at least, the twentieth century has been a tremendous success, for it has witnessed an unprecedented worldwide explosion of health, which has dramatically and nearly continuously improved survival chances for every population on the face of the earth.

The dimensions of the twentieth century's health explosion are framed by a few basic figures. At the start of our century, a rough guess would place global life expectancy at birth at about 30 years.[28] By the early 1990s, global life expectancy is thought to have risen to about 64 years, more than doubling over these nine decades. Since it is further believed that life expectancy in earlier times could not have been much lower than 20 over any long period without raising the prospect of extinction,[29] it would seem that over three-fourths of the total improvement in human longevity since the origin of our species has been achieved since 1900.

This worldwide health explosion explains the global "population explosion." Rapid population growth commenced not because human beings suddenly started breeding like rabbits but rather because they finally stopped dying like flies. As we have already seen, fertility rates in almost all regions of the world are either declining or are already below replacement. To the extent that fertility levels may have risen in recent decades in some areas, especially in sub-Saharan Africa, these increases may also speak to improvements in health. In

Higher rates of vaccination in the developing world have been a big contributor to increased global average life expectancy.

many preindustrial societies, after all, infertility was a prevalent biological condition—and was perceived locally as the most serious "population problem."[30]

Most of the improvement in global life expectancy in our century is believed to have taken place since the end of World War II. Figure 1-4 presents the United Nations's estimates and projections for the four decades from 1950/55 to 1990/95. By these calculations and assumptions, life expectancy at birth for the world as a whole increased by eighteen years, or nearly two-fifths, over these four decades, even though the share of world population comprising low-income, lower-life-expectancy societies has risen steadily between 1950 and 1995. In the world's more developed regions, life expectancy at birth is thought to have increased by less than a decade over those years; for the less developed regions as a whole, it is thought to have risen by over two decades, or more than half since the early 1950s.

In the low-income regions, improvements in life expectancy vary by country and by region. For Asia, life expectancy is thought to have increased by about twenty-three years between 1950/55 and 1990/95; for Latin America and the Caribbean, the increment is placed at about seventeen years, whereas Africa is said to have enjoyed an increase in life expectancy at birth of "only" about fifteen

FIGURE 1-4 **Estimated and Projected Life Expectancy at Birth, 1950/55–1990/95**

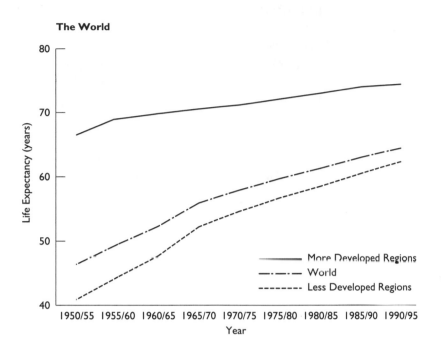

The World

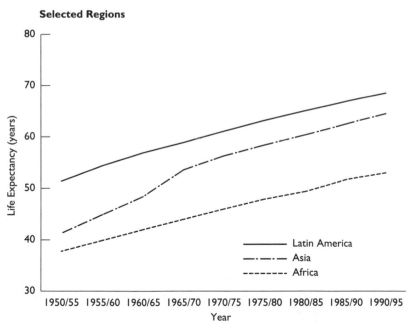

Selected Regions

Source: United Nations, *World Population Prospects*, pp. 166, 169, 175, 180, 186.

years. Of the low-income areas' most populous countries, China's life expectancy is thought to have risen by about twenty-eight years; India's, by about twenty-two; Indonesia's, by about twenty-five; Brazil's, by about fifteen; Pakistan's, by about twenty-three; Bangladesh's, by about nineteen; Nigeria's, by about fourteen. Only one region of the world appears to have experienced long-term stagnation or retrogression in health progress during the postwar era: the Soviet bloc's Warsaw Pact states. In Eastern Europe, life expectancy is thought to have risen by only one year between the late 1960s and the late 1980s; in the Soviet Union, life expectancy at birth is believed to have been slightly *lower* in the late 1980s than it had been in the early 1960s.[31] The problems of this region, it seems fairly clear, are to be explained not in terms of Malthusian mechanisms but rather by the systemic and structural defects of Soviet communist governance.[32]

Life expectancy at birth is strongly affected by the survival chances for babies. Worldwide improvements in life expectancy reflect a corresponding worldwide improvement in prospects for child and infant survival. Figure 1-5 traces these patterns over the postwar era. For the world as a whole, the infant mortality rate—that is, the number of babies dying in the first year of life for every 1,000 born—is thought to have fallen by nearly 60 percent. In the more developed regions, the drop is estimated to have exceeded 80 percent; in the less developed regions, it is thought to have exceeded 60 percent. (The reason that the world's total drop in infant mortality rates is lower than that estimated for either large region is that the less developed regions, with their much higher infant mortality rates, have accounted for a progressively larger share of the postwar world population.)

For some large populations, the drop in infant mortality rates over the postwar period has been little short of remarkable. Hong Kong, for example, is said to have witnessed a decline from 79 to 7 between the early 1950s and the early 1990s—a drop of over 90 percent. But even troubled sections within the low-income areas seem to have made tremendous strides. Over the past four decades, for example, Bangladesh is thought to have experienced a drop in infant mortality rates of 40 percent, and African infant mortality rates are believed to have been cut roughly in half.

Although activists in the international population and environmental movements do not typically comment on the significance of this international revolution in health conditions, its economic implications are major and immediate. First and foremost, better health and extended longevity constitute a self-evident improvement in the human condition—one of tremendous value to the persons concerned. Careful if necessarily speculative explorations have shown that the reduction of mortality levels confers a benefit on individuals for which they would be willing to spend considerable sums—if they could.[33] Conventional systems of

FIGURE 1-5 Estimated and Projected Infant Mortality Rates, 1950/55–1990/95

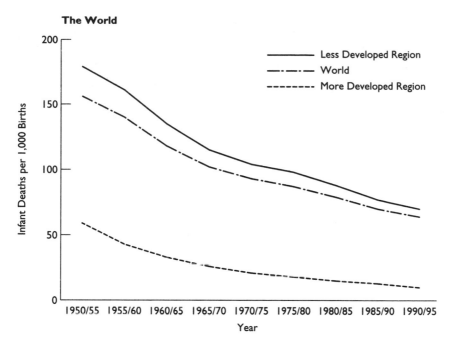

The World

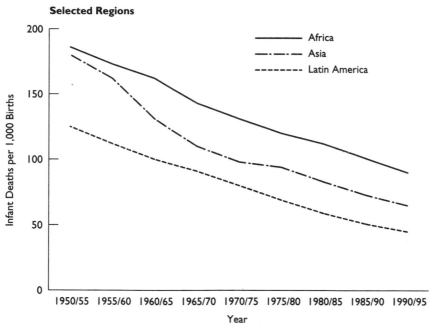

Selected Regions

Source: Ibid., pp. 206, 208, 210, 214.

national accounts, of course, do not count improvements in mortality as a positive item in the economic ledger, unless, of course, they occur among animals.

Second, it appears that reductions in mortality over the postwar period have generally been much more rapid in the less developed regions than in the more developed ones. For all the talk of a "widening gap" between rich and poor countries, it would appear that one easily measurable and preeminently important material difference has been narrowing steadily and appreciably.

Finally, improvements in health and longevity augur well for improvements in a population's productivity. It is not simply that a healthier population will be able to work harder. As Nobel economics laureate Theodore Schultz has argued, improvements in health and mortality increase the attractiveness of investments in "human capital."[34] Populations with better productivity potential, of course, may not be able to capitalize on their new capabilities for a variety of reasons. An adverse business climate or policy environment can reduce rates of return on human capital just as surely as on physical capital. Even so, the fact is that the same forces that have powered the modern "population explosion" have also eased the constraints on attaining higher levels of per capita output in contemporary Africa, Asia, and Latin America.

Food and Agriculture

Medical advances, innovations in transport and communications, and improvements in administrative capacity are among the factors that have contributed to the dramatic worldwide reductions in mortality levels of the past two generations. (Indeed, thanks to technological progress, it has become possible to "purchase" long life at an ever lower price, a truth underscored by the fact that the current global life expectancy is estimated to be only about three years shorter than was that of the United States in 1950.)[35] Improvements in worldwide health and mortality also owe much to the tremendous improvement in recent generations of the world food situation.

Other chapters in this book will examine world agricultural trends in greater detail, but since food supplies are so often viewed as the limiting constraint on the growth of human populations, a discussion is warranted here.

Estimates of agricultural output and consumption are still subject to a large margin of error, as the Appendix at the end of the book will affirm. Such uncertainties notwithstanding, there can be little doubt that the per capita food production for the world as a whole has increased markedly over the past sixty years and that per capita food availability has risen in every major region of the world. Figure 1-6 presents the time-series statistics compiled by the UN Food

FIGURE 1-6 Estimated Food Supplies for Direct Human Consumption, 1934/38–1987/89

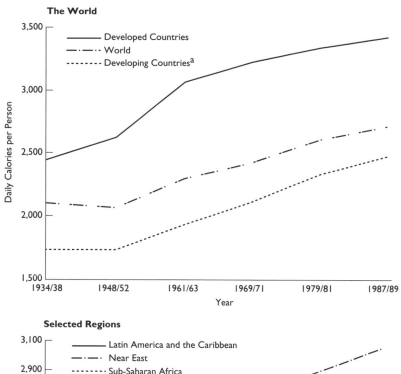

The World

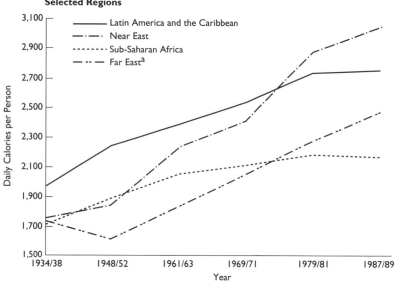

Selected Regions

[a] Excludes China for 1934/38 and 1948/52. The series links those years with more recent estimates on the basis of the index years 1962/64—a period not appreciably different from 1961/63.

Sources: FAO, *The State of Food and Agriculture 1991* (Rome: FAO, 1992), p. 14; Simon Kuznets, "Economic Capacity and Population Growth," in Richard Farmer et al., editors, *World Population: The View Ahead* (Bloomington: Indiana University Press, 1968), p. 94; P. V. Sukhatme, "The World's Food Supplies," *Journal of the Royal Statistical Society* A:129 (1966): 234.

and Agriculture Organization (FAO) and its predecessor organizations on estimated daily per capita calorie availability. By these figures, the global per capita
calorie availability rose by nearly 30 percent between the 1930s and the late
1980s. For the less developed regions of Africa, Asia, and Latin America as a
whole, per capita food supplies are thought to have risen by nearly 40 percent,
although somewhat lower or higher figures are also possible.[36] By these numbers,
however, it would seem quite likely that per capita food availability for the less
developed regions by the late 1980s had reached the level attained by the more
developed countries just before World War II.

Once again, trends in the Third World differ by region, yet even in Africa,
per capita food supplies are estimated to have been somewhat higher in the late
1980s than a generation earlier—and nearly one-fourth higher than they had
been before World War II. At least by this measure, nutritional progress in sub-
Saharan Africa was apparently rapid until the early 1960s but decelerated
sharply thereafter.

Between the early 1960s and the late 1980s, according to the FAO's estimates,
per capita food supplies rose by about 18 percent for the world as a whole and
by about 28 percent for the less developed regions. Estimates from the U.S. Department of Agriculture (USDA) differ in particulars but paint a broadly similar picture (figure 1-7). According to the USDA, between 1961 and 1990 food
supplies per person rose by 17 percent for the world as a whole and by 32 percent in the less developed regions. Like the FAO, the USDA identifies sub-
Saharan Africa as the region with the least nutritional progress between the
early 1960s and the late 1980s, although both organizations see slight improvements there over those decades.

In short, a dramatic improvement in the world's food situation has occurred,
it has taken place steadily over consecutive generations, and this period of sustained improvements coincided with the most rapid episode of global population growth in human history.

There is another noteworthy trend. In the international marketplace over
the postwar era, the real price of food grains has been gradually and significantly
declining. To be sure, international food commodity prices fluctuate unpre-
dictably, and they have at times doubled or tripled virtually without warning (as
they did during the so-called world food crisis of the early 1970s, although largely
as a response to precipitous and ill-advised governmental interventions).[37] But
there is a strong tendency evident in the long-run relative prices of food grains,
and it is unmistakably downward. Over the course of the twentieth century, according to a careful study conducted by the World Bank and published in 1988,
the relative price of food grains dropped by over 40 percent.[38] The trend in real

FIGURE 1-7   USDA Estimates of Available Food Supplies for the World, 1961–1990

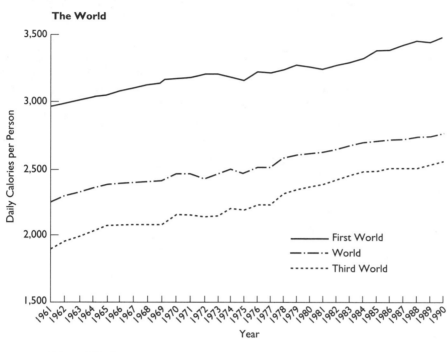

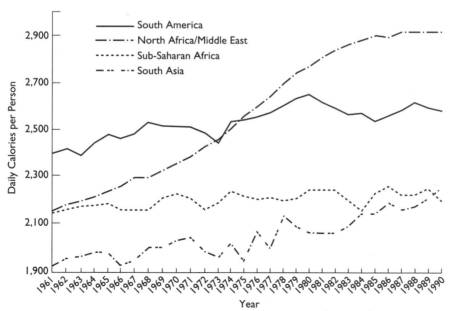

Source: U.S. Department of Agriculture, *World Agriculture: Trends and Indicators* (Washington, D.C.: USDA Economic Research Service, 1994).

Better worldwide food distribution networks have reduced the danger of famine in many poor countries. These school children in Botswana are eating a lunch of sorghum and vegetable oil.

grain prices has continued down since that study was completed. Food has been getting cheaper. By the information that prices are meant to convey, food grains have become a less scarce resource over the course of our century, despite more than a tripling of world population and an even greater increase in international demand for these commodities.

Income and Productivity

Compared with counting a nation's population or estimating a country's crop output, the task of measuring national income is fraught with difficulty. A country's production of goods and services can be totaled only through some common standard of value, and the valuation problem is much more difficult than we might suppose from the promiscuous use of economic statistics in daily news accounts, academic studies, and official policy papers. Even so, a few overarching global trends in income and productivity are unmistakable:

- Per capita output and income for the world as a whole have multiplied several-fold over the course of our century.

- Improvements in the efficiency with which resources are put to use have been a driving force in modern economic growth.

- The role of human skills has grown steadily more important in the modern production process as the role of natural resources and physical equipment has steadily diminished.

Perspective on the worldwide rise in per capita output in the modern era is provided by table 1-1. The numbers in this table were assembled by Angus Maddison, currently of the University of Groningen and for many years the director of the OECD's Development Research Centre. This sample of forty-three countries represents about three-fourths of the world's population and perhaps somewhat more of its economic output. In view of the statistical gaps for a number of the countries over the long period covered, some of these estimates should be seen as heroic, if informed, guesses; certainly these calculations were not meant to be trusted down to the last dollar. They are, however, broadly informative, and the story they tell is dramatic.

By Maddison's figures, per capita output in Western Europe, North America, and Australia increased more than thirteen-fold between 1820 and 1989. Per capita output for this group roughly doubled in the seventy years between 1820 and 1890, it more than doubled over the next sixty years, and in the four decades between 1950 and 1989 it nearly tripled.

Steady economic growth may have begun earlier in these Western European countries and overseas settlements than in the rest of the world, but Maddison's figures demonstrate that the latecomers have often availed themselves of what Alexander Gerschenkron termed the "advantages of backwardness"[39] and embarked on extremely rapid "catch-up growth." Between 1950 and 1989, for example, Spain and Portugal (countries that were widely regarded as desperately poor not so long ago) saw their per capita output levels more than quadruple. In other regions, catch-up growth has been even more dramatic. Between 1950 and 1989, for example, per capita gross domestic product is estimated to have increased in South Korea more than eightfold and in Taiwan by a factor of ten. Thailand's per capita product more than doubled in the relatively brief span between 1973 and 1989, thanks to an estimated per capita growth rate of over 5 percent per annum for those years. But even distinctly lower rates of growth, if sustained over time, can produce extraordinary transformations. Between 1913 and 1989, for example, Japan's rate of per capita growth averaged about 3.5 per-

TABLE 1-1   Estimates of Per Capita Gross Domestic Product in Forty-three Countries, 1820–1989 (1985 relative prices in dollars)

	1820	1870	1890	1913	1950	1973	1989
West European capitalist core and offshoots							
Austria	1,048	1,442	1,892	2,683	2,869	8,697	12,519
Belgium	1,025	2,089	2,654	3,267	4,229	9,417	12,875
Denmark	980	1,543	1,944	3,014	5,227	10,527	13,822
Finland	639	933	1,130	1,727	3,481	9,073	14,015
France	1,059	1,582	1,955	2,746	4,176	10,351	13,952
Germany	902	1,251	1,660	2,506	3,295	10,124	13,752
Italy	965	1,216	1,352	2,079	2,840	8,631	12,989
Netherlands	1,308	2,065	2,568	3,179	4,708	10,271	12,669
Norway	856	1,190	1,477	2,079	4,541	9,347	15,202
Sweden	1,008	1,401	1,757	2,607	5,673	11,362	14,824
United Kingdom	1,450	2,693	3,383	4,152	5,651	10,079	13,519
Australia	1,250	3,143	3,949	4,553	5,970	10,369	13,538
Canada		1,330	1,846	3,515	6,112	11,835	17,236
United States	1,219	2,244	3,101	4,846	8,605	14,093	18,282
Average	1,055	1,723	2,191	3,068	4,813	10,298	14,228
European periphery							
Czechoslovakia	836	1,153	1,515	2,075	4,365	6,980	8,538
Greece				1,211	1,456	5,781	7,564
Hungary		1,139	1,439	1,883	2,481	5,517	6,722
Ireland				2,003	2,600	5,248	8,285
Portugal		833	950	967	1,608	5,598	7,383
Spain	900	1,221	1,355	2,212	2,405	7,581	10,081
Soviet Union		792	828	1,138	2,647	5,920	6,970
Average	868	1,028	1,217	1,641	2,381	6,089	7,931
Latin America							
Argentina		1,039	1,515	2,370	3,112	4,972	4,080
Brazil	556	615	641	697	1,434	3,356	4,402
Chile			1,073	1,735	3,255	4,281	5,406
Colombia				1,078	1,876	2,996	3,979

TABLE 1-1   Continued

Mexico	584	700	762	1,121	1,594	3,202	3,728
Peru				1,099	1,809	3,160	2,601
Average	570	785	998	1,350	2,180	3,661	4,033
Asia							
Bangladesh				519	463	391	551
China	497	497	526	557	454	1,039	2,538
India	490	490	521	559	502	719	1,093
Indonesia	533	585	640	710	650	1,056	1,790
Japan	609	640	842	1,153	1,620	9,524	15,336
Korea			680	819	757	2,404	6,503
Pakistan				611	545	823	1,283
Taiwan			564	608	706	2,803	7,252
Thailand		741	801	876	874	1,794	4,008
Average	532	591	653	712	730	2,284	4,484
Africa							
Cote d'Ivoire					888	1,699	1,401
Ghana				484	733	724	575
Kenya					438	794	886
Morocco					1,105	1,293	1,844
Nigeria					608	1,040	823
South Africa				2,037	3,204	5,466	5,627
Tanzania					334	578	463
Average	400	400[a]	400[a]	580[a]	1,044[a]	1,656	1,660

[a] Rough guesses, assuming no progress in the nineteenth century.

Source: Angus Maddison, "Explaining the Economic Performance of Nations, 1820–1989," in William J. Baumol, Richard R. Nelson, and Edward N. Wolff, eds., *Convergence of Productivity: Cross-National Studies and Historical Evidence* (New York: Oxford University Press, 1994), p. 48.

cent a year, for a more than 1,200 percent increase in per capita production over the period as a whole.

By definition, of course, not all countries are success stories. For this reason, it is worthwhile examining the economic performance of some of the populations that have fared less well to date in the race toward mass affluence. Consider Latin America and the Caribbean nations. Maddison's sample includes three-quarters of that region's population. By the numbers of table 1-1, these countries appear to have tripled their per capita output between 1913 and 1989 and nearly to have doubled per capita product between 1950 and 1989. Although the region spent most of the 1980s in a depression sparked by its "debt crisis," per capita output for this grouping is said to have been higher in 1989 than it was in 1973.

Long-term economic progress is also evident in the populous low-income reaches of Asia. Poor as India may still be today, its per capita GDP is thought to have doubled between 1950 and 1989. For Pakistan, per capita output may have more than doubled over those same years, and for Indonesia it may very nearly have tripled. (In China, by these figures, per capita income would have more than quintupled; special caution is appropriate for numbers on centrally planned economies, but there can be no doubting the brisk growth in that nation, especially in more recent years.) Bangladesh stands alone in table 1-1 as a country estimated to have enjoyed no significant improvement in per capita output levels over the course of the twentieth century.

And what of today's troubled Africa? It is in the nature of economic data that reliable statistics are likely to be scarcest in precisely the regions that are of greatest humanitarian concern. The seven African countries in Maddison's sample comprise a little more than 40 percent of the continent's population and less than half of the population of the sub-Sahara. The representativeness of these estimates and guesses is a question that cannot be answered without further research. For these seven countries, nevertheless, Maddison indicates a near tripling of average per capita output between 1913 and 1989. In 1989, by his estimates, the average per capita output for these seven countries was roughly similar to that of Germany in 1890; this average, however, is strongly influenced by South Africa's performance.

According to Maddison's estimates, this sample of African countries on average stagnated over the past two decades; in four of the six sub-Saharan countries, per capita output is said to have declined between 1973 and 1989. In recent decades, by such estimates, sub-Saharan Africa would appear to have become an island of economic regression in a global sea of economic growth. This reversal is all the more puzzling and troubling in that substantial economic gains were apparently registered in Africa earlier in the century. We will return to this problem in the final section of this chapter.

Estimates by the University of Pennsylvania team provide a similar, if more detailed, portrait of global economic progress for the 1950–1990 period (figure 1-8). By their assessment, per capita GDP for the world as a whole nearly tripled over these four decades. For the less developed regions as a whole, growth is thought to have been more rapid than for the more developed regions (exclud-

FIGURE 1-8   Real GDP per Capita for Selected Regions, 1955–1992 (1985 international dollars)

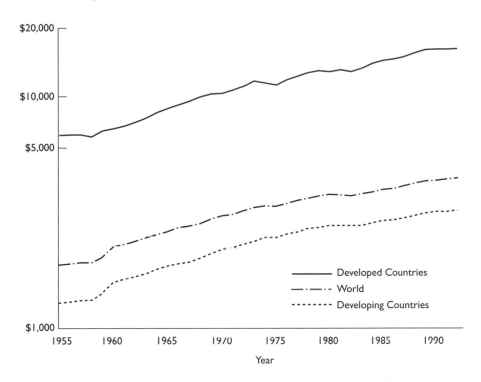

Source: Robert Summers and Alan Heston, "The Penn World Table (Mark 5.6)," unpublished.

ing the Soviet bloc, for which estimates of economic performance remain problematic). The University of Pennsylvania numbers suggest more than a doubling of per capita output in Latin America and the Caribbean between 1950 and 1990, despite severe economic setbacks in the 1980s; they depict steady and perhaps accelerating economic growth in Asia. For sub-Saharan Africa they show growth to have suddenly faltered in the 1960s, after a brisk initial ascent in the 1950s.

As the University of Pennsylvania numbers illustrate, per capita economic growth varies from one year to the next, even for large regions or entire groupings of countries. For entire regions, moreover, recessions and shocks can make for slow growth, or economic contraction, even when the trend is steadily upward. The common focus on the near term—the next harvest, the next quarterly report, the next set of monthly economic indicators—needs no justification; the purposes it serves are clear enough. But it is only with a longer look that the economic performance of the postwar world economy can be fully appreciated and that the genuine long-term problem areas can be unambiguously identified.

How have contemporary economies managed to achieve these significant and sustained increases in their levels of per capita output? Perhaps surprisingly, mobilization of labor, capital, and land—the familiar "factors of production" from classical economic treatises—are of little help in explaining the transformation. More than a generation ago Simon Kuznets demonstrated that the increase per se in hours worked and capital stock amassed could not account for the tremendous increase in output per capita in developed countries over the previous century. By his computations, something like four-fifths of their per capita growth had to be explained by other factors.[40] Subsequent exercises in growth accounting have come up with different figures for given regions and historical periods, but they have generally affirmed the great importance of qualitative improvements to the modern process of economic growth: that is, improvements in the skills of the workforce and the quality of the capital it uses; improvements in the technology utilized, the scale of operations, and the methods of organization selected; improvements in the risk environment under which potentially productive resources are allocated. In the United States between 1929 and 1982, according to one careful analysis, "total factor productivity" in the American economy more than doubled.[41] Over the twentieth century, of course, the United States has been the world's leading economy in terms of productivity levels—the economy nearest to the limits of the technologically possible. That such improvements were practicable for the United States only indicates how much greater the opportunities for technology-and-skills-driven growth were for countries at lower initial levels of productivity.

The rise toward mass affluence highlights another general trend: the tendency for human resources to matter more, and for natural resources to matter less, over the course of economic development. Land is a case in point. Around 1870, by Maddison's estimates, nearly 40 percent of the GDP of what are now the OECD countries accrued from agriculture, and farmland figured very prominently in the overall capital stock of the countries in question.[42] By the late 1980s, agriculture accounted for only about 4 percent of the OECD countries' total output, and farmland had become a relatively minor component of these nations' capital holdings. Similar tendencies can be seen in Asian and Latin American economies today.[43] Gradual diminution of the importance of natural and physical resources in the modern economy is suggested by many other changes, including the rise over time and among countries in the share of national product accruing to human beings through wages, salaries, and entrepreneurial incomes. While the twentieth century has witnessed declines in the real

High population density need not be a bar to rapid economic development, as Singapore demonstrates, but political structures are the deciding factor between dense development and dense poverty.

prices of many commodities,[44] it has seen a tremendous increase in the real price of one nonrenewable resource: human time.

If the neo-Malthusian view of the world were capable of explaining and predicting events, we should have expected the world economy to be shaped by two trends in our century: (1) a gradual global economic slowdown as poor but fast-breeding countries come to comprise an ever greater of total world population, and as rapid population growth pressed the world economy against natural resource constraints; and (2) a gradual widening of income differences within contemporary societies as poorer elements outbred more educated and affluent groups. Neither of these corollaries of the neo-Malthusian worldview corresponds with observed realities in our age. As we have seen, despite rapid world population growth, global improvements in per capita output levels have been unprecedented in the twentieth century and show no sign of stopping. As for income distribution, serious arguments today proceed about trends in particular countries over specific periods, but no serious study has ever suggested a steady worldwide "deterioration" of intracountry income distribution. In this respect as in others, contemporary neo-Malthusian doctrine can be seen as a theory still in search of facts.

Setbacks, Disasters, and Catastrophes: A Population-Resource Problem?

Despite the positive general direction of health, food, and income trends, the twentieth century has been punctuated by tragedies involving loss of life on a massive scale. It has also seen economic setbacks that have lasted decades and affected hundreds of millions of people. Do these episodes finally provide the facts that neo-Malthusian theorists have been searching for? Are the economic reversals and demographic catastrophes of our era a consequence of population-resource pressures?

To approach this question, we should examine three modern phenomena: (1) the major famines of the twentieth century,[45] (2) the recent economic slump in Latin America, and (3) sub-Saharan Africa's recent and extended economic and agricultural malaise. Before analyzing the causes of modern catastrophe, however, we should mention a seldom-noticed but nonetheless important characteristic of their aftermaths. In the modern era, recovery from catastrophes has typically been amazingly quick, as the experiences of Japan and China show (figures 1-9 and 1-10). Japan suffered devastating destruction attendant upon complete defeat in a total war, and China endured the worst famine of the twentieth century, one that may have cost as many as 30 million lives.[46] Yet the impact of these cataclysmic events on levels of national life expectancy, while terrible, was also brief. Recovery to predisaster levels of life expectancy required only a few years; thereafter, both societies were able to manage rapid improvements to levels never previously attained. Japan, in fact, enjoys the longest life expectancy at birth among contemporary societies. Note that recovery was as rapid and vigorous for a low-income society like China as for the more technologically sophisticated nation of Japan. Rapid and permanent recovery from disaster, moreover, are not characteristic of Asian societies alone. Despite the appalling

FIGURE I-9   Estimated Male and Female Life Expectancy at Birth, Japan, 1935/36–1969

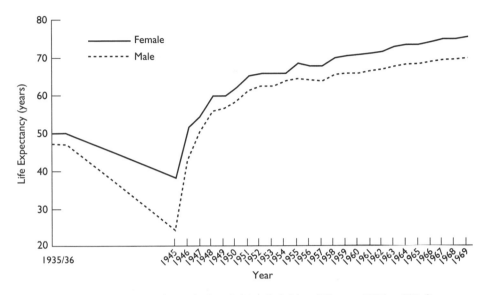

Source: United Nations, *Demographic Yearbook 1953* (New York: United Nations, 1953), p. 330; Government of Japan, Management and Coordination Agency, *Japan Statistical Yearbook* (Tokyo: Statistics Bureau), various issues.

losses from its brutal civil war, for example, Spain's health and economic progress has been rapid over the post–World War II era. By the same token, the Netherlands suffered through a serious famine in 1944/45,[47] but this evidently has not prejudiced its postwar advance.

Our century has recorded at least three famines thought each to have claimed the lives of several million persons: the Soviet famine of 1934, the Bengal famine of 1943, and China's "Three Lean Years" of 1958–1961. More recently, famine in Ethiopia is said to have taken a toll of hundreds of thousands.[48] Did these famines result from population pressures?

In a tautological sense it is true that famines represent a critical imbalance between a stricken population and the food resources available to it. In the cases noted, moreover, it is true that local fertility levels were high and rates of natural increase were generally rapid in these diverse settings just before the onset of their famines. But to describe these tragedies as demographically induced is to ignore the actual histories of each episode. In each of these tragedies, the hand of the state figured prominently; the millions who perished were killed by the direct actions of their own governments.

FIGURE 1-10   Estimated Life Expectancy at Birth, China, 1954–1984

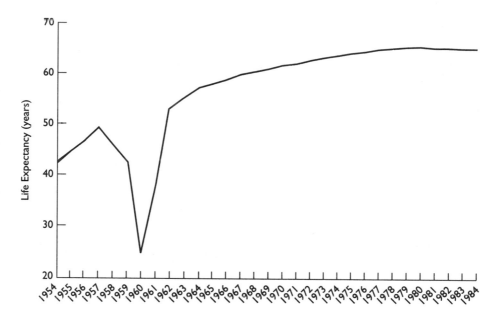

Source: Judith Banister, *China's Changing Population* (Stanford: Stanford University Press, 1987), p. 352.

The Soviet famine of 1934, for example, was the consequence of the official collectivization campaign in the Ukraine. Stalin specifically intended to use starvation as a weapon to break Ukrainian resistance to his policies, which is why the historian Robert Conquest has termed the hunger a "terror-famine."[49] The Bengal famine of 1943 took place at a time when local harvests were quite good but when British officials, fearing a possible Japanese invasion from neighboring Burma, had systematically removed local grain supplies.[50] The Chinese famine followed immediately upon Mao's Great Leap Forward, a collectivization campaign that inadvertently shattered the agricultural system in a low-income population.[51] Mass starvation erupted in Ethiopia in the 1980s after its communist government inflicted a series of harsh and injurious policies on a population whose living standard was typically only slightly above the subsistence level.[52]

In each of these instances, the reckless or intentionally punitive policies embraced by presiding governments would have been expected to result in massive loss of life, no matter what the local fertility level or population growth rate. If the famines in these diverse countries are to be described as "population problems," it is only to the extent that their problems were precipitated by their distinctly delimited populations of rulers.

What about the economic difficulties of Latin America and the Caribbean nations over the past decade? According to most available estimates, per capita output in this region was lower in the early 1990s than it had been a decade earlier. Is Latin America's prolonged economic slump an instance of population-resource pressures?

Viewed solely by its forensics, the Latin American case is not easily presented as an example of demographically induced depression. Latin America's economic slump, after all, did not take place until the early 1980s, by which time both fertility levels and population growth rates had been declining for nearly two decades. During previous decades of much higher demographic growth rates and TFRs, Latin America's pace of economic growth had been steady and quite rapid. The outlines of Latin America's slump, moreover, are hardly suggestive of an overshoot of sustainable consumption levels. Latin America's per capita output plunged in the early 1980s and has been recovering since then. The curve will be familiar to anyone who knows the phrase *business cycle*.

What prompted the economic plunge in the early 1980s? The drop can be explained without any reference to population trends. Latin America was shaken by a "debt crisis"—an economic dislocation set in motion by heavy government borrowings committed to economically questionable purposes and sparked by a sudden shift in international capital markets from negative to sharply positive real interest rates in the late 1970s and early 1980s.[53] Many East Asian states were also heavy borrowers in the 1970s, but they were generally able to weather the storm of the early 1980s because they had begun by allocating their loans to economically productive projects and had embraced disciplined programs of macroeconomic adjustment when the storm struck.[54] Performance in Latin America's economies has varied over time and among countries since the initial "debt shock"—and these distinctions have had more to do with the extent to which the degree to which sitting governments have been willing to restrain *dirigiste* impulses and unleash competitive economic forces than to any observable demographic differences among these affected societies.

What, finally, of the sub-Saharan quandary? No other region of the modern world has such a poor record of economic performance; no other region has such high fertility rates or such a rapid pace of population growth. To those for whom such fundamentals frame an airtight case, alternative explanations may seem superfluous or distracting. Yet it is difficult to see how one can examine the performance of economic systems without taking into account their general political environments or the specifics of governmental policies and practices.

> **It is difficult to see how one can examine the performance of economic systems without taking into account their general political environments or the specifics of governmental policies and practices.**

Sub-Saharan Africa's pervasive agricultural and economic malaise followed on the heels of decolonization. For the sub-Sahara, decolonization and independence happened all at once, over a relative handful of years in the late 1950s, the 1960s, and the early 1970s. Why the initial decades of self-rule should have resulted in so many failures of governance in such different cultural settings, and with separate colonial traditions, is a question that will doubtless concern historians for years to come, but the fact of this pervasive failure of governance is scarcely open to dispute. Over the past three decades civil strife and ethnic tensions have been the rule, not the exception, in sub-Saharan politics; the state has become an instrument of tribal warfare on more than a handful of occasions. To the extent that colonial rule bequeathed traditions of accountable government or rule of law—legacies admittedly uneven among the European imperialists in Africa—such heritages were commonly ignored or overturned by new rulers. The new governments in the sub-Sahara typically squeezed farmers through punitive prices and taxes and discouraged exports through perennially overvalued exchange rates. Widespread primary education was often ignored in favor of the higher schooling of elites. Budget discipline in this group of countries rapidly eroded, and accounting for official expenditures generally evolved from a routine procedure into a murky and contentious issue.[55] Under such circumstances, does one really need any demographic details to make an informed prediction about likely economic outcomes?

Does this survey demonstrate that population trends played no role whatever in the troubles of the contemporary era? Not at all. The impact of population change on social conditions and economic performance may be complex and as yet poorly understood, but it clearly entails challenges as well as opportunities. Individuals, groups, and countries do not always respond successfully to challenges. Whatever else it may portend, population growth is clearly a form of social change; nations and governments that cope poorly with change are unlikely to deal adeptly with the disequilibria that more rapid rates of population growth necessarily bring. It may be that rapid rates of population have amplified the magnitude of state-sponsored famines or the dimensions of economic failure under severe misrule. But the presumption that population growth is the determining factor in these episodes can be maintained only by resolutely ignoring the actual practices and policies of existing regimes. To hold (if implicitly) that bad governments are immutable but that their subjects' family size preferences should not be accepted as final is to enter a perilous ethical realm. More than that, it is to declare opposition to the entire purpose of modern economic development, which is, after all, the extension of human choice.

NOTES

Jonathan Tombes of the American Enterprise Institute provided expert research assistance in this effort—as always. Special thanks are due to Larry Heligman of the United Nations Population Division, Ray Nightingale of the U.S. Department of Agriculture, and Alan Heston and Valerie Mercer of the University of Pennsylvania for generously supplying unpublished results from their research; naturally, none of them is responsible for my interpretations or conclusions. Thanks (plus the standard disclaimers) are also due to C. Peter Timmer for his helpful suggestions and comments on this chapter.

1. The familiar term *Malthusian* implies that Malthus himself propounded a single, unchanging thesis over the course of his life. This was not the case. In reality, Thomas Robert Malthus offered different views of the world's population prospects in the seven editions of the famous treatise he prepared. Even between the first (1798) and the second (1803) editions of his essay, in fact, steely certainty that humanity would be unable to cope with its tendencies for population growth gave way to a guarded optimism that the population problem could indeed be managed. For more background, see Thomas Robert Malthus, *On Population*, ed. Gertrude Himmelfarb (New York: Modern Library, 1960).
2. For a brief survey of pre-Malthusian population doctrines, see United Nations, *Determinants and Consequences of Population Trends* (New York: UN Department of Social Affairs, 1973), 1: chap. 3.
3. Paul Ehrlich, *The Population Bomb* (New York: Sierra Club–Ballantine, 1968), p. i.
4. Eric Chivian et al., *Critical Condition: Human Health and the Environment* (Cambridge, Mass.: MIT Press, 1993), p. 173.
5. Lester Brown et al. *The State of the World 1994* (New York: W. W. Norton, 1994), p. 196.
6. Paul Lewis, "Curb on Population Growth Needed Urgently, U.N. Says," *New York Times*, April 30, 1992, p. A12.
7. One document that offers a thoughtful and careful review of the current state of knowledge about the relationship between population growth and economic development is National Research Council, *Population Growth and Economic Development: Policy Questions* (Washington, D.C.: National Academy Press, 1986).

8. Simon Kuznets, *Modern Economic Growth* (New Haven: Yale University Press, 1966), p. 38.

9. United Nations, *World Population Prospects: The 1994 Revision* (New York: UN Department of Economic and Social Information and Policy Analysis, 1994), pp. 11–12.

10. Kuznets, *Modern Economic Growth*, pp. 35, 38. By the figures cited here, the pace of population growth in the twentieth century would be nearly ten times as fast as the average for the eight centuries between A.D. 1000 and 1800.

11. UN, *World Population Prospects*, p. 11.

12. Kuznets, *Modern Economic Growth*, pp. 36–40.

13. For some useful background, see Dennis Hodgson, "The Ideological Origins of the Population Association of America," *Population and Development Review* 17, no. 1 (1991): 1–34.

14. See, for example, Nathan Keyfitz, "The Limits of Population Forecasting," *Population and Development Review* 7, no. 4 (December 1981): 579–593, and Ronald D. Lee, "Long-Run Global Population Forecasts: A Critical Appraisal," *Population and Development Review* 16 supplement (1990): 44–71.

15. UN, *World Population Prospects*, pp. 57–58.

16. For more details, see Nicholas Eberstadt, "Demographic Disaster: The Soviet Legacy," *National Interest*, no. 36 (Summer 1994): 53–57.

17. Data drawn from United Nations, *Demographic Yearbook 1991* (New York: United Nations, 1993), pp. 103–111.

18. Kuznets, *Modern Economic Growth*, p. 38.

19. Unlike birthrate (the number of births per thousand persons in the general population), the total fertility rate is unaffected by the age-sex structure of a population. Two populations could have identical TFRs but very different birthrates if, for example, older persons accounted for a large share of the overall population in one but not the other.

20. Among these places, according to the United Nations: Angola, Benin, Burkina Faso, Cameroon, Central African Republic, Chad, Congo, Cote d'Ivoire, Equatorial Guinea, Gabon, Guinea-Bissau, Liberia, Malawi, Mozambique, Niger, Sierra Leone, Uganda, Zaire, and the offshore islands of Comoros. UN, *World Population Prospects*, pp. 122, 126.

21. For speculations along the last line, see Allen C. Kelley and Charles E. Nobbe, "Kenya at the Demographic Turning Point? Hypotheses and a Proposed Research Agenda," *World Bank Discussion Papers*, no. 107 (1990), and Charles F. Westoff, "Age at Marriage, Age at First Birth, and Fertility in Africa," *World Bank Technical Paper Series*, no. 169 (1992).

22. UN, *World Population Prospects*, pp. 122–124.

23. Ibid., pp. 123–130.

24. Two notable exceptions are Kingsley Davis, "Population Policy: Will Current Programs Succeed?" *Science*, November 10, 1967, pp. 730–739, and Charles F. Westoff, "Is the KAP-Gap Real?" *Population and Development Review* 14, no. 2 (1988): 225–232.

25. Lant H. Pritchett and Lawrence H. Summers, "Desired Fertility and the Impact of Population Policies," *World Bank Policy Research Working Papers*, no. 1273 (March 1994).

26. John C. Caldwell. "Toward a Restatement of Demographic Transition Theory," *Population and Development Review* 2, nos. 2, 3 (1976): 321–366; Theodore W. Schultz, "The High Value

of Human Time: Population Equilibrium," *Journal of Political Economy* 82, no. 2, pt. II (1973): S2–S10.

27. Charles Tilly, "The Historical Study of Vital Processes," in Charles Tilly, ed., *Historical Studies of Changing Fertility* (Princeton, N.J.: Princeton University Press, 1978), p. 3.

28. Samuel Preston, *Mortality Patterns in National Populations* (New York: Academic Press, 1976), p. 3.

29. Alternative possibilities for stable and stationary population structures are computed in Ansley J. Coale and Paul Demeny with Barbara Vaughan, *Regional Model Life Tables and Stable Populations*, 2d ed. (New York: Academic Press, 1983). In the various "models" outlined, for example, zero population growth requires an average of six to seven live births per woman over the childbearing ages when female life expectancy at birth averages 20 years.

30. See, for example, Odile Frank, "Infertility in Sub-Saharan Africa: Estimates and Implications," *Population and Development Review* 9, no. 1 (1983): 137–144.

31. UN, *World Population Prospects: The 1992 Revision* (New York: United Nations Department of Economic and Social Development, 1993), pp. 192, 202. Though these estimates are not the UN's latest, they are used because its groupings delineating Eastern Europe and the Soviet Union conform with the boundaries of the former Soviet Bloc. The UN's 1994 revisions estimate trends for a new "Eastern Europe"; this includes the territories of the European members of the Warsaw Pact, minus East Germany, plus Belarus, Moldova, Russia, and Ukraine. For this grouping as a whole, life expectancy at birth for 1990–1995 is projected to be slightly lower than it had been in 1960–1965—thirty years earlier. UN, *World Population Prospects, 1994 Revisions*, pp. 172, 192.

32. For more details, see Nicholas Eberstadt, "Mortality and the Fate of the Communist States," *Communist Economies and Economic Transformation* 5, no. 4 (1993): 499–520.

33. See, for example, Dan Usher, *The Measurement of Economic Growth* (New York: Columbia University Press, 1980), chap. 5, and Sherwin Rosen, "The Value of Changes in Life Expectancy," *Journal of Risk and Uncertainty* 1, no. 3 (1988): 285–304.

34. Rati Ram and Theodore W. Schultz, "Life, Health, Savings and Productivity," *Economic Development and Cultural Change* 27, no. 3 (1979): 399–421.

35. According to UN projections, life expectancy at birth for the world as a whole would be just over 64 for the 1990–1995 quinquennium; for the United States in 1950, life expectancy at birth was just over 68. *World Population Prospects: 1994 Revision*, p. 180; U.S. Bureau of the Census, *Statistical Abstract of the United States 1968* (Washington, D.C.: Government Printing Office, 1968), p. 54.

36. A major lacuna involves China, whose caloric trends are not included in the FAO and predecessor series for the period from 1934/38 to the early 1960s. Figure 1-6, and the discussion preceding this note, presume that per capita caloric supplies declined over the period in question in China. If China had experienced a sharper decline than estimated in figure 1-6, per capita increases for the developing regions since the 1930s would be correspondingly lower— perhaps only on the order of 30 to 35 percent. If China' s decline is overestimated, the de-

veloping countries' per capita increases over the period could be correspondingly greater—say, 40 to 45 percent.

37. See D. Gale Johnson, *World Agriculture in Disarray*, 2d ed. (New York: St. Martin's Press, 1991).

38. Enzo R. Grilli and Maw Cheng Yang, "Primary Commodity Prices, Manufactured Goods Prices, and the Terms of Trade: What the Long Run Shows," *World Bank Economic Review* 2, no. 1 (1988): 1–47.

39. Alexander Gerschenkron, *Economic Backwardness in Historical Perspective* (Cambridge, Mass.: Harvard University Press, 1962), chap. 1.

40. Kuznets, *Modern Economic Growth*, chap. 3.

41. Edward F. Denison, *Trends in American Economic Growth, 1929–1982* (Washington, D.C.: Brookings Institution, 1985), pp. 11, 16. Denison does not refer to "total factor productivity" in his discussion; instead he estimates the sources of growth attributable to various factors. The estimate cited is the "residual" after increases in labor and capital stock are taken into account.

42. Angus Maddison, *The World Economy in the Twentieth Century* (Paris: OECD, 1989), p. 20.

43. Angus Maddison, "Explaining the Economic Performance of Nations, 1820–1989," in William J. Baumol, Richard R. Nelson, and Edward N. Wolff, eds., *Convergence of Productivity: Cross-National Studies and Historical Evidence* (New York: Oxford University Press, 1994), p. 48.

44. Grilli and Maw, *Primary Commodity Prices*.

45. Famines, of course, are not the only terrible events of the modern era to exact loss of life on a massive scale. Warfare in the twentieth century has cost tens of millions of lives and ethnic or racial hatreds many millions more. Although the proposition that population pressure provides the impetus for war and mass extermination campaigns can be encountered—sometimes in the letters section of daily newspapers—it need not be discussed here.

46. See, for example, Basil Ashton et al., "Famine in China, 1958–61," *Population and Development Review* 10, no. 1 (1984): 613–645.

47. See, for example, Nicky Hart, "Famine, Maternal Nutrition and Infant Mortality: A Re-Examination of the Dutch Hunger Winter," *Population Studies* 47, no. 1 (1993): 27–46.

48. Asmerom Kidane, "Demographic Consequences of the 1984–1985 Ethiopian Famine," *Demography* 26, no. 3 (1989): 515–522. The "Sahelian famine" of the early 1970s received considerable coverage in the popular press and in population-environment circles; subsequent examinations, however, have concluded that the drought in West Africa did not result in any significant increases in local mortality levels. See, for example, John C. Caldwell, "Desertification: Demographic Evidence," *Australian National University Occasional Paper*, no. 37 (1984).

49. Robert Conquest, *The Harvest of Sorrow: Soviet Collectivization and the Terror-Famine* (New York: Oxford University Press, 1986).

50. For more information, see Amartya K. Sen, *Poverty and Famines: An Essay on Entitlement and Deprivation* (New York: Oxford University Press, 1981).

51. For an analysis of Mao's policies and their results, see Nicholas R. Lardy, "The Chinese Economy under Stress, 1958–1965," in Roderick MacFahrquhar and John K. Fairbank, eds., *The Cambridge History of China* (New York: Cambridge University Press, 1987), 14: 360–397.

52. For background on this tragedy, see François Jean, *Ethiopie: Du bon usage de la famine* (Paris: Medicins sans frontières, 1986), and Edward W. Desmond, "Mengistu's Ethiopia: Death by Policy," *Freedom at Issue* (March–April 1986): 18–22.

53. For background, see Roland Vaubel, "The Moral Hazard of IMF Lending," *World Economy* 6, no. 3 (1983): 291–303.

54. For an exposition of the contrasts, see Jeffrey D. Sachs, "External Debt and Macroeconomic Performance in Latin America and East Asia," *Brookings Papers on Economic Activity*, no. 2 (1985), pp. 523–564.

55. For two circumspect official critiques of sub-Saharan Africa's pervasive policy problems, see World Bank, *Sub-Saharan Africa: From Crisis to Self-Sustaining Growth* (Washington, D.C.: World Bank, 1989), and *Adjustment in Africa: Reforms, Results, and the Road Ahead* (Washington, D.C.: World Bank, 1994). The latter report argues that differences in economic performance among the sub-Saharan states can be explained by variations in their official policies and business climates, although by implication these particulars are far from optimal today even in the "best" sub-Saharan economies.

Chapter 2

SAVING THE PLANET WITH PESTICIDES
Increasing Food Supplies While Preserving the Earth's Biodiversity

Dennis Avery

HIGHLIGHTS

- *Food is more abundant and cheaper today than at any other time before in history. Per capita grain supplies have increased 24 percent since 1950, while food prices have plummeted by 57 percent since 1980.*

- *Food production has outpaced population growth since the 1960s. The increase in food production in poor countries has been more than double the population growth rate in recent years.*

- *The global trouble spots for soil erosion and habitat destruction—the Amazon rain forest, parts of Africa, and other scattered locations—are experiencing problems because local agriculture has not taken advantage of high-yield technologies and efficiencies. In the case of the rain forests, problems have been compounded by government subsidies for farming and logging.*

- *Global increases in average grain yields have lagged recently because affluent countries are trying to limit their surpluses and because of chaotic conditions in the former Soviet Union. By contrast, grain yields in developing countries rose 32 percent between 1980 and 1992.*

- *Farmers are not close to reaching the point of diminishing returns in crop yields. In fact, the general trend for most crops is sustained growth, often at high and even accelerating rates of gain.*

- *Modern high-yield farming is responsible for preserving a great deal of the world's biodiversity. If crop yields had remained at levels typical of the 1950s, farmers would have had to plow down an additional 10 million square miles of wildlife habitat to raise enough food for the current world food supply.*

- *For 10,000 years soil erosion has been a major threat to humanity's sustainable existence. Traditional low-yield farming is responsible for the highest rates of soil erosion. Two modern soil-saving techniques, conservation tillage and no-till systems, cut erosion by 65 percent and 98 percent, respectively.*

- *The UN's Food and Agriculture Organization estimates that despite progress, the number of people with inadequate diets increased in the past ten years, from 540 million to 580 million. Nonetheless, it should be relatively easy to boost the vast majority over the threshold of hunger, since 90 percent of those with inadequate diets are now within 10 percent of having enough calories for good health.*

- *The risk to humans and wildlife from agrochemicals is far less than the risk of losing wildlife to low yields and the consequent plowing down of more habitat.*

Famine is a thing of the past for most of the world's people. Never before in history has food been as abundant and as cheap as it is today. Although millions remain inadequately nourished, the good news is that advances in agriculture will eliminate the remaining pockets of hunger early in the next century. This, the view of growing numbers of agricultural researchers, stands in stark contrast to the popu-

> **Famine is a thing of the past for most of the world's people. Never before in history has food been as abundant and as cheap as it is today.**

lar impression that the world's burgeoning population threatens to overwhelm humanity's ability to produce enough food to prevent massive starvation in the next century.

Alarmists point to the fact that the world's population is likely to double to more than 10 billion by 2050. Lester Brown, head of the Worldwatch Institute, warns, "The world's farmers can no longer be counted on to feed the projected additions to our numbers. Achieving a humane balance between food and people now depends more on family planners than on farmers."[1] But most researchers disagree. For example, Dr. Paul Waggoner calculates in a recent report that 10 billion people could be fed a minimal but sufficient diet right now if the farming resources already in production were used at full efficiency.[2]

The surge in world population is not the only worry of conventional environmentalists. More of the world's poor people are demanding better diets, especially more high-quality protein such as that found in meat, eggs, and milk. Without such protein, children do not grow to their full genetic stature, and many adults lack full energy. But it takes more farming resources to produce a calorie of meat, milk, or eggs than to produce a calorie of cereals—two to five times as many resources per high-quality calorie.

China's meat consumption is currently rising by 3 million tons (and 10 percent) per year. Chinese meat consumption has traditionally been very low, but rising per capita incomes are putting more meat within the reach of many more consumers. Indians do not eat much meat because of their predominant Hindu religion, but the demand for dairy products is rising by 2 million tons per year. Indonesia is Islamic, so its residents do not eat pork, and it has no extensive grazing lands for beef, but poultry consumption is rising at double-digit rates. All told, Asia's diet upgrading is the biggest surge in farm resource demand the world has ever seen.

Asia is currently consuming only about 14 grams of animal protein per capita per day, far less than the 71 grams eaten in the United States and the 55 grams eaten in Japan. Virtually all of Asia will eventually demand as much high-quality protein as Japan eats today. And each ton of added protein demand will require three to five added tons of grain and oilseed production to provide feed to produce the additional animal protein.

In short, to meet world demand for better diets, agricultural output must triple by the middle of the next century. Can humanity meet this challenge? Agricultural researchers are increasingly confident that we can. Success depends on continued research in plant breeding, fertilization, pest control, and other high-yield agricultural systems, and free trade in farm products so we can use the world's best and safest land to meet food needs with fewer acres and less soil erosion. Research and trade should enable farmers to feed 8 billion to 10 billion people comfortably from the land and water already used in agriculture. In fact, high yields obtained with conservation tillage and no-till systems and precision farming should make food production more sustainable than ever before in history.

Humanity's Gains against Hunger

In 1970, the Nobel Peace Prize was awarded to Norman Borlaug, the plant breeder who developed the famous Green Revolution wheat varieties. The Green Revolution's "miracle" wheat and rice varieties are one of the outstanding achievements in human history; in a matter of a few decades, these new varieties increased cereal production dramatically.

More recently, some environmentalists, including biologist Paul Ehrlich and Lester Brown, have asked whether it was a good idea to save 500 million people from famine. Might not these additional millions put more strain on the earth's ecosystems, crowd out some of the world's disappearing wildlife, destroy irreplaceable natural resources, and be victims of an even greater famine? Yet despite these concerns, the Green Revolution in farming is continuing.

Let's look at what happened in the first phase of the Green Revolution in the 1960s and 1970s. In 1950, the world produced 692 million tons of grain, the key foodstuff for a population then at 2.5 billion people.[3] The director-general of the United Nations's Food and Agriculture Organization (FAO) estimated that two-thirds of the world's population lacked adequate food. The world had just witnessed a major famine in Bengal (1943), and another, in China, was near. But by 1992, the world produced 1,952 million tons of grain for 5.7 billion people, a 24 percent gain per person in grain supplies[4] and enormous progress in putting billions of poor people over the threshold from hunger to food sufficiency. Per capita food supplies in the world's poor countries have increased faster than in the First World.

TABLE 2-1   Worldwide Calories per Capita per Day

	1963	1992	Percentage Increase
World	2,287	2,697	+18
United States	3,067	3,642	+19
Third World	1,940	2,473	+27
Asia	1,888	2,494	+3
Latin America	2,363	2,690	+8
Africa	2,155	2,348	+2

Source: FAO Production Yearbooks (Rome: FAO, various years).

Per capita calories have soared in the years since the Green Revolution (table 2-1). In 1950, Americans were eating 3,200 calories per day. The average resident of China was getting about 2,100 calories per day. In India, the average was estimated at 1,700. In Indonesia, the caloric average was 1,750, and virtually the entire population lived in abject poverty. Bolivians were eating 1,760 calories daily. No one was even looking at the calories available in sub-Saharan Africa.[5] Per capita calories in the Third World have increased 27 percent since 1963, and probably a full one-third since 1950 (table 2-1).[6] In addition, according to the World Resources Institute, the global average price of food has plummeted by an astonishing 57 percent just since 1980 (figure 2-1).[7] Consequently, the vast majority of people living in poor countries are getting adequate calories most of the time.

Despite these enormous improvements, many people still have inadequate diets. In fact, the FAO estimates a slight increase in the numbers of the developing world's underfed, from around 540 million in 1979/81 to about 580 million in 1989/90.[8] But in the World Bank and FAO hunger numbers, roughly 90 percent of the "hungry" are within 10 percent of having adequate calories for good health.[9] Also the world's population grew by 23 percent during those years—nearly 1 billion people—so the increase in per capita food supplies is astonishing.[10] On a percentage basis at least, the world's underfed population is declining.

Today, severe hunger probably stalks no more than 5 percent of the world's population, except when Africa is having one of its periodic major droughts. Fortunately, better seeds and high-yield farming systems are reaching into such remote regions as Ethiopia, Ecuador, and Mongolia.

FIGURE 2-1 Total Food, World Commodity Price Index (1979–1981 = 100)

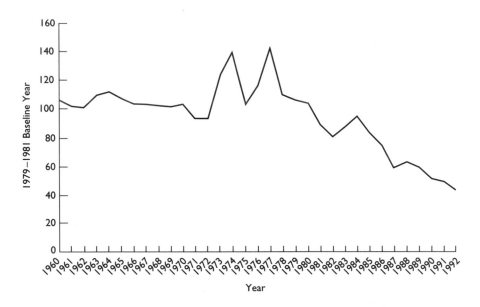

Sources: World Resources 1988–89 (New York: Basic Books, World Resources Institute [WRI] and International Institute for Environmental Development, with UN Environment Programme [UNEP], 1988), table 14.3; *World Resources 1992–1993* (WRI with UNEP and UN *Development Programme* [UNDP], Oxford University Press, 1992); *World Resources 1994–95* (WRI with UNEP and UNDP, Oxford University Press, 1994). Data for 1990–1992 indexed by Competitive Enterprise Institute.

Causes of Recent Famines

Most of the 5 percent of the world's people who are at greatest risk of starvation lack food because of a war or the policies of their own governments. That was true in China under Mao and more recently in Somalia, Ethiopia, Liberia, Angola, and most of Africa's other recent famines, which tend to be smaller than those of the past. In short, as economist Amartya Sen has pointed out, our belief that famine is caused by drought is wrong.[11] Rather, famines come about when political systems fail to encourage agriculture and distribution successfully. And those political failures have a pattern: they occur in centralized, authoritarian systems. Free-market economics do not produce famines.

The fact that modern famines are relatively small does not make them unimportant, but we must not let our anguish over suffering blind us to the fact that humanity has found successful strategies to stop large-scale famines. Famine caused 20 to 25 million deaths in the last quarter of the nineteenth century. For today's larger population, a comparable number of famine deaths for the current 1975–2000 quarter of the century would be about 50 million people, yet the famine death toll for 1975–2000 is likely to be 2 million or fewer.[12]

Recent per Capita Food Gains

Now that total fertility rates are tapering off, progress against malnutrition is picking up speed. The numbers in table 2-2 clearly show that the rate of increase in food production in poor countries has been more than double their average population growth rate. The World Bank projects that the average population growth rate in developing countries over the next decade will decline from 1.6 percent. Meanwhile, there has been no slackening in the rate of food production gain, contrary to reports issued by the Worldwatch Institute and other environmental groups. Clearly the world's food production (figure 2-2) is not being overwhelmed by population growth. Per capita food supplies have continued to grow as well (figure 2-3), except in Africa, the only at-risk region (table 2-3). But note that sub-Saharan Africa contains only 7 percent of the world population. Asia, where gains against hunger have been concentrated, represents the vast majority of the developing countries' population and 59 percent of the total world population. Fortunately it appears that Africa will soon have its turn at increased food production (box 2-1).

Prices and Food Abundance

One of the key measures of scarcity is always price, and the real prices of food have been coming down rapidly as modern agriculture has boosted crop productivity. For example, the U.S. Department of Agriculture (USDA) reports that season average wheat prices in constant dollars dropped from $256 per ton in 1950 to as little as $90 per ton in the 1990s. The only anomaly during the

TABLE 2-2   Third World Food versus Population Growth

	Food Production	Population Growth	Rate of Gain Differential
1960–1965	+2.4%	+2.6%	−9%
1970–1975	+3.0	+2.4	+25
1987–1992	+4.4	+1.9	+131

Sources: Ibid.; Index of Total Food Production, All Developing Countries, in *FAO Production Yearbook, 1992* (Rome: FAO, 1993), pp. 43–44.

FIGURE 2-2   World Index of Food Production, Total

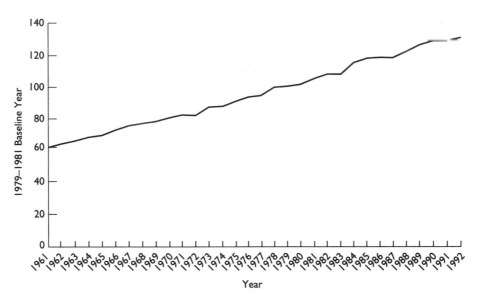

Source: World Resources Database, World Resources Institute/DSC Dataservices Inc., Washington, D.C. 1994.

FIGURE 2-3   World Index of Food Production, Per Capita

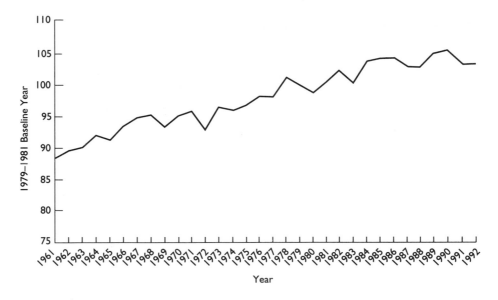

Source: World Resources Database, World Resources Institute/DSC Dataservices Inc., Washington, D.C. 1994.

TABLE 2-3   Index of Food Production (1979–1981 = 100)

	Total Food Production						Total Food Production per Capita					
	1978	1981	1984	1987	1990	1992	1978	1981	1984	1987	1990	1992
All developing countries	99	104	116	125	141	147	99	102	105	108	114	114
Africa	103	102	106	118	132	134	103	99	94	96	98	93
Latin America	100	104	109	117	127	132	100	102	100	101	103	103
Asia	99	104	119	128	145	152	99	102	110	112	120	121

Source: FAO Production Yearbooks, various years.

forty years was a three-year surge in all commodity prices triggered by the OPEC oil cartel in the mid-1970s.

Nor is wheat an exception. The season average prices for corn, which tend to be 20 percent lower per ton than wheat, paralleled the fall in wheat prices.[13] The real price of rice dropped by 50 percent between 1966 and 1991.[14] The decline in the real prices of inputs such as fertilizer was a factor in the drop in global food prices. The real price of ammonia (nitrogen) fertilizer in constant dollars fell from $1.75 per nitrogen unit in 1965 to $1.50 in 1985.[15]

The Secret of the Green Revolution

The secret of the Green Revolution has been to make two blades of grass grow where one grew before. It took 611 million hectares of cropland to produce 692 million tons of grain in 1950 because the world's average grain yield was only 1.1 tons per hectare.[16] In 1992, it took only 700 million hectares to produce 1,920 million tons, more than twice as much grain, because the yields had much more than doubled to 2.8 tons per hectare (table 2-4).[17] The key to these high yields is science. Plant breeding, chemical fertilizers, sophisticated pest control, artificial insemination for animals, and a whole host of high-yield food production technologies have been developed over the past sixty years.

Plant breeders like Norman Borlaug have turned out a profusion of new plant varieties that start earlier, grow faster, resist pests, tolerate drought stress, and preserve more of the harvest. Many have shorter stalks so more of their energy goes into producing grain, and the sturdier stems support bigger seed heads. Hybridization offers faster gains in yield potential and greater ease in maintaining

BOX 2-1   Africa's Turn Is Coming

It was no accident that Asia got the first and biggest gains from the Green Revolution. In the 1950s, hundreds of millions of people were at immediate risk of famine, so Asian farming was specifically targeted by agricultural researchers.

Africa today is even less well off than the African calorie averages would suggest. Africa's well-fed populations are concentrated in Egypt (2,568 calories per capita per day) and South Africa (3,133 calories). Meanwhile caloric averages in such countries as Somalia (1,874) and Mozambique (1,805) are clearly on the borderline of nutritional risk. Fortunately, Africa is only a latecomer to the world's food production success, not the pattern for any future failure of the world food system.

When Green Revolution research began, Africa was a relatively small and sparsely populated continent by comparison with Asia and Latin America. By the time that the revolution seeds and farming systems began to bear fruit, Africa also had a weak set of newly independent governments. Few of them put high priority on agricultural research. Most put price ceilings on foodstuffs instead, to placate the urban populations who could surround the presidential palaces at any moment. Most of Africa's governments also adopted socialist policies. Few offered farmers incentives for higher output. Africa's many civil wars also drastically disrupted already shaky farming and food sectors. Food has been openly used as a weapon in Africa. In the 1980s, during the Ethiopian civil war, the Mengistu government tried to prevent U.S. food aid from reaching rebellious provinces during droughts. More recently, the Islamic government of northern Sudan has blocked food aid to the Christian and animist rebels in southern Sudan.

Today, however, even Africa is on the way to more adequate food supplies from high-yield agriculture. In much of central Africa, the average corn yield is about 0.8 ton per hectare, and the potential is 7 tons.[1] New farming systems such as tied ridges and alley cropping promise much higher yields with only modest use of commercial fertilizer. A new sorghum hybrid for Sudan will yield more in the midst of a drought than the local varieties do now in a good year. In good years, it doubles and triples the current yields. Plantings of the new cereal have been severely hampered by Sudan's ongoing civil war.[2]

One plant breeder working for just six years at the international research center in Nigeria produced cassava varieties resistant to several endemic pests.

These new varieties produce three to five times as much food per hectare as past varieties did.[3] In addition, biologists have beaten back attacks on Africa's cassava crop by the green spider mite and the cassava mealy bug. The pests are immigrants from Latin America, so the researchers imported several predator insects from their home territories. It is the biggest success for biological pest control in history.[4]

Plant breeders have developed banana and plantain varieties that resist the black Sigatoka disease, which has been spreading from the Fiji Islands and could have devastated one of Africa's key food crops.[5]

Many of Africa's governments are just beginning to shift their farm policies from socialist regulation to market incentives. Recently several African governments achieved astonishing results by liberalizing their agricultural policies. In Ghana corn production has tripled, in Togo cotton production doubled, and Zambia saw a 20 percent increase in agricultural output. "If all black African leaders were to . . . lift price controls to permit their peasants to sell their produce in open, free markets, there would be no food crisis in the continent," says Ghanaian economist George Ayittey.[6] Better economic policies and better crop varieties should help ease Africa's food deficit in the future.

[1] U.S. Department of Agriculture, Foreign Agricultural Service, *Ivory Coast Grain and Feed Annual*, report from the U.S. agricultural attaché, Ivory Coast (1991).

[2] Gebisa Ejeta, "Development and Spread of Hageen Dura I, the First Commercial Sorghum in Sudan," *Journal of Applied Agricultural Research* 3 (1988).

[3] International Institute of Tropical Research, *IITA Strategic Plan 1989–2000* (Ibadan, Nigeria, 1988), pp. 59–61.

[4] International Institute of Tropical Agriculture, *IITA Annual Report 1989/90* (Ibadan, Nigeria, 1990), pp. 17–19. See also Yaniek, Onzo and Ojo, "Continent-Wide Releases of Neotropical Phytoseiids against Exotic Cassava Green Mite in Africa," *IIT Research*, no. 8 (March 1994): 14–17.

[5] Black Sigatoka Resistant Plantains on Trial," *International Institute of Tropical Agriculture Annual Report 1992* (Ibadan, Nigeria, 1992), pp. 10–12.

[6] Cited by David Osterfeld, "Hope for the World Food Situation," *Journal of Economic Growth* (Winter 1989–1990): 42–43.

Soybean yields have been increasing, and double cropping is now possible.

the vigor of seed breeding lines. The new crop plants permit higher yields and more double-cropping, and they protect the land more effectively than the old "land-race," or farmer-bred, varieties.

Third World grain yields rose 32 percent just between 1980 and 1992, from 1.9 tons per hectare to 2.5 tons. Meanwhile, affluent countries are struggling with surpluses and deliberately trying to limit food production gains. Their low rate of increase, plus the troubles in the agriculture of the former Soviet Union, kept the world average of grain production increase down to less than 1 percent per year between 1986 and 1991. This is a statistic often mistakenly quoted by environmental doomsayers as evidence for impending global famine.[18]

Genetic research has also paid off in livestock, poultry, and fish. For instance, the amount of feed needed to produce a pound of chicken meat has been cut in half by improved breeding. Artificial insemination has been increasing the yield of milk per dairy cow by about 2 percent annually since the 1940s. Today's good American herds may produce over 15 tons of milk per cow annually instead of the 5.4 tons, which was considered good in 1950.

Donald Plucknett, a former senior science adviser to the Consultative Group on International Agricultural Research, which is the network of research cen-

TABLE 2-4   Gains in Key Crop Yields, 1950–1990 (metric tons per hectare)

	1950	1990
U.S. corn	2.7	6.8
South African corn	1.0	2.7
Argentine corn	1.6	2.7
Chilean corn	1.6	8.4
Chinese rice	2.5	5.7
Indonesian rice	1.7	4.4
French wheat	2.2	6.7
Mexican wheat	1.1	4.2
Indian wheat	.7	2.3
Indonesian cassava	7.8	12.4
Chilean tomatoes	13.4	36.0
Canadian rapeseed	.9	1.32
U.S. soybeans	1.4	2.3
Australian cotton	1.8	3.7

Source: Ibid.

ters that nurtured the Green Revolution, sees no slowing of yield increases on the horizon (figure 2-4). He notes:

> A fascinating aspect of yield analysis is that many yield levels do not slow down as yields rise. Indeed, the data indicate that as yield takeoff proceeds above the 2,000 kilograms per hectare (kg/ha) level, it takes less time to attain each new 1,000 kg/ha of yield. . . . For countries that today attain national average yields of 6,000 to 7,000 or even 8,000 kg/ha, average annual yield gains are often very high.
>
> Most countries are well below their potentially-attainable, practical yields. . . . The high rates of yield gain being achieved by some of the highest-yielding countries indicate there is room yet for further yield improvement. . . . Yield performance . . . does not indicate significant environmental degradation that might affect yields. The general trend for most crops and countries is sustained growth, often at high and even accelerating rates of gain. The data present no environmental warning signals.[19]

FIGURE 2-4   World Production of Cereals

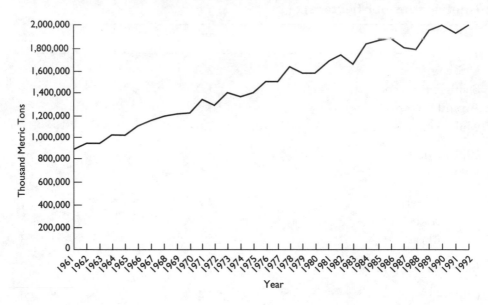

Source: World Resources Database, World Resources Institute/DSC Dataservices Inc., Washington, D.C. 1994.

His point is that the critical moment for any country's agriculture is when its farmers stop trying to push their crops out onto frontiers or marginal land and decide to raise more per acre. That becomes the point of yield take-off. For example, farmers can raise beans at any level of intensity. They can scatter a few seeds on a weed-covered hillside and come back weeks later to see if they have any beans. Or they can grow the best disease-resistant varieties climbing up poles on prime soils kept free of weeds. The second approach—the intensive one— yields one hundred times more beans per acre. Many farmers in the developing countries are still farming at low intensity. Much of Africa is getting about 0.8 ton of corn per hectare, when hybrid seeds and fertilizer would yield 7 tons.

No country that has ever achieved yield take-off has been stopped more than temporarily in its quest for higher yields. Even the first pioneers and the highest-yielding farms are still raising their yields.

A group of Dutch scientists at Wageningen University in the 1970s tried to estimate "the absolute maximum food production . . . the upper limit" on crop yields. Their conclusion was about 11 tons per hectare or 400 bushels of grain per acre. Nobody has yet reached 400 bushels per acre, but when they do, the odds are overwhelming that the potential will have moved to 500 or 700.

The success of high-yield farming has been so broad and rapid that its continuing important role in keeping food production ahead of population growth seems unquestioned. Yet critics abound. Some object to the use of farm chemicals, though billions of dollars in research have failed to find the negative impacts they fear for human health, soils, or wildlife. Others claim that the high-yield technologies favor big farmers and make the poor worse off. It is hard, of course, to argue that famine and hunger make anyone better off, but the income distribution question is also important. Time has shown, however, that both big and small farmers benefit from high-yield farming systems; small farmers simply let big farmers take the risk of trying new technologies first. History also suggests that the number of farms in any country tends to fall (and their size increases) as the value of an off-farm job increases. In most cases, it is hoped, workers are being attracted away from the farm rather than forced off. But modern economies do not have 50 percent of their labor forces on farms, nor will they ever again.

A potentially more serious claim is that humanity is running out of technology—that the research results that worked in the past have been virtually used up, and few new ones will be available to prevent global famine and population crash. Lester Brown claimed in 1974 that farmers could not count on agricultural research because it had not produced higher-yielding soybeans or cows that would give birth to twin calves.[20] Ironically, research has since done both, along with raising the world's grain yields from 1.1 to 2.8 tons per hectare.[21] Now Brown is once again asserting that "the backlog of unused agricultural technology is shrinking in industrial and developing countries alike, slowing the rise in cropland productivity."[22]

The question nevertheless remains a good one: Are we running out of farming technology? The evidence says clearly that we are not. Past investments in agricultural research have paid off handsomely, permitting humanity to get more food and better diets at a declining real cost. Researchers continue to achieve major gains from cross-breeding, chemical fertilizers, and the other established research approaches. More nitrogen still combines with higher-yield plant genes to produce more yield, even if the field is already producing 150 bushels of grain per acre.

Corn

American farmers more or less launched high-yield farming when they began planting hybrid corn in the 1930s, in the first broad, practical success for plant breeding. At the time, U.S. corn yields averaged about 25 bushels per acre. Hybrids helped them gain a very rapid 4 percent annually in the 1930s. Still, that added up to only one-half bushel of yield gain per year.[23] In the 1990s, U.S. corn yields are gaining less than 1 percent per year, but the national yield average is now above 138 bushels per acre, and the yield gains are mounting at 2 bushels per acre annually. As this book goes to press, in fact, it appears that America set new records in 1994 for both corn and soybeans. And there is no end in sight:

- Corn yield contest winners are getting over 300 bushels of corn per acre.[24]

- Because new corn varieties are more drought tolerant, 40 percent more yield can be saved in dry years.[25]

- New high-protein corn has 90 percent of the food value of nonfat dry milk and 2 percent more feed value for livestock.[26]

- Genetically engineering natural pesticide into corn may raise yields another 15 percent in the United States by controlling the corn borer. This innovation could help raise Third World corn yields even more sharply because poor farmers often cannot afford enough pesticides.

- Eventually biotechnology will probably create corn with the nitrogen-fixing properties of legume crops, so it will provide much of its own fertilizer.

Rice

The International Rice Research Institute (IRRI) has recently released a new type of rice plant with 25 percent higher yield potential. It puts 10 percent more of its energy into the grain head, supported by fewer but larger stalks. Next, IRRI plans to hybridize the new rice through two separate systems to open the potential for the same kind of continuing-yield gains that hybrid corn keeps demonstrating. Third, the new rice will be genetically engineered to resist the tungro virus, the white-backed planthopper, the yellow stem borer, drought, and salt. The result in 2030, says IRRI, should be rice varieties with 50 to 75 percent higher yields than the world's farmers get today.[27]

Cross-Breeding

The new International Center for Living Aquatic Resources Management (ICLARM) in the Philippines has produced a "super-tilapia" for fish farming.

Work continues year-round in the experimental rice fields of the International Rice Research Institute at Los Baños in the Philippines. The IRRI has developed many of the varieties of high-yielding "miracle" rice that have fueled the Green Revolution.

The tilapias are a prized group of freshwater fish from Africa that can thrive in intensive fish ponds or a flooded rice paddy. As we reach the limits of hunting and gathering for wild fish stocks, species like tilapia are the major future source for expanding fish harvests.

The tilapia was in trouble in the 1980s. Poorly conducted breeding programs and even "introgressive hybridization" with subpar species had robbed the fish of much of its productivity. No one had set up the kind of carefully designed and controlled breeding programs that had produced high-yield results in plants, birds and mammals. Then in 1988, ICLARM established its tilapia breeding program. It went back to the African sources for breeding stock, put them through quarantine, and started its tilapia renewal. In on-farm trials, ICLARM's new synthetic breed of tilapia was soon growing 60 percent faster than existing farm stocks, with a 50 percent higher survival rate.

The Philippines began national distribution of the "super-tilapia" in 1993. China, India, Bangladesh, the Ivory Coast, Egypt, Ghana, Indonesia, Malawi, Thailand, and Vietnam are all collaborating with the tilapia research, hoping to generate major gains in fish production and efficiency.[28]

Underused Resources

The world has large tracts of underused cropland. The United States, for example, has been diverting about 60 million acres of reasonably good cropland in most nonwar years since 1933. It is not America's best land, but USDA and other experts say it is about 80 percent as productive as the average American cropland. Since this land is already cleared, and most of it has periodically been used for crops, it can be brought back into production without loss of significant wildlife habitat. Sixty million acres times 80 percent of the average U.S. cereal yield (2.2 tons per acre) would produce about 105 million tons of additional grain per year.

Argentina too has a great deal of good, underused cropland. Due to high taxes and pervasive trade barriers overseas, Argentine farmers have been minimizing their out-of-pocket costs rather than farming for high yields They have been pasturing cattle on about 75 million acres of the Pampas, cropland as fine as any other in the world. They have been planting old-style low-yield flint corn without fertilizer or pest protection.

If Argentina achieved U.S.-average yields on its Pampas cattle pasture, that would produce another 165 million tons of grain per year. We could also expect Argentina to match U.S. average yields on its other grain lands, adding another ton per hectare of production on roughly 25 million acres. That adds up to roughly 190 million additional tons of Argentine grain.

If the U.S. and Argentine land were all planted with corn—perhaps to supply feed grain demand in China and India—then the yields would be higher yet. The United States currently averages 2.8 tons per acre, potentially adding nearly 350 million additional tons of corn per year.

Brazil has more than 100 million acres of unplanted arable land in its Cerrados plateau, most of it idle because the soil is too acid for most crop plants. This has been tropic wasteland, covered with stunted brush and coarse grass. But now Brazil's researchers have produced acid-tolerant corn, which can yield 3.4 tons per acre. The Cerrados can add hundreds of millions of tons of additional grain per year. If acid-tolerant corn grows well in Brazil, it should grow equally well in Zaire (in southern Africa), which has acid savannas comparable in size and agroclimate to those of Brazil.[29]

The Promise of Biotechnology

Biotechnology is already making important breakthroughs in medicine; soon, it is likely to make an enormous contribution to raising crop and forest yields, essentially by helping farmers to "manufacture" more land. The many criticisms of biotechnology applied to animals echo in concerns for plants, but it is important to realize that biotechnology is not a radical departure from the plant and animal genetics research scientists have been doing. For years, agricultural researchers have cross-bred and selectively bred all kinds of crops.

Biotechnology speeds the pace of this research. It allows researchers to target their breeding work much more directly than old-fashioned cross-breeding. Researchers also gain access to far more of the world's genes through cross-breeding crop varieties that were previously impossible and through interspecies crosses. Does this mean that researchers are "playing God," and irreversibly altering the "natural" world to their whims? Not necessarily. Biotechnological advances often use genes from wild species that have evolved successfully through countless adaptations to the ever-changing world. Thus, biotechnology makes the world's wild genes more valuable. Until now, the odds were heavily against most of them being useful in research and technology. Biotechnology will help create additional, valuable biodiversity. It may recreate the American elm and chestnut trees, driven virtually to extinction by natural diseases. It may create powerful new bacteria that could remove the heavy metals that contaminate urban sewage sludge and limit its utility in agriculture.

> **Does this mean that researchers are "playing God," and irreversibly altering the "natural" world to their whims?**

Already biotechnology has produced:

- Rice plants that resist the tungro virus and thus will produce another 7 million tons of rice per year.[30]

- Wheat plants with the strongest resistance yet to the pervasive rust diseases, one of the worst pests that attack wheat crops all over the world. The rust resistance comes from a wild cousin of the wheat plant and would be useless without biotechnology because it is too distant a relative to cross-breed.[31]

- A genetically engineered copy of the natural pork growth hormone that produces hogs with half as much body fat, which means more healthful pork, and raised with one-fourth less feed grain. Think of pork growth hormone as the equivalent of producing millions of extra tons of feed corn per year from laboratory bacteria instead of ploughed-down wildlife habitat.[32]

- Cloned and tissue-cultured Georgia yellow pine, planted in Brazil, which can produce sixteen times as much pulpwood per hectare per year as a Swedish natural forest. Each acre of the high-yielding trees can protect fifteen wild acres from being logged, let alone cleared.[33]

Removing Constraints, Not Diminishing Returns

Environmentalist Lester Brown regularly predicts that farmers are about to arrive at the point of diminishing returns in agricultural productivity. For example, he recently claimed that extra fertilizer is not yielding more grain.[34] In contrast Paul Waggoner, former director of the Connecticut Agricultural Experiment Station, sees no diminishing returns. His research indicates that farmers are in fact engaged in the serial elimination of constraints. As they deal with each constraint in turn, their yields keep going up.[35]

Weeds

Farmers traditionally have dealt with weeds by fallowing land (letting the weeds come up and killing them mechanically), plowing (burying them too far down to germinate), and most recently by applying chemical weed killers. Each of these solutions has successively raised productivity per acre. (If farmers still had to leave room for mechanical cultivators, they would be back to 10,000 corn plants per acre instead of today's 18,000 to 20,000, and that would be reflected in lower yields).

Water

Where water is scarce, farmers have terraced land, irrigated, and raised crops (like sorghum and millet) that need less rainfall. More recently researchers have developed irrigation systems using center-pivot sprinklers and drip tubes that raise water efficiency from perhaps 30 percent (flood irrigation on unleveled Third World fields) to 85 to 95 percent. These high-efficiency irrigation systems can be used on more and more of the world's irrigated land in the future as water increases in value. They both boost the efficiency of scarce water supplies and can help virtually to eliminate waterlogging and salt buildup in irrigated soils.[36]

Insects and Diseases

Insect and disease control will be increasingly important as both populations and incomes rise in the tropics. Insects continue to be voracious and adaptable predators of crops and livestock all over the globe, and especially in warm climates, where winters are too mild to kill back their numbers.

The Green Revolution might not have succeeded without modern pest control strategies. Farmers started with DDT and have now progressed to narrow-toxicity, low-volume, rapidly degrading pesticides and Integrated Pest Management. In the future, plant breeders and biotechnicians hope to build pest control into the crops. Recently Mississippi State University announced a pest-resistant soybean for the southern United States that yields 45 bushels per acre, compared to 21 bushels for current varieties.[37] Biotechnology is already inserting the toxin gene from *Bacillus thuringiensis*, which acts as a natural pesticide in both corn and cotton. In the near future, biotechnology should offer much more potential for built-in pest control than standard breeding strategies.

Plant Nutrients

Perhaps the key constraint to increasing yield is plant nutrients. Farmers deal with this constraint effectively by supplying a complete menu of nitrogen, phosphate, potash, and trace minerals. Whole regions (North Africa, Turkey) have soils short of phosphate, boron, zinc, or some other key menu item, and yields can be boosted sharply by eliminating that constraint. In addition, intensive cropping causes even the best fields' supplies of soil nutrients to fall behind the rate of natural replenishment. Farmers must either accept lower yields or add more nutrients, in the form of green manure crops such as clover and alfalfa, animal manure, or chemical fertilizer.

The organic food movement leaders believe that organic sources of plant nutrients are far superior to chemical sources. Nitrogen, however, is a basic element. Neither science nor plant tests have found any chemical distinction between the nitrogen from clover and the nitrogen from a factory. Moreover, the world cannot realistically expect organic farming to grow the same amount of food produced by modern agrochemical farming, let alone tripling production for the future. The reason is that crop yields from organic farms are typically half or less of those of high-yield mainstream farmers.[38] Organic farms get both lower yields and a lower cropping intensity. They keep more land in green manure crops because nitrogen is a key constraint on their production. The United States has only about one-third of the organic nitrogen needed to support current crop output.[39] The rest of the world may not have even 20 percent of the organic nitrogen needed. Organic farms also suffer heavier pest losses. A recent review of organic farming in seven European countries indicates that giving up pesticides would by itself cut the yields for wheat by 35 percent, for potatoes by 40 percent, for sunflower by 30 percent, and for carrots by 40 percent. Other crop yields would be cut comparably.[40]

Supporters of organic farming are fond of saying that Americans are wasting lots of organic fertilizer potential, and surely it must seem that way. The United States has millions of tons of biomass that could, at a cost, be turned into fertilizer and compost. But the quantities needed to replace chemical fertilizer are enormous. American farms currently use about 2.4 million tons of "pure" nitrogen per year. Sewage sludge contains nitrogen but only at low levels. If all of America's sewage sludge could be turned into fertilizer, it would make up for only 2 percent of the chemical nitrogen fertilizer we currently use on crops.[41]

The only practical way to get huge new supplies of organic nitrogen would be to plow much more land and grow much more clover, alfalfa, and other legume crops. Until and unless supporters of organic farming can clearly demonstrate the advantages of organic nitrogen and compost, giving up chemical fertilizer would cost the world dearly in lost food supplies.

Besides feeding humanity's burgeoning numbers, modern high-yield agriculture has another great advantage: it preserves the earth's biodiversity. Some of the world's leading naturalists and biologists are worried about the possibility of major losses in global biodiversity over the next fifty years. Professional science journals are filled with their anticipatory despair. Paul Ehrlich of Stanford University and E. O. Wilson of Harvard University project that up to half the world's wildlife species—and they claim there may be as many as 100 million species— might go extinct because of population growth and affluence.[42] Such a loss of biodiversity would mean a real loss in quality of life for the people on the planet. More important, it could mean the loss of creatures that play key roles in many vital ecosystems. The lost species would take with them millions of irreplaceable and potentially valuable genes, just as we are acquiring the knowledge to use such genes in resolving critical problems of human and environmental health.

The main threat to the world's wildlife is the destruction of habitat. Continued dependence on low-yield farming in the developing nations would mean the plowing of additional acres of wildlife habitat to grow food for their increasing populations. A more populous world that also wants room for wildlife has no room for low-yield farming.

Hungry people will almost certainly find ways to feed themselves and their children. The question is whether they will be fed from a few high-yielding acres that leave space for wildlife. If poor people in developing countries are not given the opportunity to adopt high-yield farming, they will feed themselves by extending traditional, organic, low-yield farming—and crowd wildlife out.

> **The plain fact is that the Third World cannot hire enough game wardens to protect wildlife if people are hungry.**

The plain fact is that the Third World cannot hire enough game wardens to protect wildlife if people are hungry.

High yields let farmers concentrate food production on the best and safest cropland, producing the same amount of food with fewer acres and cutting down both wildlife habitat loss and soil erosion. Biologists have also discovered that there is less biodiversity on good than on poor land. On the good land, it is apparently easy for a few species to dominate—as bison, antelope, wolves, and prairie dogs dominated the mammalian life of the Great Plains. In such difficult environments as the Peruvian rain forests, biologists have found dozens of new species in every square mile. Thus farming the best land is less of a threat to biodiversity.[43]

The naturalists who are concerned about losing wildlife species are not worried that farm chemicals will make them extinct but that the expanded population will mean too much wildlife habitat will be lost. It is estimated that the cities of the future will take only 3.5 percent of the earth's land area. If the cities treat their sewage and invest in clean energy, they offer little threat to wildlife, but low-yield agriculture does.

Saving Wildlife Habitat

The world is currently cropping or farming about 5.8 million square miles of land, roughly a land area the size of South America (figure 2-5).[44] Humanity might already be farming three times that much land if farmers were still getting the low yields typical of the 1950s, before most of the world began using fertilizer and pesticides. That would have meant plowing down the land equiv-

FIGURE 2-5   Cereals: World Area Harvested

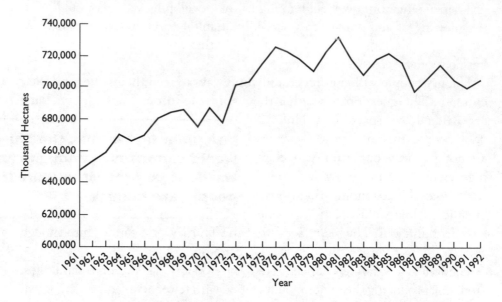

Source: World Resources Database, World Resources Institute/DSC Dataservices Inc., Washington, D.C. 1994.

alent of North America in addition to South America. Thus, in a real sense, high-yield farming is saving 10 million square miles of wildlife habitat right now. Without it, we would already have lost a whole continent's worth of wildlife. There is no chance that wildlife will be saved if people are going hungry. Consider the plight of elephants and gorillas in Africa, where these species are rapidly losing their habitats. Local people are dealing with rising human population by extending low-yield farming.[45] As areas are converted to farmland to feed the hungry people, elephants and gorillas lose out.

High-yield agriculture also deserves environmental credit for the big tracts of steep, rocky, drought-prone, and sandy soils that have been put back into grass and trees in the United States since farmers began boosting crop yields. The steep slopes of New England, West Virginia, and my own Shenandoah Valley in Virginia no longer court erosion with row crops. Big tracts of low-quality land in the Carolinas and gulf states are planted with fast-growing pines for timber and pulpwood rather than erosion- and drought-prone corn. In 1880, there were 68 million acres of farmland in the Northeast. Today it has dropped by two-thirds, to 23 million acres. Fewer than 10 million acres are urbanized or used for transportation. The result has been a dramatic expansion of wildlife habitat in the region since the turn of the century.[46]

America has lost only four of its bird species since the landing of the Pilgrims, all before the introduction of DDT.[47] One pesticide (Furadan 15G) was recently banned in the state of Virginia. It was supposed to be incorporated in the soil, but granules sometimes remained on the surface, where some birds mistook them for seeds. The chemical caused hundreds of bird deaths officially—thousands according to the Audubon Society—before the ban. The manufacturer withdrew the compound from the entire U.S. market after the Virginia ban, but without alternative pesticides, this solution would *worsen* the problem. Virginia has 44,000 square miles and millions of birds. Several hundred bird deaths per year is unfortunate, but how many more birds would Virginia lose if it had to accept a 50 percent cut in crop yields—and make up for it by plowing down more bird habitat? A 50 percent cut in crop yields would mean plowing down 1.5 million extra acres.

Conquering Soil Erosion

Yields on the best land are often twice as high as yields on poorer land. That means half as much soil erosion from the start. Additionally, steeper land can be left in grass and trees, making soil erosion gains still bigger. Less erosion means that more crop productivity can be preserved for the future, and less freshwater and marine life is threatened downstream by silt and nutrient runoff.

Some observers concede that high-yield farming has achieved enormous gains to date but claim that they have been achieved by "mining" soil and groundwater. They contend that world farm production, far from continuing to rise, is poised to collapse.[48] There are, in fact, farmers in parts of the world who permit too much soil erosion, mine their groundwater, and steal too much crop biomass from their fields for the long-term good of their soils. However, the most serious soil erosion, the most serious groundwater and irrigation problems, and the most serious "soil mining" are occurring in the areas being farmed with traditional methods, not the areas using modern high-yield farming:

- Traditional farmers are clearing the upper slopes of Nepal's highlands to grow corn—and causing serious flooding in Bangladesh downstream. Nepal could double the crop output of its Churai lowlands with a better irrigation system, and in fact, it has some of the world's best undeveloped dam sites for doing just that. Then there would be no need to clear fragile mountain slopes.

- Africa's per capita food supplies have been declining, but Africa has always fed itself on low-cost slash-and-burn farming. Now Africa's population is burgeoning, and it must turn to higher-yielding farming systems or lose its forests and wildlife cultivation. African farmers will have to find ways to buy fertilizer and to produce plant nutrients with mulch cropping and alley cropping. (Mulch cropping grows nitrogen-fixing plants as a ground cover between the rows of crop plants; they provide extra nitrogen for the crop, suppress competing weeds, and reduce erosion. Alley cropping grows crops between rows of leguminous trees. The trees reach deeper into the soil for moisture and nutrients, and help shade the young seedlings from the harsh tropic sun. Later, the trees can be pruned, and their leaves and branches used as a mulch to help nourish the crop.) Otherwise the farmers will strip and degrade most of the continent's wildlife habitat.

- Farmers have cleared too much of the tropical rain forests, although these can rarely be farmed successfully using traditional methods. Most heinous, the clearing has been abetted by government subsidies for farming and logging.

Saving Soil with Agrochemicals

Soil erosion is the real long-term threat to sustaining the food supplies for even the current level of human population. Contrary to what many conventional environmentalists want to believe, the fact is that traditional and organic farmers suffer the highest rates of soil erosion per ton of food output.

First, consider that doubling the crop yields on the best land cuts soil erosion per ton of food produced by more than half. Twice as much food comes from each acre of soil exposed to the elements. In addition, the higher yields allow farmers to focus crop production on the best and safest land. Thus, higher yields in themselves radically reduce soil erosion.

In addition to their yield advantage, modern chemically assisted farmers now have the soil-safest farming systems ever devised. Conservation tillage and no-till farming use herbicides to kill weeds instead of plowing and other bare-earth farming methods. They cut soil erosion by 50 to 98 percent on farms that were already getting high yields and thus already had relatively low erosion rates.

Since biblical times farmers have left half of their fields bare to wind and water for the whole year, so that the weed seeds would sprout and be killed by hand hoeing. As recently as the 1960s, even the best modern farmers plowed their fields in the fall and left them open to wind and water for the harsher half of the year. That was the only way to get the plowing done soon enough so the crops could be planted on time in the spring. In the 1970s, however, farmers were suddenly offered a much wider range of more effective weed killers. For the first time, they had an alternative to plowing for weed control.

The first of the soil-saving crop systems is called conservation tillage. Farmers using conservation tillage do not plow under the crop residue in their fields and leave a clean, bare soil surface, which allows too much erosion. Instead, conservation tillage mixes at least 30 percent of the crop residue (corn stalks, wheat stubble, etc.) into the top three or four inches of the soil surface. The residue makes little dams and half-buried points of resistance for wind and water. It slows runoff, and soil erosion with it.

Done properly, conservation tillage cuts soil erosion by about 65 percent.[49] The tillage is done with a disc plow that works only three to four inches deep instead of eight to twelve inches deep using a regular plow. And there is another advantage: disc plowing can be done far more rapidly and with far less tractor fuel.

No-till is the second stage of the tillage revolution. In no-till, the soil is rarely left open to erosion. The farmer starts with a sod cover over the field, perhaps

No-till agriculture in which the soil is never left bare to wind and water is spreading rapidly in the United States. No-till farming reduces soil erosion up to 98 percent.

planted at the end of the previous growing season. At planting time, the sod is killed with an herbicide, and the seeds are planted with a special planter that slices open the dead sod. Once the new food crop has a good start, the farmer plants another cover crop to provide winter protection after the harvest of the food crop. No-till can cut erosion by 98 percent.[50]

These two forms of tillage are expanding rapidly in both temperate and tropical countries.[51] Already, more than 100 million acres of farmland in the United States use one of these methods. The USDA estimates that by the year 2000, nearly 70 percent of all U.S. cropland will use these new types of tillage.[52]

Other Elements of Sustainability

Preserving Traditional Seeds and Animals

With the increasing reliance on high-yield hybrid crop varieties, there is real danger of losing the genetics packaged in the old seed varieties and breeds of livestock and poultry developed by centuries of on-farm selective breeding. These old gene packages are a problem because they cannot survive in the wild. They were developed on farms where they were protected from many kinds of wild competition. Yet some of their genes could be very valuable for future breeding programs. Thus, some governments, private individuals, and the international agricultural research networks are setting up gene banks to preserve the landrace seeds and their wild cousins. Tens of thousands of plant selections have been gathered and identified, and they are being maintained in growth-viable status.[53]

The old strains of livestock and poultry are more difficult to sustain because they cannot be stored over the long term in a climate-controlled drawer. Some preservationists and hobbyists have filled part of the void by raising antique breeds, but more of this live preservation effort is probably needed.

Maintaining Water Quality

High-yield agriculture has recently been portrayed as "America's biggest pollution problem." It is not. The average American water well contains only 5 to 7 parts per million (ppm) of nitrate (from legume plants, manure, and/or chemical fertilizer).[54] It takes at least 100 ppm to trigger the famous "blue baby" syndrome, a type of cyanosis that can occur in children under 1 year old which is the only health threat ever linked to nitrates in water. The United States has had only one blue-baby death in more than twenty years, caused by a huge fertilizer spill in the mid-1980s near a water well in North Dakota.

The fact is that the field application of pesticides is not contaminating drinking water. The only pesticide residues exceeding good health standards in recent EPA and state surveys have been from known pollution sites: accidental spills, old rinse sites, and pesticide distributorships dating back before the current requirements for impermeable dikes, covered handling facilities, and rinsate lagoons.

The most widely cited "risk chemical" in our water has been atrazine, a widely used corn herbicide. But atrazine has just had its safety rating (known as the "reference dose") raised sevenfold by the EPA.[55] The maker of atrazine is suing the EPA to force the agency to raise its water quality limits accordingly.

The effluent from intensive livestock and poultry operations is now handled under tight regulation and poses little, if any, threat to the water supplies or aquatic ecosystems. The "contamination" from livestock grazing beside streams is trivial and gone within a few feet downstream.

Overfertilization of lakes, rivers, and estuaries by nitrogen and phosphate is a problem. This overfertilization can be traced primarily to inadequate treatment of urban sewage and storm sewer runoff—multibillion-dollar problems for urban taxpayers that few officials want to address.

Agriculture does, however, play a role in overfertilizing inland surface waters. The biggest problem caused by agricultural runoff is the eutrophication of lakes and streams. But the combination of conservation tillage and precision farming should radically lower the modest current impact of fertilizer on most aquatic ecosystems. It will be much harder to deal

The fact is that the field application of pesticides is not contaminating drinking water.

with the nitrogen deposited by urban auto exhausts and washed into streams through storm sewers.

Using the Best and Safest Farmland

Traditionally, the nations of the world have wanted to be self-sufficient in food production. Garrett Hardin and some bioregionalists in the environmental movement have come to favor this notion as well.[56] Environmentally, this is a very bad idea because the yields on the best and safest croplands are two to three times higher than on marginal lands, cost less, suffer less soil erosion, and tend naturally to harbor less biodiversity.

The world's good soils are not well distributed for the twenty-first century. A high proportion of the best soils are found in a few places that are not heavily populated, such as America's Cornbelt and Great Plains, Argentina's rich Pam-

Researchers from the developing world are learning how to boost their countries' wheat yields by studying techniques developed at the International Maize and Wheat Improvement Center in Mexico.

pas, and the plains of France and Hungary. Additionally, some soils can be made much more productive through new technologies or capital investments. Brazil's Cerrados plateau, with highly acid soils, can be productive cropland because of the acid-tolerant crops developed by that country's researchers. Turkey is creating a replica of California's hugely productive Central Valley by building twenty-one dams to irrigate 4 million acres of semiarid land in the Upper Euphrates Valley. Most of the additional food demand, on the other hand, will come in Asia, already about six times as densely populated as the Western Hemisphere per acre of farmland and thought to be perhaps nine times as densely populated by 2050.

We can readily overcome the maldistribution of cropland by producing food where it grows best and moving it to consumers after harvest. Grain, meat, and produce can be shipped anywhere in the world where they are needed. If people in the developing countries are blocked from getting cheap food produced by high-yield farming, they will be forced to plow down their wildlands in order to feed themselves. Thus global free trade in food increases both food security and preserves the natural environment.

Agricultural research continues to raise food production and per acre productivity faster than population growth, and there is no end in sight. The widely predicted famines of the future are not likely to occur. High-yield farming has not only been our best defense against famine but is humanity's greatest environmental triumph. Where we need to turn in order to save biodiversity and prevent famine is toward the political and technological problems that stand in the way. If we could end all authoritarian regimes, we would come close to ending all chance of famine. If we could speed the adoption of high-yield technology in underdeveloped regions, we could stop clear-cutting of important habitats. These are the real challenges for food production in the late twentieth century.

NOTES

1. Lester Brown et al., *State of the World 1994: A Worldwatch Institute Report on Progress toward a Sustainable Society* (New York: W. W. Norton, 1994), p. 178.

2. Paul Waggoner, *How Much Land Can Ten Billion Spare for Nature?* Task Force Report 121 (Ames, Iowa: Council for Agricultural Science and Technology and the Rockefeller University, February 1994).

3. Food production statistics from *FAO Annual Production Yearbook Series* (Rome: FAO). Population statistics are from Francis Urban and Ray Nightingale, *World Population by Country and Region, 1950–1990 and Projections to 2050* (Washington, D.C.: U.S. Department of Agriculture, Economic Research Service, 1993).

4. *FAO Annual Production Yearbook Series*, 1970 and 1992.

5. *FAO Annual Production Yearbook* (Rome: FAO, 1969).

6. "Calories per Capita," in *FAO Annual Production Yearbook 1992* (Rome: FAO, 1992).

7. World Resources Institute, *World Resources 1994–95* (New York: Oxford University Press, 1994), p. 262.

8. UN Administrative Committee on Coordination, Subcommittee on Nutrition, and International Food Policy Research Institute, *Second Report on the World Nutrition Situation* (Geneva: World Health Organization, 1992).

9. Thomas Poleman, *Quantifying the Nutrition Situation in Developing Countries*, Cornell Food Research Institute Studies, vol. 18, no. 1 (1981). Also see S. Bhalla, "Measurement of Poverty Issues and Methods," unpublished preliminary paper for *World Development Report*, no. 3 (Washington, D.C.: World Bank, 1980).

10. Francis Urban and Ray Trueblood, *World Population by Country and Region, 1950–2050* (Washington, D.C.: U.S. Department of Agriculture, 1990).

11. Amartya Sen, *Poverty and Famines: An Essay on Entitlement and Deprivation* (New York: Oxford University Press, 1981).

12. D. Gale Johnson, *World Food Problems and Prospects*, Foreign Affairs Study, no. 20 (Washington, D.C.: American Enterprise Institute, 1975), p. 17.

13. U.S. Department of Agriculture, Economic Research Service, *Feed Situation and Outlook*, FDS-330 (Washington, D.C.: Government Printing Office, 1994).

14. *World Development Report 1991*, p. 74.

15. National Fertilizer Development Center, Tennessee Valley Authority, *Fertilizer Trends: Production, Consumption and Trade* (October 1986).

16. *FAO Production Yearbook 1970* (Rome: FAO, 1970).

17. *FAO Production Yearbook 1992*.

18. Brown et al., *State of the World 1994*, p. 184.

19. Donald Plucknett, "Science and Agricultural Transformation" (lecture to the International Food Policy Research Institute, Washington, D.C., September 9, 1993).

20. Lester Brown, *By Bread Alone* (New York: Praeger, 1974), p. 7.

21. "Total Cereals, Area, Yield and Production," *FAO Annual Production Yearbooks 1994*.

22. Brown et al., *State of the World 1994*, p. 177.

23. D. Gale Johnson and Robert L. Gustafson, *Grain Yields and the American Food Supply* (Chicago: University of Chicago Press, 1962).

24. Cited in *Tabulation of Winners of the 1992 Corn Yield Contest* (St. Louis, Mo.: National Corn Growers Association, 1993).

25. Boyce Rensberger, "Building a Better Ear," *Washington Post*, June 27, 1994, p. A3.

26. Board on Agriculture, National Research Council, *Quality Protein Maize* (Washington, D.C.: NRC, 1988).

27. International Rice Research Institute, *Rice Research in a Time of Change* (Manila: IRRI, 1993).

28. International Center for Living Aquatic Resources Management, *ICLARM Report 1993* (Manila: ICLARM, 1993), pp. 7–11.

29. Pedro Sanchez, executive director, TROPSOILS, Soil Science Department, North Carolina State University, personal communication, 1992.

30. Robert Herdt, "The Potential Role of Biotechnology in Solving Food Production and Environmental Problems in Developing Countries" (paper presented at the ASA-CSSA-SSSA Annual Meetings, Cincinnati, November, 1993).

31. "Wild Genes May Tame Leaf Rust," *Successful Farming* (August 1992).

32. R. D. Goodband et al., "The Effects of Procine Somatotropin and Dietary Lysine on Growth Performance and Carcass Characteristics of Finishing Swine," *Journal of Animal Science* 68 (1990): 3261–3276.

33. Roger Sedjo, "The Competition for Albertan Forest Products" (speech to the Conference on International Competitiveness of Canadian Forest Products, Center for Business Studies, University of Alberta, Edmonton, October 23, 1991).

34. Brown et al., *State of the World 1994*, p. 184.

35. Waggoner, "How Much Land Can Ten Billion Spare for Nature?"

36. Ibid.

37. U.S. Department of Agriculture, Agricultural Research Service, *Quarterly Report of Selected Research Projects* (Washington, D.C.: Government Printing Office, October–December 1993).

38. National Research Council, "Economic Evaluation of Alternative Farming Systems," in *Alternative Agriculture* (Washington, D.C.: National Academy Press, 1989). See also Landell Mills Research Group, *Organic Farming: Summary of Findings from a Study of Seven European Countries* (Brussels: European Crop Protection Association, 1992).

39. Donald L. Van Dyne and Conrad B. Gilterson, *Estimating U.S. Livestock and Poultry Manure and Nutrient Production*, U.S. Department of Agriculture, ESCS-12 (Washington, D.C.: Government Printing Office, 1978).

40. Landell Mills Research Group, *Organic Farming*.

41. Calculation by Steven Graef, director of technical services, Western Carolina Regional Sewage Authority, Greenville, S.C., based on "Estimated Mass of Sewage Sludge Disposed Annually," *Federal Register*, February 19, 1993, p. 9257.

42. Paul Ehrlich and E. O. Wilson. "Biodiversity Studies: Science and Policy," *Science*, August 16, 1991, pp. 758 761. See also E. O. Wilson, *The Diversity of Life* (Cambridge: Harvard University Press, 1992).

43. Michael Huston, "Biological Diversity, Soils, and Economics," *Science*, December 10, 1993, pp. 1676–1679.

44. John F. Richards, *The Earth as Transformed by Human Action* (Cambridge: Cambridge University Press, 1990).

45. "Cereals, Total Area; Roots and Tubers, Total Area; and Pulses, Total Area," in *FAO Production Yearbook 1992*.

46. Indur Goklany and Merritt Sprague, *An Alternative Approach to Sustainable Development: Conserving Forests, Habitat and Biological Diversity by Increasing the Efficiency and Productivity of Land Utilization* (Washington, D.C.: U.S. Department of the Interior, April 4, 1994), pp. 12–15.

47. Samuel Florman, "Progress for the Birds," *Technology Review* (July 1993): 63.

48. Brown et al., *State of the World 1994*.

49. Conservation Technology Information Center, National Association of Conservation Districts, West Lafayette, Indiana.

50. W. T. Flinchum, "Producing Soybeans under a High-Erodable Situation," in *Proceedings of the 1993 Southern Soybean Conference*, American Soybean Association, St. Louis, Missouri.

51. Peter Oram and Behjat Hojjati, "The Growth Potential of Existing Agricultural Technology" (paper presented at the Roundtable Meeting on Population and Food in the 21st Century, Washington, D.C., February 14–16, 1994), p. 7.

52. Ibid.

53. International Board for Plant Genetic Resources (IBPGR), *Partners in Conservation: Plant Genetic Resources and the CGIAR System* (Rome: Consultative Group on International Agricultural Research, 1990). See also Erich Hoyt, *Conserving the Wild Relatives of Crops* (Rome: IBPGR, 1988).

54. U.S. Environmental Protection Agency, *Another Look: National Survey of Pesticides in Drinking Water Wells, Phase 2 Report*, EPA 570-9-91-020 (Washington, D.C.: Government Printing Office, 1992).

55. J. M. Addiscott et al., *Farming, Fertilizers and the Nitrate Problem* (Wallingford, U.K.: CAB International, 1991).

56. Kirkpatrick Sale, "Bioregionalism—A New Way to Treat the Land," *Ecologist* 14 (1984): 167–173; Garrett Hardin, *Living within Limits: Ecology, Economics, and Population Taboos* (New York: Oxford University Press, 1992).

Chapter 3

GLOBAL WARMING:
Messy Models, Decent Data, and Pointless Policy

Robert C. Balling, Jr.

HIGHLIGHTS

- *The earth's atmosphere has actually cooled by 0.13 degrees Celsius since 1979 according to highly accurate satellite-based atmospheric temperature measurements. By contrast, computer climate models predicted that the globe should have warmed by an easily detectable 0.4°C over the last fifteen years.*

- *The scientific evidence argues against the existence of a greenhouse crisis, against the notion that realistic policies could achieve any meaningful climatic impact, and against the claim that we must act now if we are to reduce the greenhouse threat.*

- *Current computer climate models are incapable of coupling the oceans and atmosphere; misrepresent the role of sea ice, snow caps, localized storms, and biological systems; and fail to account accurately for the effects of clouds.*

- *Temperature records reveal that predictive models are off by a factor of two when applied retroactively in projecting the change in global temperature for this century.*

- *The amount of warming from 1881 to 1993 is 0.54°C. Nearly 70 percent of the warming of the entire time period—0.37°C—occurred in the first half of the record—before the period of the greatest buildup of greenhouse gases.*

- *Accuracy in land-based measurements of global temperatures is frustrated by the dearth of stations, frequent station relocations, and changes in how oceangoing ships make measurements.*

- *Although all of the greenhouse computer models predict that the greatest warming will occur in the Arctic region of the Northern Hemisphere, temperature records indicate that the Arctic has actually cooled by 0.88°C over the past fifty years.*

- *Corrective environmental policies would have a minuscule impact on the climate. According to its own projections, the Intergovernmental Panel on Climate Change's own plan would spare the earth only a few hundredths of a degree of warming by the middle of the next century.*

Many environmentalists, politicians, and reporters believe that global warming represents a significant threat to the future of humanity and the planet. They think that the continued buildup of greenhouse gases will cause the world to warm substantially, sea level will rise, ice caps will melt, droughts will become more frequent and more intense in the breadbaskets of the continents, severe storms will batter the coastlines with greater intensity, and wildfire activity will increase in forests. This apocalyptic vision of the greenhouse effect has been presented through a barrage of printed stories and editorials, scientific documentaries, and even several full-length films. The greenhouse advocates have been extremely effective in selling this message to the public, and as result, the global warming issue continues to remain high in the environmental conscientiousness.

Three underlying and fundamental messages are heard regularly:

1. The continued buildup of greenhouse gases will produce substantial warming, thereby generating a variety of undesirable outcomes.

2. Realistic policy options are available that can reduce or even eliminate the greenhouse threat.

3. We must act now. Waiting for action is too dangerous given the severity of the global warming crisis.

From a climatological perspective (as opposed to social, economic, or political perspectives), I believe that each of these three messages regarding the greenhouse effect is seriously flawed. The professional literature in climatology is full of articles that demonstrate the weaknesses in these three statements, and yet this literature often is overlooked in presenting the greenhouse catastrophe to the public and policymakers. Anyone seriously interested in the greenhouse issue must become aware of the debate surrounding global warming projections, and they must come to understand that credible scientists are producing results, and publishing these results in leading peer-reviewed professional journals, that do not support the highly popularized vision of a greenhouse disaster.

> **Credible scientists are producing results, and publishing these results in leading peer-reviewed professional journals, that do not support the highly popularized vision of a greenhouse disaster.**

Origins of the Greenhouse Scare

Nearly a century ago, Arrhenius published an article in the *Philosophical Magazine* presenting calculations showing that a doubling of the atmospheric concentration of carbon dioxide (CO_2) would lead to a planetary temperature rise of 6°C.[1] Arrhenius understood that certain gases, like CO_2, trap heat energy that otherwise would escape more quickly into space. (The analogy is made to a greenhouse; hence, gases with this radiative property have become known as greenhouse gases.) Many of these greenhouse gases, including water vapor, are naturally occurring and help to maintain the "normal" temperature of the earth. However, due to a variety of human activities that involve fossil fuel burning and deforestation, the atmospheric concentration of CO_2 has been increasing measurably since the beginning of the Industrial Revolution (figure 3-1). Over the past two centuries, atmospheric concentrations of CO_2 have increased by 25 percent (from approximately 285 ppm to 356 ppm), virtually all in the past 100 years.

During the same period of industrial development, many other gases with the potential to produce greenhouse warming have been added to the atmosphere. Methane (CH_4) concentrations were near 0.75 ppm in 1800; recent measurements show levels to be near 1.70 ppm, and the increase is related largely to various agricultural activities, most notably, rice paddy agriculture. Nitrous oxide (N_2O) is another naturally occurring greenhouse gas that has increased in atmospheric concentration due to deforestation, fossil fuel burning, and the use of some fertilizers. Atmospheric concentrations of N_2O have risen from about 285 parts per billion (ppb) for pre–Industrial Revolution levels to approximately 310 ppb in 1990.

Carbon dioxide, methane, and nitrous oxide occur naturally in the atmosphere. Unlike these greenhouse gases, the pre–Industrial Revolution atmos-

FIGURE 3-1   Atmospheric Carbon Dioxide Concentration

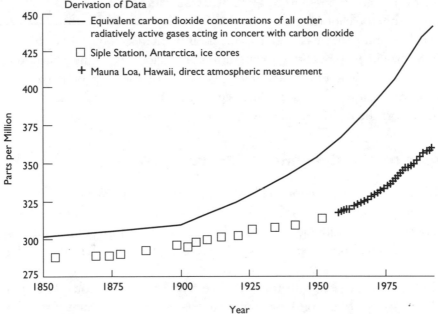

Note: For a description of the data, see R. C. Balling, Jr., *The Heated Debate: Greenhouse Predictions Versus Climate Reality* (San Francisco: Pacific Research Institute, 1992).

pheric concentrations of the chlorofluorocarbons (CFCs) were essentially zero. These CFCs are very powerful greenhouse gases, and despite having concentrations that are measured in parts per trillion, they add significantly to the overall greenhouse effect. They destroy some ozone in the stratosphere, and because ozone also operates as a greenhouse gas, the destruction of ozone by the CFCs may ultimately minimize the total greenhouse contribution of the CFC molecules.[2]

The overall radiative effects of these many greenhouse gases may be approximated by "equivalent carbon dioxide" values. The resultant value gives an indication of how much CO_2 would be required to produce the same greenhouse effect as other trace gases found in the atmosphere. Equivalent CO_2 levels were approximately 290 ppm at the beginning of the Industrial Revolution, 310 ppm in 1900, and nearly 440 ppm in 1994 (figure 3-1). Since the beginning of the Industrial Revolution, equivalent CO_2 has increased by 50 percent, and in the past 100 years, the increase has been fully 40 percent.

There has been a tendency to believe that the trends in atmospheric concentrations in these gases would continue without any great surprises into the next century. In the late 1980s and early 1990s, however, the growth rate in the concentration of many greenhouse gases fell well below expected levels. Some, such as methane and carbon monoxide, have leveled off or even declined.[3] Others, such as CO_2, have shown a substantial reduction in the rate of increase; the atmospheric concentration is still increasing, but the rate of increase is far below previous levels. These findings have surprised many climatologists, who are now groping for an explanation. An unusually long-lived El Niño/southern oscillation event, a reduction in biomass burning in tropical savannas, the eruption of Mount Pinatubo, enhanced growth in the biosphere, and even major repairs of large pipelines have all been suggested as possible causes of the trends in various greenhouse gases. Finally, although the rate of increase has slowed in the most recent few years, many scientists continue to expect the equivalent CO_2 value to double by the middle of the next century with respect to pre–Industrial Revolution values and pass the 600 ppm level.

Given the observed and projected rise in the atmospheric concentration of these greenhouse gases, many scientists have readdressed the calculations Arrhenius made a century ago. But unlike the limitations he faced, today's scientists have a battery of powerful tools for analyzing the sensitivity of climate to the buildup of greenhouse gases. The most powerful of these tools are enormous computer programs designed to simulate the physics principles that govern the earth-ocean-atmosphere system. These general circulation models have been independently designed throughout the world and have been used repeatedly to examine the warming that could be associated with a doubling of equivalent

CO_2. The popularized projections for massive warming, sea level rise, glacier melting, droughts, severe storms, and wildfires ultimately come from these numerical climate models. In essence, these theoretical, mathematical, simulation models are the basis for the catastrophic vision of global warming.

The Modern Crystal Balls?

Numerical models of climate represent an enormous achievement in atmospheric physics and applied mathematics. Many of the world's best numerical climate models have been developed at large centers employing full teams of Ph.D.s who have devoted their professional careers to the construction of such models. As a result of their efforts, we have more than a dozen very sophisticated, complex, and intricate models (that nevertheless are far from perfect representations of reality and referred to by many modelers as crude or primitive) that can be used to test the climate response to a doubling of carbon dioxide concentrations ($2 \times CO_2$).[4] The world's largest and fastest computers are pushed to the limit by these models; in many respects, the computing capabilities often restrict the modeling efforts.

In order to conduct a $2 \times CO_2$ simulation experiment, the value of atmospheric CO_2 in the numerical climate model changes from 300 ppm to 600 ppm. In the equilibrium experiments, the model is run for 300 ppm and then run for 600 ppm, and the differences in the simulated climate are determined. In more difficult but more realistic time-dependent experiments, the model's equivalent CO_2 levels are raised gradually and realistically from 300 ppm to the 600 ppm level, and the resultant climate is determined along the time continuum. These are somewhat different approaches to the problem, but in gross terms, they tend to produce similar results.

All climate models predict a global increase in annual average near-surface air temperatures as a doubling of equivalent CO_2 occurs. The range of projected global changes is from 1°C to 5°C, with a mean near 3.5°C. The tropics are predicted to warm the least and the high latitudes to warm the most; more warming is predicted in winter than in summer. The models also predict an increase in globally averaged precipitation due to an intensification of the hydrological cycle and the ability of the warmer atmosphere to hold more water. Irrespective of the changes in precipitation, the predicted warming should act to increase the rate at which water can be evaporated into the atmosphere. The rise in evaporation and transpiration is predicted to overwhelm any increases in precipitation, and soil moisture levels are expected to decrease over many continental areas. Climate predictions about wildfires, drought frequencies, or severe storm outbreaks are derived from the model results, but they are not direct outputs from these models.

Many people have argued that the best models in the world are basically all predicting something of a climate disaster for a doubling of the atmospheric concentration of greenhouse gases.[5] Given that these models represent the best tools for peering into the climate future, they argue that we should pay close attention to the warning. They very fairly asked how we can neglect this warning, or they asked how will we ever explain to future generations that we knew about the coming disaster and did nothing to stop it. If we place much faith in the models, then global warming must be viewed as a very real threat.

The same scientists who work on these models, however, are among the first to point out that the models are far from perfect representations of reality and probably are not advanced enough for direct use in policy implementation. Major weaknesses remain in the models; in particular, the role of the ocean in absorbing CO_2 and storing and transporting heat is not adequately included. Coupling the best ocean models to the best climate models is very tricky business computationally, but a necessary step in building more reliable forecast tools.

Clouds play a critical role in maintaining the energy balance of the earth, and cloud representations are particularly questionable in the models.[6] If the greenhouse enhancement acts to produce an increase in high clouds, the feedback could be positive and the warming accentuated. But if the greenhouse effect yields an increase in low clouds, the feedback could be negative, and the cloud increases could counter and even cancel the greenhouse warming.[7] These and many other complexities regarding the clouds are not adequately handled in the models, and relatively small adjustments in the representation of these feedbacks can have a profound impact on the calculated climate response to a $2\times CO_2$ condition.

Other widely acknowledged problems in the models are related to the representation and treatment of sea ice, snow cover, localized storms, and biological systems. We have so far totally neglected the fact that other nongreenhouse gases are being added to the atmosphere that can have strong local and regional cooling effects.[8] Sulfur dioxide emissions have increased at the global scale, and sulfur dioxide is known to have a cooling effect via the production of aerosols that reflect incoming sunlight, brighten clouds, and extend the lifetime of existing clouds. Simply doubling CO_2 in a model to represent future conditions is an interesting and valuable academic exercise, but in order to simulate the future realistically, changes in other gases, such as sulfur dioxide, must be considered in the modeling experiments.

So although the numerical models for testing the sensitivity of climate to various changes in atmospheric chemistry are powerful and complex, they are crude and deficient in many of the critical representations of the climate system. The models are getting better all the time, and advances in computing technology, along with advances in the atmospheric sciences, will produce the next generation of models. Therefore, I suspect climate model outputs of the immediate future will look quite different from today's results as this critical coupling of ocean, cloud, and other effects is achieved.

These existing models may be viewed as crystal balls for peering into the climatic future, and we are free to believe in as much or as little of their predictions as we like. I believe that as we learn more about the models, less confidence will be placed on their projections for the future. Many of the best modelers are ill at ease with any reference to their model outputs as forecasts or projections; they are far more comfortable using the phrase *sensitivity experiments* to describe their work. They understand that their climate models are not very good crystal balls. Nevertheless, many greenhouse proponents place remarkable confidence in the model outputs and demand policy changes based on their climatic crystal balls.

Lessons from the Recent Climatic Past

One obvious test of the models involves their ability to simulate accurately the observed climate changes that have occurred during the past century when equivalent CO_2 levels rose by 40 percent. Here we begin to see the role of the observational climate record in the overall greenhouse question. What is called the greenhouse "debate" often refers to the tension that exists between scientists who build and perfect the numerical models and the scientists who analyze the actual changes that have been observed in the climate system. Here theoretical calculations meet empirical evidence, and the debate is at its most heated.

The temperature records from around the world prior to approximately 1880 are especially incomplete and unable to pass selected homogeneity tests commonly used in climatology.[9] If we begin with the mean annual temperature of the earth in 1881 and complete the record to the near present, we see that the thermometer network of the world shows a linear increase in temperature of 0.54°C (figure 3-2), seemingly empirical support for the greenhouse projections: the models are all predicting warming for an increase in greenhouse gases, and the earth appears to have warmed during the period of historical records. However, there is a lot more to this story, and the connection between the warming and the increase in greenhouse gases may not be so straightforward.

FIGURE 3–2   Mean Annual Global Near-Surface Air Temperature Anomalies, 1881–1993

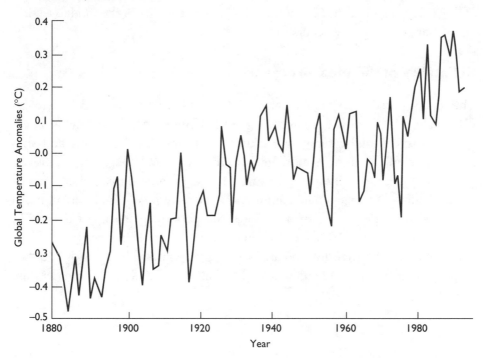

Source: P. D. Jones, T. M. L. Wigley, and P. B. Wright, "Global Temperature Variations Between 1861 and 1984," *Nature* 322 (1986): 430–434.

To begin, we can examine the timing of the warming over the past 113 years of record. While the amount of warming from 1881 to 1993 is 0.54°C, the warming during the first half of the record is 0.37°C. Nearly 70 percent of the warming of the entire time period occurred in the first half of the record; the bulk of the greenhouse gas buildup clearly occurred in the second half of the record. Much of the warming of the past century preceded the large increase in greenhouse gas concentrations.

Given the approximate 40 percent increase in atmospheric concentration of equivalent CO_2 over this 1881-to-1993 time period, the observed rise in temperature is low given numerical model predictions for a doubling of equivalent CO_2.[10] Given this buildup in equivalent CO_2, the models suggest that we should have observed at least 1.0°C of warming over the same period. Even if all of the 0.54°C can be ascribed to the effect of greenhouse gases, the models appear to be off by a factor of two. Noted climate modeler Stephen Schneider acknowledged this point and concluded that "the twofold discrepancy . . . is still fairly small."[11] However, alternative explanations for the observed warming, the timing of the warming, and/or the geography of the warming may increase the size of the discrepancy and cast even more doubt on the ability of the models to simulate changes in climate accurately.

Reliability of the Measurements

The global near-surface air temperature data presented in figure 3-2 were developed by the Climatic Research Unit at the University of East Anglia. These scientists screened the temperature data from land-based stations and ship records to eliminate as many nonclimatic errors as possible (e.g., station moves, instrument changes). Efforts were made to avoid large cities, where urban-induced climate changes could contaminate the record. As with the land-based records, extensive quality control analyses were performed on the ocean records. A 5-degree latitude by 10-degree longitude grid was established, and the station and ship data were then interpolated to the grid points. The scientists identified two significant problems with the reliability of this global temperature record.

Researchers prepare to drill ice cores in Antarctica. Ice cores help researchers learn more about past climates and the composition of the atmosphere.

First, station relocations produce changes in exposure, elevation, and topography that can change the recorded temperature and create a discontinuity in the record.[12] If the station move is well documented, some of the effects of the relocation can be statistically removed from the record. In addition to potential shifts in the station location, the time of observation may change from one observer to the next, and the temperature record is altered. Also, the instruments themselves, along with the recommended exposure to the sun, have changed through time.

Second, the marine air temperature measurements are prone to similar problems through time. Possibly most significant, ships are getting larger, and the thermometers used to measure the air temperature are getting higher above the ocean surface. This change in height, along with other onboard changes, make the marine air temperature records difficult to adjust to some baseline level.

All of these problems influence the long-term temperature record; despite every effort to remove or minimize their effect, the record remains contaminated by these uncertainties. Additionally, the geographic distribution of the stations and ship records creates yet another difficulty for the "global" temperature record. Some areas of the world are well sampled with the existing network; others are virtually unmeasured. Some of these problems in the record will tend to cancel out, but they absolutely increase the uncertainty in the 0.54°C trend of the past 113 years.

Contaminants to the Records

The problems already described cannot be overlooked in the search for any greenhouse signal, yet the potential impact on the temperature record caused by the urban heat island effect represents a major contaminant to many of the temperature records. Recognizing that cities tend to warm their local environments, a number of scientists have attempted to quantify the urban heat island effect in the historical land-based temperature records of the globe.[13] A variety of schemes have been used in these analyses, and from this research, it would appear that the global temperature data set has a global urban warming bias somewhere between 0.01°C and 0.10°C per century, with the most likely value near 0.05°C.[14]

The urban effect creates a localized warming signal that is not representative of the surrounding area. Recently it has been discovered that overgrazing and desertification may be producing a large-scale warming signal that is clearly not related to the greenhouse gases. The role of desertification in changing the regional temperature was strongly debated following a landmark article suggesting that overgrazing in arid and semiarid lands would increase the albedo (reflectivity) by removing the dark-colored vegetation.[15] The increased albedo would reflect more of the sun's energy, less solar energy would be absorbed by the surface, and surface and air temperatures would drop. Soon after the introduction of this hypothesis, others argued that removal of vegetation would reduce evapotranspiration rates; less solar energy would be consumed in evaporating and transpiring water, leaving more solar energy to warm the surface and the air.[16] Most empirical data[17] and recent theoretical findings[18] support the notion that overgrazing and desertification would act to warm, not cool, the surface and air temperatures.

Because the overgrazing, resultant desertification, and landscape degradation occur over decades, it is reasonable to expect a relative warming trend for the areas of the earth that have experienced substantial desertification. Warming signals have been found in the global temperature records that appear to be related to this nongreenhouse forcing; the desertification warming signal in the global temperature record, like the urban heat island effect, accounts for between 0.01°C and 0.10°C of the global warming trend of the past century.[19]

The Role of Volcanic Dust

In addition to the buildup of the greenhouse gases, many other phenomena could be causing the upward trend in the surface air temperature data. There remains the strong possibility that some of the trend shown in figure 3-2 could be explained by some external forcing of the climate system.

Many climatologists believe that volcanic dust in the stratosphere can act to cool the earth, particularly when the eruption emits large amounts of sulfur.[20] The resultant dust and sulfuric acid particles in the stratosphere may increase the reflection of incoming radiation and ultimately act to cool the planet. So the question remains, Is some of the trend in the global temperatures of the past century related to volcanic dust in the stratosphere?

To answer this question, we need a long-term record of dust measurements in the stratosphere, and fortunately, a data set has been produced that can be used in this regard.[21] The statistical relation between the stratospheric dust index and global temperatures is relatively easy to establish, and when the effect of stratospheric dust is removed from the global temperatures, fully 0.18°C of the total trend is eliminated. Nearly one-third of the global temperature trend of the past 113 years disappears when the stratospheric dust index is considered.

Effects of a Variable Sun

Obviously the total energy output of the sun could play a major role in governing the planetary temperature. For many years, some scientists have argued strongly in favor of this mechanism as a primary control of planetary temperature,[22] while others have rejected the idea that small variations in solar output can explain much of the trend of the past century.[23] Recently two researchers have found that the length of the solar sunspot cycle is related strongly to the fluctuations in temperatures on the earth.[24] Although the physical mechanism responsible for the linkage remains elusive, it is noteworthy that over 75 percent of the observed global warming in this century can be statistically explained by the variations in the length of the solar sunspot cycle.

Global Temperatures from Satellites

Satellite-based lower-tropospheric atmospheric temperature measurements are available for 2.5-degree latitude by 2.5-degree longitude grid cells on the monthly basis for the period 1979 to the present.[25] These temperature measurements are made by a passive microwave sensor system by the 53.74 gigahertz channel

Despite all the talk about global warming during the 1980s, the buildup of greenhouse gases during the 1979 to 1994 time period, and the anticipated 0.3° per decade warming, the highly accurate satellite-based global temperature measurements not only show no warming but show very real cooling.

that detects thermal emission of molecular oxygen in the middle and lower troposphere. The measurement is not particularly affected by changes in water vapor, cloud variations, or changes at the surface. In addition, the temperature changes occurring in the stratosphere do not significantly affect the microwave data.[26] When areally averaged for the world as a whole, the resultant global temperature is accurate to within plus or minus 0.01°C at the monthly time scale.

A plot of the satellite-based monthly temperatures from January 1979 to April 1994 is presented in figure 3-3. These data reveal a statistically significant cooling of 0.13° C over the 184-month period. Despite all the talk about global warming during the 1980s, the buildup of greenhouse gases during the

FIGURE 3–3 Satellite-Based Monthly Global Temperatures, January 1979–April 1994

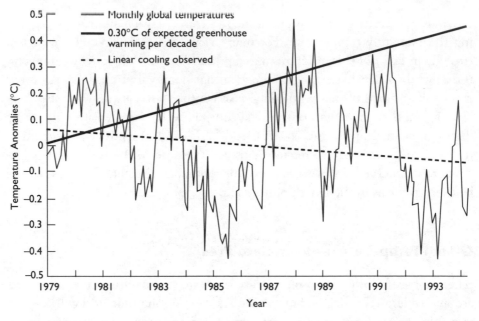

Source: R. W. Spencer, J. R. Christy, and N. C. Grody, "Global Atmospheric Temperature Monitoring with Satellite Microwave Measurements: Method and Results, 1979–1984," *Journal of Climate* 3 (1990): 1111–1128.

1979 to 1994 time period, and the anticipated 0.3° per decade warming, the highly accurate satellite-based global temperature measurements not only show no warming but show very real cooling. The eruption of Mount Pinatubo in June 1991 undoubtedly contributed to this cooling pattern; however, Christy and McNider controlled for such volcanic eruptions as well as El Niño/southern oscillation events, and they calculated a statistically insignificant warming of 0.09°C per decade using the satellite-based global temperature measurements.[27]

Geography of the Warming

It should be clear that the amount of warming at the global scale over the past decade or century is not particularly supportive of the apocalyptic global warming projections. Although the regional predictions made by the models are typically viewed with considerable skepticism, it is noteworthy that all of the models predict the greatest warming to occur in the Arctic region of the Northern Hemisphere.[28] Given this prediction of warming in the high latitudes, many sci-

FIGURE 3-4 **Arctic-Area Annual Near-Surface Air Temperature Anomalies, 1890–1993**

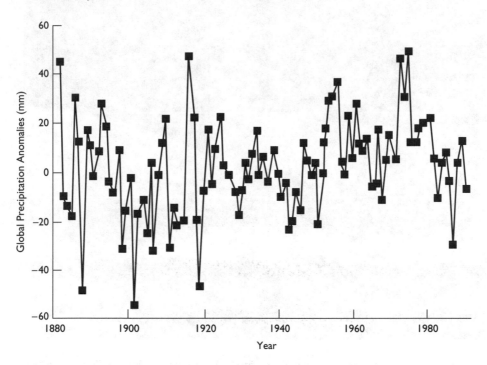

Sources: Jones, Wigley, and Wright, "Global Temperature Variations"; Spencer, Christy, and Grody, "Global Atmospheric Temperature Monitoring."

Note: Anomalies for the period 1979–1993 are satellite-based Arctic area measurements.

entists have attempted to determine the temperature trends in this part of the world.

Figure 3-4 presents updated Arctic-area annual near-surface air temperature anomalies over the past century. The amount of linear warming over the past century in the Arctic is 1.03°C, and at first glance, this trend appears to be supportive of the projections from the models. However, over the most recent fifty years of the record and during the time of the greatest greenhouse gas increase, Arctic temperatures have actually *cooled* by 0.88°C. Virtually all of the Arctic warming of this century occurred in and around the 1920s. Furthermore, the satellite-based temperature measurements show Arctic cooling of 0.21°C during the 1979–1993 period (figure 3-4). Additionally, others found no warming over the past forty years in near-surface air temperatures as measured from balloons launched daily throughout the Arctic.[29] The high latitudes of the Northern Hemisphere should

Temperature measurements by highly accurate instruments on this satellite show that the earth's average temperature has cooled by more than one-tenth of a degree Centigrade over the past fifteen years.

be warming quickly according to the $2 \times CO_2$ model projections, and yet the best regional data sets show no warming, or even cooling, in this part of the earth.

Day versus Night Warming

Detailed studies of the climate record have uncovered a particularly interesting and important pattern in the temperature data: the diurnal temperature range (the difference between the daily maximum and minimum temperatures) has declined significantly over the past half-century in many locations around the world.[30] The decline has been well documented in North America using a variety of data sets and analytical procedures.[31] Similar decreases in the diurnal temperature range have been identified in Europe, Australia, Asia, and Africa.[32] In contrast, no evidence is found for a decrease in New Zealand and surrounding islands,[33] and an increase in the diurnal temperature range has occurred in

India.[34] In order to explain these observed trends, investigators have proposed many interrelated mechanisms, including changes in cloud cover, precipitation, snow cover, atmospheric sulfate levels, and greenhouse gas concentrations.

It is noteworthy that a decrease in the diurnal temperature range has been reported from several $2\times CO_2$ numerical climate model experiments.[35] However, the Intergovernmental Panel on Climate Change (IPCC) concluded that "there is no compelling evidence for a general reduction in the amplitude of the diurnal cycle."[36] Several years later, IPCC further concluded that although a decrease in diurnal temperature range was observed over 25 percent of the global land area, "the reasons for this change, which is largely a result of an increase in minimum temperatures, are not yet clear."[37]

As noted by Michaels and Stooksbury, among others, the trend in the diurnal temperature range is critical in determining the severity of the greenhouse threat.[38] A lower diurnal temperature range would not allow daytime evaporation rates to climb, thereby generating the predicted increases in droughts; growing seasons would be longer; plants would experience less thermal stress; and polar melting would be reduced. In many respects, the decrease in the diurnal temperature range could be beneficial to a substantial portion of the global ecosystem. It should be clear that the timing of any temperature change (day versus night) is critical in assessing the impact of the change on other elements of the ecosystem.

Clouds and Precipitation

All of the discussion thus far has centered on temperature measurements compared to temperature predictions from the $2\times CO_2$ numerical experiments. And just as all of the models are predicting an increase in temperature for the buildup of greenhouse gases, they also predict increases in cloud cover and precipitation across the globe. Not surprisingly, scientists have been assembling data on these variables and examining trends over the period of historical records.

The global precipitation index (available from the World Meteorological Organization), which has been established for over 5,000 stations around the world, shows departures from the average based on a 1951–1970 "normal" period. The index reveals an upward trend of 16.55 millimeters over the period 1882 to 1990 (figure 3-5). In the broadest terms, this general increase in precipitation is consistent with predictions from the $2\times CO_2$ numerical climate experiments.

Given the observed increase in precipitation, we would expect an increase in cloudiness over the past century, and in fact, such an increase has been observed. Results by many scientists suggest that global cloudiness has increased between

FIGURE 3-5   Global Precipitation Anomalies, 1882–1990

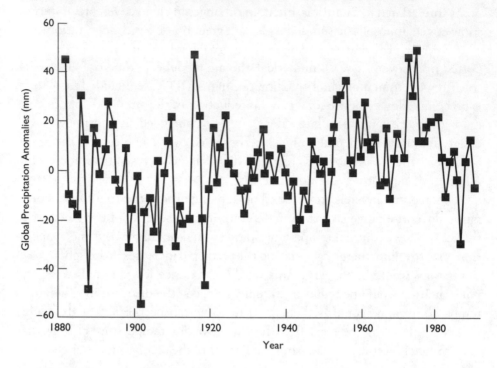

Source: The Global Climate System (Geneva: World Meteorological Organization, 1991), p. 25.

5 and 10 percent over the past century over land areas; data presented by others also show a total cloud cover increase over the oceans during the past fifty years.[39]

A Benign Greenhouse Effect?

What emerges from this discussion is a greenhouse effect of slightly higher temperatures, a reduction in the diurnal temperature range, and an increase in cloudiness and precipitation. This view of the greenhouse effect is consistent with the observational record of the past century and reasonably consistent with the model simulation studies, particularly when the climate effects of aerosol sulfates are included in the modeling experiments. It is not consistent with the popularized vision of a global warming catastrophe. Although we rarely hear about greenhouse benefits, it is clear that nighttime warming would lengthen growing seasons, and the lack of warming during the daytime would not force upward potential evaporation rates that could cause an increase in droughts.

The earth is becoming increasingly cloudy. Increased cloud cover may be offsetting any increases in temperature that computer climate models predict will occur as atmospheric carbon dioxide levels rise, owing to the burning of fossil fuels

More clouds and more rain should generally increase soil moisture levels and alleviate moisture stress to plants. No one would argue that all greenhouse effects are bound to be beneficial, but in an environment of thinking only of greenhouse costs, potential benefits must be examined.

Stopping the Crisis?

Global warming is almost always presented as an environmental crisis that can be stopped or minimized with appropriate policy actions. Policymakers can debate the impact and the cost-effectiveness of their policies forever, but from a straight climatological perspective, the evidence suggests that realistic policies are likely to have a minimal climatic impact.

Figure 3-6 was derived directly from the 1990 IPCC report. The uppermost line represents the IPCC "business-as-usual" trend in global temperature to the

FIGURE 3-6   IPCC Projected Global Warming

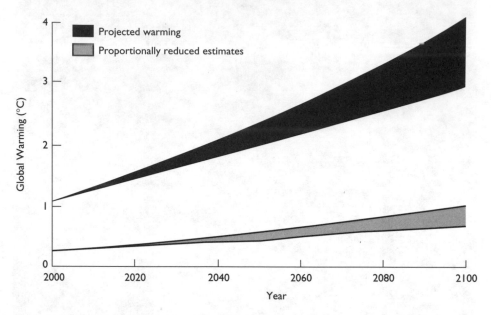

Source: J. T. Houghton, B. A. Callander, and S. K. Varney, eds., *Climate Change 1992: The Supplementary Report to the IPCC Scientific Assessment* (Cambridge: At the Press, 1992).

year 2100. According to that scenario, the earth will warm by approximately 4°C over the natural, background planetary temperature by the end of the next century. If that were to occur, many elements of the greenhouse disaster would become reality. If we adopt IPCC's "Scenario B"—moving to lower carbon-based fuels, achieving large efficiency increases, controlling carbon monoxide, reversing deforestation, and implementing the Montreal Protocol (dealing with chlorofluorocarbon controls) with full participation—the panel projects that the earth would warm according to the line at the bottom of the black area: still nearly 3°C of warming. Scenario B spares the earth very little warming (the black area); by the year 2050, the policies of this IPCC scenario have spared the earth only 0.3°C of warming. These policies do not stop global warming at all; indeed, they barely slow it.

Furthermore, the climatic impact of any policy is directly dependent on the amount of warming predicted over the next century. Figure 3-6 also shows the impact of the IPCC scenario assuming a business-as-usual 1°C temperature increase, which is much more consistent with the historical record. The Scenario B policies spare the planet less than 0.3°C by the year 2100, and by 2050, they would have spared the earth something near 0.07°C. As scientists lower their estimate of temperature rise for the next century, they also reduce the potential climate impact of any corrective policies. In a very recent and important study, scientists performed a numerical modeling study and concluded that it will take seventy to one hundred years to detect any climatic difference between the business-as-usual scenario and the most draconian measures proposed by the IPCC.[40]

Must We Act Now?

Despite all the material presented to this point, many scientists and policy-makers continue to argue that we must act immediately to avoid a greenhouse disaster; waiting, they say, is too dangerous given the threat. Several scientists, however, have evaluated the climate difference between acting immediately and waiting a decade or more to implement selected policies. Schlesinger and Jiang used a numerical model to simulate the impact of realistic policies hypothetically adopted in 1990, and they calculated the global temperature for the middle of the next century.[41] They then simulated the impact of waiting a decade to implement the same policies and found that the temperature of the earth by the middle of the next century was not affected by the delay. Their results, which support the unpopular view that we do not need to rush into policy regarding the greenhouse issue, obviously generated a tremendous debate in the scientific and policy arenas.

Nonetheless, many nations seem impatient with the progress made on meeting the many commitments signed at the 1990 Earth Summit in Rio de Janeiro. Accordingly, some nations, including the United States, may be pressured into adopting stronger measures to control greenhouse gas emissions. These measures may increase the costs to the nation and seem doomed to failure. Rather, the evidence argues that there is no need for urgent action, and the suggested measures will probably have only a trivial effect on the global greenhouse gas concentrations.

The Bottom Line

The greenhouse debate will be with us for many years to come. The climate models are getting better all the time, and their predictions for a doubling of atmospheric CO_2 will continue to capture professional and public interest. As the climate effects of other gases are included in the model simulations (e.g., sulfur dioxide), the projected temperature rise will likely be lowered, and the threat of global warming will weaken further. Equally important, climate databases on a variety of variables will continue to be expanded in both time and space. The state-of-the-art satellite-based planetary temperature measurements continue to show no warming at all. As the satellite database grows, I suspect global warming will be even more difficult to prove on empirical grounds. As models and climate databases improve, there is little doubt that the scientists will develop an even better understanding of how the climate system will respond to future changes in atmospheric chemistry.

It seems equally obvious that the greenhouse issue will continue to be a major force in the policy arena. Government agencies, from worldwide to regional organizations, have been developed to deal with the policy concerns raised by the greenhouse issue. Many of these government groups have expanded in recent years, and their very fate is tied to the perception that global warming represents a significant threat to the planet. Bureaucratic inertia virtually guarantees that the greenhouse question will remain high on the list of environmental policy priorities. Despite the evidence reviewed in this chapter, the upcoming versions of the various IPCC reports will continue to trumpet the threat of global warming. Press releases will probably emphasize the risks of the greenhouse effect while remaining silent about the potential benefits of the greenhouse world.

I have become increasingly concerned that a wide gap has opened between many of the policy-oriented groups and the science-oriented groups. Too often policymakers appear to neglect the enormous evidence that argues against the greenhouse disaster and freely accept and promote the scientific evidence in favor of the crisis. It is imperative that the policies developed for the global warming issue be built on the best science available, not on the extreme viewpoints that seem to satisfy and justify the policymakers. The scientific evidence argues against the existence of a greenhouse crisis, against the notion that realistic policies could achieve any meaningful climatic impact, and against the claim that we must act now if we are to reduce the greenhouse threat.

> **Too often policymakers appear to neglect the enormous evidence that argues against the greenhouse disaster and freely accept and promote the scientific evidence in favor of the crisis.**

NOTES

1. S. Arrhenius, "On the Influence of Carbonic Acid in the Air upon the Temperature of the Ground," *Philosophical Magazine* 41 (1896): 237–276.
2. J. T. Houghton, B. A. Callander, and S. K. Varney, eds., *Climate Change 1992: The Supplementary Report to the IPCC Scientific Assessment* (Cambridge: Cambridge University Press, 1992).

3. M. A. K. Khalil and R. A. Rasmussen, "Global Decrease in Atmospheric Carbon Monoxide Concentration," *Nature* 370 (1994): 639–641.

4. Houghton, Callander, and Varney, *Climate Change 1992*, pp. 97–134.

5. Ibid.

6. R. D. Cess et al., "Interpretation of Cloud-Climate Feedback as Produced by 14 Atmospheric General Circulation Models," *Science* 245 (1989): 513–516.

7. D. L. Hartmann and D. Doelling, "On the Net Radiative Effectiveness of Clouds," *Journal of Geophysical Research* 96 (1991): 869–891.

8. R. J. Charlson et al., "Climate Forcing by Anthropogenic Aerosols," *Science* 255 (1992): 423–430.

9. C. D. Schönwiese, "Moving Spectral Variance and Coherence Analysis and Some Applications on Long Air Temperature Series," *Journal of Climate and Applied Meteorology* 26 (1987): 1723–1730.

10. Houghton, Callander, and Varney, *Climate Change 1992*, pp. 135–170.

11. S. H. Schneider, *Global Warming: Are We Entering the Greenhouse Century?* (San Francisco: Sierra Club Books, 1989), p. 118.

12. T. R. Karl et al. "The Recent Climate Record: What It Can and Cannot Tell Us," *Reviews in Geophysics* 27 (1989): 405–430.

13. T. R. Karl, H. F. Diaz, and G. Kukla, "Urbanization: Its Detection and Effect in the United States Climatic Record," *Journal of Climate* 1 (1988): 1099–1123.

14. P. D. Jones et al., "Assessment of Urbanization Effects in Time Series of Surface Air Temperatures over Land," *Nature* 347 (1990): 169–172.

15. J. G. Charney, "Dynamics of Deserts and Drought in the Sahel," *Quarterly Journal of the Royal Meteorological Society* 101 (1975): 193–202.

16. R. D. Jackson and S. B. Idso, "Surface Albedo and Desertification," *Science* 189 (1975): 1012–1013.

17. R. C. Balling, Jr., "Impact of Desertification on Regional and Global Warming," *Bulletin of the American Meteorological Society* 72 (1991): 232–234; H. A. Nasrallah and R. C. Balling, Jr., "Spatial and Temporal Analysis of Middle Eastern Temperature Changes," *Climatic Change* 25 (1993): 153–161.

18. S. H. Franchito and V. B. Rao, "Climatic Change Due to Land Surface Alterations," *Climatic Change* 22 (1992): 1–34.

19. Balling, "Impact of Desertification," pp. 232–234; H. A. Nasrallah and R. C. Balling, Jr., "The Effect of Overgrazing on Historical Temperature Trends," *Agricultural and Forest Meteorology* 71 (1994): 425–430.

20. C. F. Mass and D. A. Portman, "Major Volcanic Eruptions and Climate: A Critical Evaluation," *Journal of Climate* 2 (1989): 566–593.

21. Z. Wu, R. E. Newell, and J. Hsiung, "Possible Factors Controlling Global Marine Temperature Variations over the Past Century," *Journal of Geophysical Research* 95 (1990): 11799–11810.

22. F. Seitz, R. Jastrow, and W. A. Nierenberg, *Scientific Perspectives on the Greenhouse Problem* (Washington, D.C.: George C. Marshall Institute, 1989).

23. T. M. L. Wigley and S. C. B. Raper, "Natural Variability of the Climate System and Detection of the Greenhouse Effect," *Nature* 344 (1990): 324–327.

24. E. Friis-Christensen and K. Lassen, "Length of the Solar Cycle: An Indicator of Solar Activity Closely Associated with Climate," *Science* 254 (1991): 698–700.

25. R. W. Spencer, J. R. Christy, and N. C. Grody, "Global Atmospheric Temperature Monitoring with Satellite Microwave Measurements: Method and Results 1979–84," *Journal of Climate* 3 (1990): 1111–1128.

26. B. L. Gary and S. J. Keihm, "Microwave Sounding Units and Global Warming," *Science* 251 (1991): 316–317.

27. J. R. Christy and R. T. McNider, "Satellite Greenhouse Signal," *Nature* 367 (1994): 325.

28. Houghton, Callandar, and Varney, *Climate Change 1992*, pp. 135–170.

29. J. D. W. Kahl et al., "Tropospheric Temperature Trends in the Arctic: 1958–1986," *Journal of Geophysical Research* 98 (1993): 12,825–12,838.

30. T. R. Karl et al. "Asymmetric Trends of Daily Maximum and Minimum Temperature," *Bulletin of the American Meteorological Society* 74 (1993): 1007–1023.

31. T. R. Karl, G. Kukla, and J. Gavin, "Decreasing Diurnal Temperature Range in the United States and Canada, 1941 through 1980," *Journal of Climate and Applied Meteorology* 23 (1984): 1489–1504; R. C. Balling, Jr., and S. B. Idso, "Decreasing Diurnal Temperature Range: CO_2 Greenhouse Effect or SO_2 Energy Balance Effect?" *Atmospheric Research* 26 (1991): 455–459; M. S. Plantico et al., "Is the Recent Climate Change across the United States Related to Rising Levels of Anthropogenic Greenhouse Gases?" *Journal of Geophysical Research* 95 (D10) (1990): 16617–16637; D. P. Lettenmaier, E. F. Wood, and J. R. Wallis, "Hydro-Climatological Trends in the Continental United States, 1948–88," *Journal of Climate* 7 (1994): 586–607.

32. Karl et al., "Asymmetric Trends."

33. M. J. Salinger et al., "Southwest Pacific Temperatures: Diurnal and Seasonal Trends," *Geophysical Research Letters* 20 (1993): 935–938.

34. K. R. Kumar, K. K. Kumar, and G. B. Pant, "Diurnal Asymmetry of Surface Temperature Trends over India," *Geophysical Research Letters* 21 (1994): 677–680.

35. D. Rind, R. Goldberg, and R. Ruedy, "Change in Climate Variability in the 21st Century," *Climatic Change* 14 (1989): 5–37; H. X. Cao, J. F. B. Mitchell, and J. R. Lavery, "Simulated Diurnal Range and Variability of Surface Temperature in a Global Climate Model for Present and Doubled CO_2 Climate," *Journal of Climate* 5 (1992): 920–943; J. Hansen et al., "How Sensitive Is the World's Climate?" *Research and Exploration* 9 (1993): 143–158.

36. J. T. Houghton, G. J. Jenkins, and J. J. Ephraums, eds., *Climate Change: The IPCC Scientific Assessment* (Cambridge: Cambridge University Press, 1990), p. 153.

37. Houghton, Callandar, and Varney, *Climate Change 1992*, p. 119.

38. P. J. Michaels and D. E. Stooksbury, "Global Warming: A Reduced Threat?" *Bulletin of the American Meteorological Society* 73 (1992): 1563–1577.

39. A. Henderson-Sellers, "Cloud Cover Changes in a Warmer Europe," *Climatic Change* 8 (1986): 25–52, "Increasing Cloud in a Warming World," *Climatic Change* 9 (1986): 267–309, and "North American Total Cloud Amount Variation This Century," *Global Planetary Change* 1 (1989): 175–194; K. McGuffie and A. Henderson-Sellers, "Is Canadian Cloudiness In-

creasing?" *Atmosphere-Ocean* 26 (1988): 608–633; S. G. Warren et al., *1988: Global Distribution of Total Cloud Cover and Cloud Type Amounts over the Ocean* (Washington, D.C.: U.S. Department of Energy and National Center for Atmospheric Research, 1988); F. Parungo et al., "Trends in Marine Cloudiness and Anthropogenic Sulfate," *Journal of Climate* 7 (1994): 434–440.

40. B. D. Santer et al., "Signal-to-Noise Analysis of Time-Dependent Greenhouse Warming Experiments. Part 1: Pattern Analysis," *Climate Dynamics* 9 (1994): 267–285.

41. M. E. Schlesinger and X. Jiang, "Revised Projection of Future Greenhouse Warming," *Nature* 350 (1991): 219–221.

Chapter 4

THE COMING AGE OF ABUNDANCE

Stephen Moore

HIGHLIGHTS

- *The objective, scientific evidence available today shows that the prophets of doom were wrong in virtually every prediction they made in the 1960s and 1970s when they forecast increasing natural resource scarcity and rising commodity prices.*

- *Every measurable trend of the past century suggests that humanity will soon be entering an age of increasing and unprecedented natural resources abundance.*

- *Technological improvements and advances in productivity have continually outpaced our consumption of natural resources and have led to the net creation of more resources available to future generations.*

- *Today, natural resources are about half as expensive relative to wages as they were in 1980, about three times less expensive than they were fifty years ago, and roughly eight times less costly than they were in 1900.*

- *In spite of Paul and Anne Ehrlich's projection that mineral supplies would be largely depleted by 1985, proven reserves of virtually all important minerals have skyrocketed since 1950.*

- *The continuing discovery of new mines, technological innovations in mining techniques, and introductions of less expensive or superior substitutes for the use of some minerals have caused most minerals to become less scarce rather than more scarce over the past 100 years.*

- *Although Paul Ehrlich refers to the 1980s as a "catastrophic decade" in terms of consumption, the period witnessed tremendous increases in the supply of almost all raw materials.*

- *A well-publicized book,* The Energy Crisis, *projected in 1972 that the earth held thirty years left of gas reserves and twenty years left of oil reserves. Since 1950, however, proven reserves for oil and gas have climbed by over 700 percent.*

- *The very concept of "finite natural resources" embraced by geologists is a flawed way of thinking about the earth and nature. "Natural resources" have value only when humanity invents a use for them.*

The twenty-first century may be the first era in the history of humanity that natural resource scarcity ceases to act as a significant constraint on economic growth. Every measurable trend of the past century suggests that humanity will soon be entering an age of increasing and unprecedented natural resource abundance.

This is admittedly an extraordinary and controversial prediction. It contradicts virtually everything that we read in the newspapers, view on television, and are told by many prestigious academic scholars. We are routinely deluged with precisely the opposite message: that humanity is on the brink of an era of severe, and perhaps catastrophic, physical limits to growth.

The prediction that we may soon be entering an age of resource abundance also seems to collide with pure common sense. There are already 5.6 billion people on earth, each of whom consumes vast quantities of the earth's natural resources every day. That number may double by the year 2050 alone, and a larger population means

For at least the past one hundred years, virtually every natural resource has experienced declining prices. A drop in price is a market signal of less, not more, scarcity.

more consumption of the earth's energy sources, minerals, and other natural resources. As Herman Daly of the World Bank has insisted, more people using more resources can continue indefinitely only if scientists discover "a way to widen the diameter of the earth."[1] In other words, we seem to be speeding toward a head-on collision with the geological reality of resource scarcity.

Yet history proves that the pundits, as well as our intuition, are wrong. For at least the past one hundred years, virtually every natural resource has experienced declining prices. A drop in price is a market signal of less, not more, scarcity.

Figure 4-1 shows how the composite price of thirteen major minerals, metals, and energy sources has changed since 1900.[2] Because the prices are indexed to the wage rate, we can compare for any two points in time the number of hours a worker had to work to purchase a given quantity of these resources. The results confirm the conclusion of a long-term trend toward increasing resource abundance:

- Natural resources today are about half as expensive relative to wages as they were in 1980.

- Natural resources are three times less expensive today than they were fifty years ago.

- Natural resources are roughly eight times less costly than they were in 1900.

How is it possible that even as we use ever increasing amounts of seemingly finite amounts of natural resources, they still decline over time in price? Part of the explanation lies in the fact that technological improvements and advances in productivity continually outpace our consumption of resources. New mining techniques, for example, have opened the doors to vast new quantities of minerals that were never formerly thought to be recoverable. Fifty years ago it would have been inconceivable that oil could be mined from the bottom of the ocean

FIGURE 4-1   Index of Mineral and Energy Prices Relative to Wages

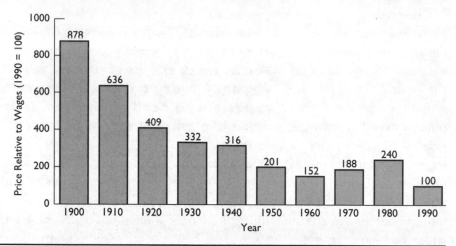

	Average	Aluminum	Antimony	Copper	Lead	Mercury	Nickel	Silver	Platinum	Tin	Tungsten	Zinc	Coal	Oil
1900	878	2,990	690	944	651	1,238	789	1,570	89	400	645	407	633	371
1910	636	1,570	380	568	513	1,025	489	832	360	541	952	406	490	148
1920	409	811	170	278	324	619	171	537	428	478	344	204	687	260
1930	332	629	161	217	236	923	183	377	181	237	564	131	366	107
1940	316	412	230	158	184	1,184	152	161	123	166	802	150	305	77
1950	201	177	221	135	215	248	69	134	119	217	500	148	347	86
1960	152	166	150	129	123	409	77	87	82	128	252	87	221	63
1970	188	123	470	155	109	532	96	117	89	148	307	67	179	47
1980	240	107	228	125	136	235	103	621	209	332	466	75	343	143
1990	100	100	100	100	100	100	100	100	100	100	100	100	100	100

Note: Prices relative to wages (1990 = 100).

Sources: U.S. Department of the Interior, Bureau of Mines, *Mineral Commodity Summaries* (1992); Bureau of Mines, *Nonferrous Metal Prices in the United States Through 1988* (1989); U.S. Department of Energy, Energy Information Administration, *Annual Energy Review* (1989).

or be produced from sand (or shale). Today we have the know-how to obtain huge quantities from each of these sources.

But, one might ask, might this be only a temporary benefit? Aren't there, in fact, only limited quantities of oil and other resources, which may be larger than we previously realized but still not infinite? Yes—and no. The history of progress is one of human ability to produce increasing levels of output and wealth with decreasing amounts of natural resources and human time and effort. Moreover, the pace of this progress throughout the past century has persistently accelerated rather than slowed, and despite the increase in population.

As we progress further into the current information age, the notion of finite physical resources is becoming all the more outmoded. At an unparalleled pace, human ingenuity is unlocking ever more spectacular advances in technology and scientific knowledge that are advancing our mastery over the finite physical universe.

This point is most vividly illustrated by the breathtakingly rapid developments now taking place in the computer microchip industry. The microchip is slightly larger than the size of a man's thumbnail. It is made of silicon, or sand—a natural resource that is in great abundance and has virtually no monetary value. Yet the combination of a few grains of this sand and the infinite inventiveness of the human mind has led to the creation of a machine that will both create trillions of dollars of added wealth for the inhabitants of the earth in the next century and will do so with incomprehensibly vast savings in physical labor and natural resources. Today a single microchip is capable of retaining as much information as is contained in all the books in a small community library. By the end of this century, the computer industry could potentially release a microchip with the capability of retaining and processing all of the information contained in every publication in the U.S. Library of Congress. Even if the world's population were to double every fifty years, the microchip is doubling in capacity roughly every four years, a geometrical rate of growth that is several magnitudes larger than population growth.

Herein lies the explanation for the seeming paradox of the abundance of natural resources today as measured by continuously declining prices: human ingenuity has led to the net creation over time of more resources available to us and future generations, not less.

The Modern Doomsday Mind-set

Throughout the 1990s Americans have been inundated with stories grimly predicting the demise of the planet. Never was this more apparent than in September 1994 when the international environmental community gathered in Cairo for the World Population Conference. The delegates from over one hundred nations signed a document urging that the world community join together to work toward a policy of population stabilization. "What is needed," the report stated, "is a sustainable balance between human numbers and the resources of the planet."[3] Among world statesmen, including those from the United States, there was little dissent from this opinion.

In large part these concerns of political leaders reflect the consensus opinion of most academics—particularly demographers, biologists, geologists, and environmental scientists. One of the most widely quoted academics on population issues, biologist Paul Ehrlich of Stanford University, recently noted: "We are literally using up in a few generations the biological and mineral wealth of the Earth that took millions and millions of years to create."[4] He ridiculed the idea that economic progress can solve the ecological problems of the globe as "growth mania," which he described as "a fixation on growth for growth's sake. . . . It is the creed of the cancer cell." Lester Brown, editor of the *State of the World* annual report, concurs with this antieconomic growth vision as the only way to save the planet; he writes that "current notions of economic growth . . . are at the root of so much of the earth's ecological deterioration." Another highly influential environmentalist, Barry Commoner of the Center for Biology of Natural Systems, added his voice to the choir when he recently lamented: "If we do nothing and just continue the way we are, using 95 percent of our energy from fossil fuels, it will create an economic collapse long before 50 years from now."[5]

Even the vice president of the United States has become a fervent and influential disciple of the doomsday logic. In his best-selling book, *Earth in the Balance*, Al Gore laments that future generations of Americans "will look back on 1990, at the kind of ecological destruction we have, and will wonder, as they shake their heads, how people could have thought in ways that allowed them to condone that kind of activity."[6]

These doomsday outlooks and predictions are anything but new. Just over twenty years ago, for instance, the Club of Rome released its highly influential *Limits to Growth* report, which predicted that over the next thirty years unbridled population growth would lead to mass famines and severe shortages of energy, minerals, trees, and other precious resources.[7] The book's famous conclusion was that "shortages of natural resources will lead to a dismal and depleted existence by the beginning of the next century." *Limits to Growth* was recently updated with a new title, *Beyond the Limits*. The new version attempts to validate the earlier conclusions and then glumly reports that resource and environmental trends are much worse today than even previously suspected.

In 1980 this apocalyptic vision of the future received the official sanction of the U.S. government when the Carter administration released its frightening *Global 2000* report.[8] The $1 million report, sponsored by the State Department and twelve other federal agencies, predicted that most resources—energy, minerals, food, and forests—would be in severe shortage by the year 2000 "if present trends continued" and that "the world's people would be much poorer than today [1980]." These terrifying predictions commanded headlines throughout the nation.[9] I will reserve judgment for the accuracy of these predictions for later; my point here is that this apocalyptic outlook was the prevailing view of the U.S. government at the end of the 1970s.

In the Reagan years of the 1980s, the shift toward a free market–based governing philosophy led to the abandonment of these apocalyptic views. Indeed, most of the calls by Carter administration officials for price controls, population stabilization policies, the development of synthetic fuels, government-funded conservation programs, command-and-control–type energy rationing, and stricter natural resource regulatory policies were not only ignored, but preexisting controls were in many cases lifted.

The Clinton administration, not surprisingly given Vice President Gore's environmental philosophy, has reembraced many of the environmental groups' concerns over world population growth and natural resource scarcity. In sum, the U.S. government has returned full circle to promoting the "limits to growth" governing philosophy and policy outlook that was first popularized two decades ago by the Club of Rome's work.

How much stock should we put in these fears?

Taking Inventory of the Planet

In late 1973, as the United States found itself in the grip of an energy crisis, *Newsweek* magazine dedicated its cover story to the haunting question: "Are We Running Out of Everything?"[10] What we really need to ask is, How do we even go about objectively answering that question?

One way to measure the stock of natural resources available is to examine the quantity of known reserves. Environmental analyst Jerry Taylor of the Cato Institute recently compiled the statistics for reserves of twelve strategic natural resources in 1950 and 1990 and found that "proven reserves of virtually all important minerals have skyrocketed since 1950" (table 4-1).[11] Only reserves of tin fell over the past forty years. Perhaps most surprising of these trends is that "proved reserves" for oil and gas have climbed by over 700 percent since the 1950s, roughly equivalent to a five-hundred-year supply of known reserves of fossil fuels. How can reserves have gone up? The answer is they have been newly discovered or newly made feasibly accessible. The information in table 4-1 not only calls into question the idea that we face imminent scarcity of the earth's resources, it also contradicts the standard notion that for every barrel of oil humanity uses, there is one less available to us, our children, and our grandchildren.

Yet the truth is that the somewhat encouraging data in table 4-1 prove very little about the long-term availability of natural resources. The table shows that

TABLE 4-1   Proven Reserves of Various Resources, 1950–1990 (million metric tons)

Resource	1950	1990	Change (%)
Bauxite	1,400	21,500	1,436
Chromium	70	420	500
Copper	100	350	250
Iron ore	19,000	145,000	663
Lead	40	70	75
Manganese	500	980	96
Nickel	17	59	247
Oil and gas[a]	30	250	733
Coal[a]	450	570	27
Tin	6.0	4.2	–30
Zinc	70	145	107

Sources: Jerry Taylor, *Market Liberalism* (Washington, D.C.: Cato Institute, 1992); *World Bank Development Report 1992* (Washington, D.C.: World Bank, 1993). Used with permission.
[a] Billion tons of oil equivalent.

as of 1990 we had 350 million metric tons of known reserves of copper. Does this mean that as the earth's growing population uses up this 350-million-ton reserve that the earth's inventory gauge will then hit empty? No. Proven reserves merely measure the amount of a mineral that can be extracted economically at any given point in time given current prices and current technological capabilities. As Taylor emphasizes, "Proven reserves are a function of economics and know-how, not geological abundance."[12] Indeed, current trends in industry show that firms are moving aggressively toward finding ways to hold lower inventories of goods and services, through such practices as the just-in-time inventory methods. Yet we do not fear that we are running out of the supply of milk because the grocery store holds only a three-days' supply. Similarly, we should not expect to run out of copper simply because copper mining companies calculate that they only have a certain number of years of reserves. When they use up those reserves, they will have a renewed incentive to locate new sources of supply. As a result of these defects, the best, and perhaps the only reliable, measure of a resource's supply is the change in its market price.[13] Prices are the most objective way that economists have of measuring the relative scarcity of goods and services. A rising price of a commodity, good, or service is a signal that demand is outstripping supply (or is expected to outstrip supply in the future) and that a shortage may emerge.[14]

This same law of supply and demand applies to natural resources. If there were an impending shortage of coal, copper, rubber, or tin, then buyers and sellers would consistently bid up their price. Conversely, if a huge new reserve of oil were discovered, or demand for oil were expected to drop because of the sudden introduction of an alternative energy source, buyers and sellers would consistently bid down oil prices. In sum, a rising price of a resource indicates increasing supply relative to demand; a falling price indicates declining supply relative to demand.[15]

It is noteworthy that even the doomsayers acknowledge that increased scarcity necessarily translates into higher prices of the resource in question. One of the consistent and dire predictions of *Limits to Growth* was that with less availability of resources, we can expect ever-increasing prices. Similarly, Barry Commoner has written that "each barrel of oil drawn from the earth causes the next one to be more difficult to obtain. . . . The economic consequence is that it *causes the price to increase continuously*" (emphasis added).[16] A final example comes from the *Global 2000 Report*. It predicted that nonfuel mineral prices would rise by 5 percent per year through the year 2000 as a consequence of impending scarcity. So it would seem to be a generally accepted proposition, from doomsayers and resource optimists alike, that if we are entering an age of resource scarcity, we *must* observe rising prices. Falling real prices would be incompatible with the concept of resource exhaustion.

Some do argue, however, that the price of a resource or raw material may reflect the available supply today, not the relative abundance or scarcity in future years. If world population growth were to place enormous new demand pressures on resources, then even if the prices had been falling in the past, they may suddenly and sharply rise in the future. This common argument reveals a deep misunderstanding of how the modern price system operates.

Today's price of oil reflects not only the availability of that resource today relative to demand, but its expected availability in the future. An asset's value is determined by the discounted present value of its future return. If the marketplace believed that oil was going to be in short supply in ten years, then owners would bid up prices today. And any market analyst who firmly believed that oil prices will soar in the future could buy oil at today's prices, hold on to it, and sell it in the future at the expected higher price. Indeed one would not even need to take physical possession of the resource. There are now futures markets for most resources that allow traders to purchase the future resale rights of oil, or pork bellies, or other commodities. Thousands of speculators do this; they

Copper produced by this mine in Chile continues, as do most other basic minerals, to fall in price as mining technologies improve.

hoard gold, copper, chickens, farmland, or whatever the resource that is expected to grow scarce might be.

In sum, the price of a natural resource today not only equilibrates current supply and demand but the market's best estimate about the future levels of supply and demand. There isno law of nature that says that the market is always setting the right price. Speculators who bought oil futures in the early 1970s and then sold them in the late 1970s made a fortune. Speculators who bought oil futures in the 1980s lost fortunes. But the market price does have the virtue of incorporating all of the best and most relevant information we have available.

Resource Price Trends over the Long Term

Historical price data for natural resources in the United States are readily available from standard government sources. For some of the resources these figures date back as far as 1800.[17] The following sections examine the price trends for three categories of resources: energy, minerals, and raw materials.

Because minerals are commonly thought to be "nonrenewable" resources, most analysts believe that they must be becoming scarcer over time—almost by definition. Yet the long-term price trends for the thirteen minerals in table 4-2 reveal that relative to the consumer price index, all but four of the minerals have declined in real price over the past 100 to 150 years. That is, the price of most minerals has declined relative to the price of other goods and services. Antimony, platinum, and tin are notable exceptions.

Relative to wages, however, all but platinum have fallen steeply in price (table 4-3). Most of the minerals at the turn of the century were five to ten times more expensive than today in terms of the numbers of hours of work needed to purchase them. As an example, the decline in lead prices relative to wages since 1820 is shown in figure 4-2.

The steady decline in mineral prices is attributable to several factors: the constant discovery of new mines, technological innovations in mining techniques that lower the cost of resource recovery and allow mining from areas where excavation was previously technologically and economically infeasible, and the introduction of less expensive or superior substitutes for the use of some minerals, such as the use of more efficient fiber-optic cables in place of copper telecommunication cables, that lowered demand for some of these commodities.

Price data for basic raw materials are available only back to 1960. But as tables 4-4 and 4-5 show, the prices of cement, glass, metals, and rubber have

TABLE 4-2   Mineral Prices Relative to the Consumer Price Index (1990 = 100)

	Aluminum	Antinomy	Copper	Lead	Magnesium	Manganese	Mercury	Nickel	Silver	Platinum	Tin	Tungsten	Zinc
1800			325										
1810			296										
1820			224	128									
1830			223	99									
1840			265	138				511					
1850			279	162			627	685	413				
1860			275	177			312	400	412				108
1870			181	139			238	308	272				98
1880			241	147			169	311	328	11	74		101
1890			190	140			307	234	323	13	82		109
1900	682		216	149	191		283	180	190	20	124	148	93
1910	414		150	135	119		270	129	142	95	126	252	107
1920	284	60	97	114	667		218	60	132	150	83	121	72
1930	248	64	86	93	250		365	72	63	72	65	224	52
1940	232	132	89	104	179		669	86	76	70	122	454	85
1950	128	160	97	156	104		179	50	99	86	137	363	107
1960	150	140	119	114	102	106	377	71	81	76	118	233	80
1970	130	490	162	114	77	46	557	101	123	93	154	323	70
1980	160	242	133	145	131	65	251	110	661	223	353	499	80
1990	100	100	100	100	100	100	100	100	100	100	100	100	100

Sources: U.S. Department of the Interior, Bureau of Mines, *Mineral Commodity Summaries* (1992); Bureau of Mines, *Nonferrous Metal Prices in the United States through 1988* (1989).

TABLE 4-3   Mineral Prices Relative to U.S. Wages (1990 = 100)

	Aluminum	Antinomy	Copper	Lead	Magnesium	Manganese	Mercury	Nickel	Silver	Platinum	Tin	Tungsten	Zinc
1800			7,250										
1810			5,340										
1820			3,670	2,090									
1830			2,820	1,250									
1840			3,060	1,580				5,890					
1850			2,540	1,470			5,699	6,230					
1860			2,170	1,390			2,455	3,150	3,750				850
1870			1,310	1,010			1,723	2,240	3,230				700
1880			1,490	904			1,039	1,920	1,965	69	620		620
1890			928	683			1,496	1,140	2,020	61	457		530
1900	2,990	690	944	651			1,238	789	1,570	89	400	645	407
1910	1,570	380	568	513			1,025	489	832	360	541	952	406
1920	811	170	278	324	1,883		619	171	537	428	478	344	204
1930	629	161	217	236	627		923	183	377	181	237	564	131
1940	412	230	158	184	314		1,184	152	161	123	166	802	150
1950	177	221	135	215	143		248	69	134	119	217	500	148
1960	166	150	129	123	109	116	409	77	87	82	128	252	87
1970	123	470	155	109	73	45	532	96	117	89	148	307	67
1980	107	228	125	136	122	65	235	103	621	209	332	466	75
1990	100	100	100	100	100	100	100	100	100	100	100	100	100

Sources: Ibid.

FIGURE 4-2   Lead Prices Indexed by Wages

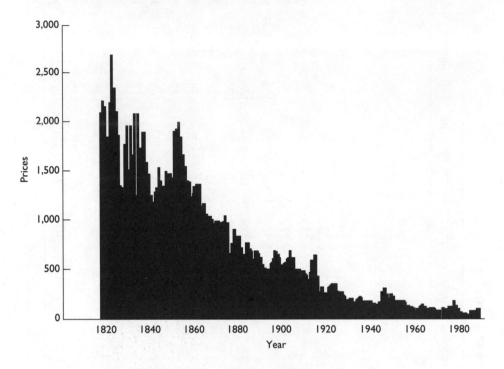

Sources: Bureau of Mines, *Mineral Commodity Summaries* and *Nonferrous Metal Prices.*

declined relative to prices and wages since then. The worldwide price trends for these raw materials are much the same as for those in the United States. There is no conceivable case for predicting any world shortage of basic raw materials.

Over the past century energy prices in the United States have fluctuated substantially. Unlike many other natural resources, most energy prices, indexed by consumer prices, are no lower today than in the early 1900s (table 4-6). The exception is the price of electricity, which today is only one-half to one-third its level in the early part of the century. Relative to wages, however, all forms of energy have experienced price reductions. In 1920, for example, the wage-indexed price of oil was roughly twice its current level, the price of electricity was six times above its current level, and coal was almost seven times more expensive (table 4-7). The data for the long-term trend in oil prices indexed by wages are shown in figure 4-3. Like all other natural resources, energy is not becoming scarcer over time; it is becoming more plentiful, though not as rapidly as other commodities.

Scientists continue to refine their techniques for finding new sources of energy. Here a researcher uses a computer model to study the formation of natural gas fields.

So we see that for almost all basic natural resources, the long-term trends are almost universally in the direction of lower prices and greater affordability. Now we arrive at the core question of this chapter: Have these trends continued in recent years? Or to put it another way, how have the predictions of Paul Ehrlich, Lester Brown, *Global 2000, Limits to Growth*, and other leading declinists compared with the actual evidence over recent years? Let's examine each area separately.

Minerals

In 1976 Paul and Anne Ehrlich projected that "before 1985 mankind will enter a genuine age of scarcity . . . in which the accessible supplies of many key minerals will be facing depletion."[18] A few years prior to that, the authors of

TABLE 4-4   Raw Material Prices Relative to the Consumer Price Index (1990 = 100)

	Cement	Glass	Metals	Rubber
1960	155		112	
1970	125	467	107	
1980	167	154	123	167
1990	100	100	100	100

Source: U.S. Department of the Interior, Bureau of Mines, *Mineral Commodity Summaries* (1991).

Limits to Growth wrote: "Even taking into account such economic factors as increased prices with decreasing availability, it would appear at present that the quantities of gold, zinc, and lead are not sufficient to meet demands. At the present rate of expansion . . . silver, tin, and uranium may be in short supply at higher prices by the turn of the century."[19] Virtually all of those predictions can now be proved wildly inaccurate. Figure 4-4 shows that from 1980 to 1990:

- Of thirteen minerals examined, eleven declined in price. The two exceptions were manganese and zinc.

- The real prices of antimony, mercury, platinum, silver, tin, and tungsten declined by more than 50 percent.

- Copper, lead, and magnesium prices dropped by 20 percent.

Contrary to the predictions of *Limits to Growth*, silver, tin, uranium, and lead have lower prices today than in 1972, not higher prices. Only zinc has risen in price as predicted. As for the Ehrlichs' claim that "by 1985 . . . many

TABLE 4-5   Raw Material Prices Relative to Wages (1990 = 100)

	Cement	Glass	Metals	Rubber
1960	168		121	
1970	120	447	103	
1980	157	144	115	150
1990	100	100	100	100

Source: Ibid.

TABLE 4-6 Energy Prices Relative to the Consumer Price Index (1990 = 100)

	Coal	Electricity	Natural Gas	Oil
1870				183
1880				58
1890				51
1900	76			85
1910	68			39
1920	126	213		91
1930	76	324	97	42
1940	90	299	115	43
1950	131	153	67	62
1960	107	125	78	58
1970	98	88	65	49
1980	191	117	103	153
1990	100	100	100	100

Sources: U.S. Department of Energy, Energy Information Administration, *Annual Energy Review* (1989). More recent data from Energy Information Administration, unpublished data.

TABLE 4-7 Energy Prices Relative to Wages (1990 = 100)

	Coal	Electricity	Natural Gas	Oil
1870				1,320
1880				356
1890				248
1900	633			371
1910	490			148
1920	687	608		260
1930	366	822	244	107
1940	305	530	203	77
1950	347	212	92	86
1960	221	136	84	63
1970	179	84	62	47
1980	343	110	96	143
1990	100	100	100	100

Source: Ibid.

FIGURE 4-3   Oil Prices Indexed by Wages

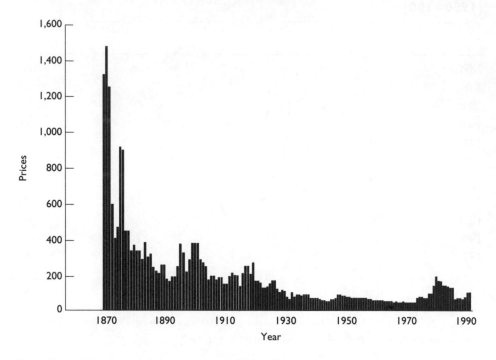

Source: Energy Information Administration, *Annual Energy Review.*

minerals will be facing depletion," we now know that it is plainly and absurdly false.

Raw Materials

Recently Anita Gordon and David Suzuki lamented that "our insatiable appetite for more of the planet's resources has finally caught up with us."[20] Meanwhile Paul Ehrlich recently wrote of the past ten years that we cannot afford "another catastrophic decade like the 1980s."[21] This is the kind of public information that is disseminated about raw material usage in the 1980s. Are we really running up against limits?

No. In the 1980s almost all raw materials declined in price. Figure 4-5 shows the drop in the prices of cement, glass, rubber, and a composite index of all metals between 1980 and 1990: glass prices fell by 34 percent, cement prices by 39 percent, metals prices by 15 percent, and rubber prices by 40 percent. In fact,

FIGURE 4-4   Real Change in Mineral Prices, 1980–1990

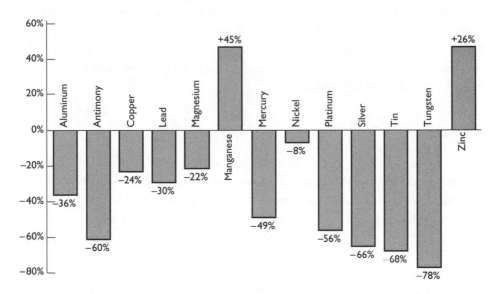

Source: Bureau of Mines, *Mineral Commodity Summaries.*

FIGURE 4-5   Real Change in Raw Material Prices, 1980–1990

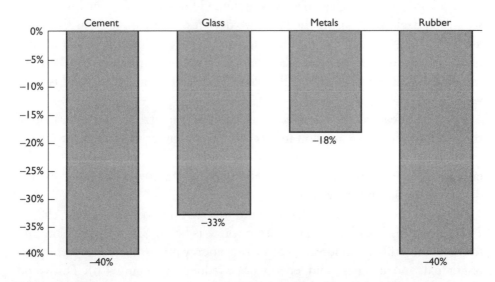

Source: Energy Information Administration, *Annual Energy Review,* and unpublished data.

contrary to the unending rhetoric about the overconsumption of the 1980s, this was a decade of enormous increases in the supply of resources. Economist G. F. Ray recently reported that the 1980s "will go down in economic history as a period when [the price] of primary products hit rock bottom, whatever method is used to illustrate their real value of purchasing power. There have been hardly any exceptions to this decline."[22]

Energy

Throughout the 1970s, as America was held hostage to the OPEC oil embargo, it became commonplace to see newspaper stories and television reports predicting massive and worsening energy shortages as the world entered a new era of scarcity. Oil prices were expected to reach $50 a gallon or more by the year 2000. Lengthy waiting lines at gas stations were expected to become routine. Energy conservation and price controls were the public policy prescriptions of the day. Jimmy Carter urged stores and public buildings to turn the thermostat to what *Newsweek* described as "a chilly 65 degrees in the winter and no lower than a sweaty 80 in summer."[23] The national Democratic party endorsed gas rationing as a way to slow the impending crisis.

Everyone was caught up in the hysteria. President Carter gloomily predicted in 1977 that "we could use up all of the proven reserves of oil in the entire world by the end of the next decade."[24] Two environmental scientists, Lawrence Rocks and Richard Runyon, released a well-publicized book in 1972, *The Energy Crisis*, which declared: "During the next two decades, severe oil and gas shortages are inevitable! We shall be powerless to infuse energy sources on a sufficiently massive scale to meet the demands of our industrial life-support system."[25] They said that at current rates of use, the earth held thirty years' of gas and twenty years' of oil.

Almost all federal officials agreed with this assessment. The vice chairman of the Federal Power Commission in 1971, John A. Carver, described the energy crisis ahead as "endemic and incurable. . . . We can anticipate that before the end of this century energy supplies will become so restricted as to halt economic development around the world."[26]

How did these predictions fare? Today, in 1995, oil does not sell for $50, or $40, or even $30 per barrel. The current price is below $20 a barrel. Contrary to the doom-and-gloom forecasts of soaring energy prices, the 1980s were years of gradual, and at times rapid, energy price reductions (figure 4-6). This trend reversed the upward spiral of the 1970s: oil prices fell by 35 percent, electricity

FIGURE 4-6   Real Change in Energy Prices, 1980–1990

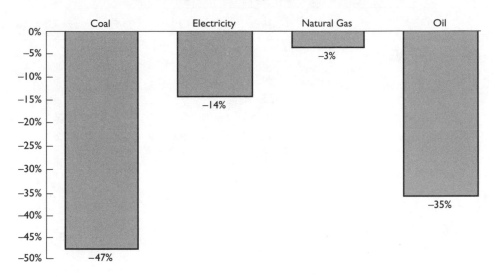

Source: Ibid.

prices by 14 percent, coal prices by 47 percent, and natural gas prices fell by 4 percent. (Natural gas was still under price controls in the 1980s.)

A July 1991 *New York Times* story buried in the Business section contrasted the actual energy situation in the 1990s with the predictions made ten to twenty years earlier. It reported that "adjusting the price of gasoline at the pump to reflect changes in the consumer price index, the average price in the first five months of this year, $1.15, was 6 percent lower than in 1972, the year before the OPEC oil embargo."[27] Gasoline prices may seem high today, but that is only because of the general inflation of the past twenty years and because taxes on gasoline have surged to over thirty-five cents a gallon in many states. Yet the pretax price is lower today than at any other time since 1947. "The good old days when it comes to oil prices, are now," recently exalted Daniel Yergin, energy historian at Cambridge Energy Research Associates.[28]

Worldwide Resource Price Trends

The reductions in prices in the United States in the 1980s were enjoyed by consumers of other nations as well. The World Resources Institute recently com-

TABLE 4-8   World Commodity Prices and Index, 1980 and 1992

	1980	1992	Percentage Change
Commodity indexes[a]			
Petroleum	199	76	−61.8
Beverages	234	74	−68.4
Cereals	183	94	−48.6
Fats and oils	205	105	−48.8
Timber	129	112	−13.2
Metals and minerals	134	83	−38.1
Commodity prices[b]			
Cocoa (kg)	3.62	1.03	−71.5
Coffee (kg)	4.82	1.32	−72.6
Tea (kg)	3.10	1.88	−39.4
Rice (mt)	603.0	269.7	−55.3
Grain sorghum (mt)	179.1	96.4	−46.2
Maize (mt)	174.1	97.8	−43.8
Wheat (mt)	265.1	166.1	−37.3
Sugar (kg)	0.88	0.19	−78.4
Beef (kg)	34.83	2.30	−40.0
Lamb (kg)	4.01	2.49	−37.9
Bananas (mt)	526.6	443.9	−15.7
Oranges (mt)	542.7	459.0	−15.4
Copra (mt)	629.0	356.9	−43.3
Coconut oil (mt)	936.2	541.9	−42.1
Groundnut meal (mt)	333.9	146.1	−56.2
Groundnut oil (mt)	1,193.6	572.3	−52.1
Linseed (mt)	487.6	198.0	−59.4
Linseed oil (mt)	968.6	372.7	−61.5
Palm kernels (mt)	479.5	215.8	−55.0
Palm oil (mt)	810.8	369.2	−54.5
Soybeans (mt)	411.6	221.0	−46.3
Soybean oil (mt)	828.8	402.4	−51.4
Soybean meal (mt)	364.0	191.8	−47.3
Fish meal (mt)	700.3	451.8	−35.5
Cotton (kg)	2.84	1.20	−57.7

TABLE 4-8   Continued

Burlap (mt)	0.50	0.27	−46.0
Jute (mt)	428.0	299.9	−29.9
Sisal (mt)	1,062.7	474.4	−55.4
Wool (kg)	6.39	3.69	−42.3
Rubber (kg)	2.26	0.96	−57.5
Logs (cm)	271.6	234.9	−13.5
Plywood (sheet)	3.80	3.57	− 6.1
Sawnwood (cm)	507.3	481.4	− 5.1
Tobacco (mt)	3,195.8	1,848.4	−42.2
Coal (mt)	59.89	38.09	−36.4
Crude petroleum (bbl)	42.38	16.23	−61.7
Gasoline (mt)	497.4	n/a	n/a
Fuel oil (mt)	235.9	n/a	n/a
Aluminum (mt)	2,023.1	1,176.9	−41.8
Bauxite (mt)	57.25	30.02	−47.6
Copper (mt)	3,031.8	2,140.3	−29.4
Lead (kg)	1.26	0.51	−59.5
Tin (kg)	22.84	5.62	−75.4
Zinc (kg)	1.06	1.16	+ 9.4
Iron ore (mt Fe)	39.03	29.65	−24.0
Manganese ore (10 kg)	2.18	n/a	n/a
Nickel (mt)	9,057.5	6,569.0	−27.5
Phosphate rock (mt)	64.9	39.2	−39.6
Diammonium phosphate (mt)	308.7	136.2	−55.9
Potassium chloride (mt)	161.0	105.2	−34.7
Triple superphosphate (mt)	250.1	113.2	−34.7
Urea (mt)	308.6	131.6	−57.4

The following calculations do not include gas, fuel oil, and manganese ore because of incomplete data: (49 commodities)

Average 1980 price[c] = $640.11

Average 1992 price[c] = $397.22

Average commodity price change[c] = −37.95%

Source: World Resources Institute, *World Resources 1993–1994* (New York: Basic Books, 1994), p. 262.

[a] Based on constant prices, with 1990 = 100.

[b] In constant U.S. dollars.

[c] Excluding gas, fuel oil, and manganese ore because of incomplete data.

piled worldwide indexes for more than fifty commodities, ranging from Peruvian fish to Malaysian aluminum (table 4-8). All but two of the resources declined in real price. For consumers all over the world, natural resources have become more affordable. Between 1980 and 1982, the worldwide real price of food declined by 49 percent, the worldwide real price of petroleum by 62 percent, and the worldwide real price of timber by 13. Also, the worldwide real price of metals and minerals fell by 38 percent from 1980 to 1988.

In sum, the objective, scientific evidence available today reveals that the prophets of doom were wrong in most predictions they made in the 1960s and 1970s when they forecast increasing natural resource scarcity and rising prices of commodities. In the 1980s in the United States and abroad, natural resources became more, not less, plentiful. Those trends have generally continued in the early 1990s.

Why the Doomsayers Got It Wrong

The persistent price declines of the 1980s thoroughly confounded the doomsayers, who had predicted precisely the opposite course of events. Yet it is important to emphasize that a vocal minority of scientists and economists, such as Julian Simon and Herman Kahn, correctly forecast the price trends of the past fifteen years. At the time, these scholars were summarily dismissed by critics as holding views that are outside the mainstream and irresponsible.[29]

In 1980, economist Julian Simon of the University of Maryland bet Paul Ehrlich $10,000 that the real price of five natural resources of Ehrlich's own choosing would be less expensive in 1990 than in 1980. All five of the resources Ehrlich chose—copper, chrome, nickel, tin, and tungsten—fell in price. Ehrlich lost the bet.[30]

In the face of a mountain of evidence today that the environmental predictions of the doomsayers were widely off-target, many of these pundits now maintain that their reports issued valuable warnings and led to the implementation of policy prescriptions that helped avert the ecological disaster they had predicted. The truth is that the limits-to-growth corrective measures that had been urged were rejected outright by the Reagan administration, which preferred to focus on economic expansion and wealth creation. If *Limits to Growth* and *Global 2000* had been correct, the predicted path toward severe scarcity should have accelerated in the 1980s, not reversed. Price and production controls were lifted rather than expanded. Mandatory conservation programs were abandoned rather than extended. The windfall profits tax was repealed. The doomsayers routinely attacked these policies as wrongheaded and dangerous.

Indeed, we now have fairly compelling evidence that it was precisely the government interventionist resource policies of the 1970s that exacerbated and lengthened the energy crisis of that unfortunate era. The experience with oil price controls provides an eye-opening example. The

Indeed, we now have fairly compelling evidence that it was precisely the government interventionist resource policies of the 1970s that exacerbated and lengthened the energy crisis of that unfortunate era.

objective of these price ceilings, implemented in 1971, was to protect U.S. consumers by preventing U.S. oil companies from collecting "windfall profits" as a result of the higher prices charged by OPEC. Instead the oil price controls had three effects (see figure 4-7):

1. *Price controls reduced domestic oil production.* U.S. oil production fell from 11 million to 9 million barrels a day from 1971 to 1980. Price controls made it unprofitable for domestic producers to increase output through more expensive processes, such as drilling deeper wells, fracturing, steam or water injection, and offshore drilling. Oil price controls brought new drilling in the United States to a standstill.

2. *Price controls increased U.S. oil consumption.* Oil consumption by U.S. households and industry rose from 15 to 19 million barrels of oil a day from 1970 through 1978 because price controls kept oil prices artificially lower than they would have otherwise been in the short term. This discouraged conservation. In large part because of price controls from 1975 to 1978, the real price of oil fell. These artificially low prices contributed directly to the wrenching 1978 to 1981 energy crisis.

3. *Price controls increased reliance on foreign oil.* U.S. imports of foreign oil rose from 4 million to 8.5 million barrels of oil a day from 1970 to 1977 as demand for oil rose and domestic production fell.

In 1981 Ronald Reagan's first act as president was to lift all oil price controls. The impact was immediate:

1. *Energy efficiency and conservation increased.* In 1981 the price of crude oil rose from $29 to $36 per barrel after lifting price controls. By the end of 1981, oil consumption in the United States dropped by 20 percent.

2. *Oil imports declined and domestic production rose.* OPEC commanded almost 60 percent of the U.S. market in the 1970s. By the end of 1981, after the lifting of price controls, the imported share was down to 45 percent (today it is about 50 percent). By January 1982 U.S. oil production rose by 50 percent above 1980 levels and by 100 percent above 1979 levels.

3. *Oil prices eventually declined, and OPEC was defeated.* The increased U.S. production of oil led to a glut in the market by late 1982. As always happens with an increase in supply, eventually prices tumbled from their high of $35 per barrel in 1981 to a low of $13 per barrel in 1986. Two years after the elimination of energy price controls OPEC had been crushed. A *Time* magazine caption read: "Down, Down, Down: OPEC Finds That It Is a Crude, Crude, World." *Newsweek* was more succinct: "OPEC: From Cartel to Chaos!"[31]

In sum the energy crisis in the 1970s was attributable mainly to government folly. Congress prevented the price system from operating properly. When the market system for energy was permitted to function in the 1980s, the crisis quickly subsided, and American consumers enjoyed cheaper and abundant oil.

A case could be made that the very concept of "finite natural resources" embraced by geologists is a flawed way of thinking about the earth and nature. Natural resources have market value only when human beings invent a use for them. In the eighteenth century,

> **Jimmy Carter's Global 2000 report, for example, was a case study in the abuse of forecasting as science.**

people worried about shortages of the major transportation form of the day, horses, and of the "fuel" that made them run, rangeland for grazing. In that same era, the discovery of a black gooey substance known as oil underneath one's property depreciated the value of that property.[32] No one had invented a use for it. Now it is valued as one of our most vital resources, and grazing land is relatively abundant and inexpensive.

Another explanation for why we are often confronted with persistently wrong environmental predictions about the future of the planet's resources is that the traditional model for forecasting the future availability of resources is irreparably damaged. Jimmy Carter's *Global 2000* report, for example, was a case study in the abuse of forecasting as science. Its dismal view of the future was based on forecasts that assumed that "current trends will continue." It is such analyses that lead to the conclusion that the earth will hold some 12 billion people by

Companies respond to an increase in the price of a resource by seeking more of it or by developing cheaper substitutes. The rise in the price of oil in the 1970s led both to the opening of new oil fields like this one in the North Sea and to the consequent steep drop in oil prices as more supplies came on line.

the middle of the next century or that the last barrel of oil will be drilled from the earth early in the next century. This type of extrapolation analysis often leads to bizarre conclusions.[33] For instance:

- Economists Charles Maurice and Charles Smithson of Texas A&M computed that if one were to extrapolate publishing trends from the 1970s, there would be 14 million doomsday books published by the year 2000—or more than half as many books as there are in the entire Library of Congress today.[34]

- At the end of World War II, economists extrapolating from the declining birth figures from the 1930s predicted a declining U.S. population in the post–World War II era and thus a period of "secular stagnation." These predictions were made only a few years prior to the baby boom and two decades of economic prosperity.[35]

- The *Economist* magazine recently demonstrated that by using "present trend" analysis, and depending on the years one starts and ends the data collection to develop the trend, very different conclusions can be forecast. Trend analysis could predict that by the year 2000 Arnold Schwarzenegger's earnings would top $360 million. But if different starting and ending points are chosen, using the same estimation technique one could predict Schwarzenegger's income to be −$130 million in 2000.[36]

As nonsensical as these estimates may seem, none of them is any less defensible than projecting that resources will run out or that population will explode in the next century. The reason that the analyses are wrong is that short-term trends almost never continue because human beings adapt their behavior and change their environment in response to changing conditions. For example, although economic output in 1989 was roughly 30 percent higher than in 1979 and although the U.S. population rose by 10 percent over this period, energy consumption was up only roughly 3 percent. In other words, btu's per dollar of GNP were roughly 20 percent lower in 1989 than in 1979. Inventive minds responded to the economic incentives of the "energy crisis" by finding ways to make industry and consumers more energy efficient. Trend analysis cannot account for such behavioral changes. For this reason, reports such as *Global 2000* have very little

For this reason, reports such as Global 2000 have very little scientific value.

scientific value. Their "contribution" is to scare needlessly people and divert resources from real problems.

Conclusion

Natural resources have been growing more plentiful over the course of the past
century, as measured by their price. The 1980s, contrary to popular belief, was
no exception to this long-term trend. Indeed, thanks to the steep across-the-
board declines in natural resource prices in the 1980s, many of the earth's resources today are at their lowest price ever in recorded history. Even as a growing
population and a more economically developed society uses more resources than
ever before, the introduction of new technologies and innovations, which make
us more efficient in consuming and producing natural resources, has meant that
the earth's resources have continually become less of a limit to growth over time
rather than more so.

Even the U.S. government now apparently recognizes the errors of its judg-
ments in the past. Reversing the forecasts of studies such as *Global 2000* in 1988,
the Office of Technology Assessment concluded: "The nation's future has probably never been less constrained by the cost of natural resources."[37]

There is no inevitability of declining prices of natural resources. Unwise government intervention policies as were experimented with in the 1970s can often
have economically and ecologically debilitating consequences. But if politicians
can resist the ever-present temptation to intervene in natural resource markets,
America and the rest of the world face a surprisingly rich resource future in the
twenty-first century.

NOTES

1. Herman Daly, "The Steady State Economy: Alternative to Growthmania," in *Population-
   Environment Balance Report* (Washington, D.C., 1987).
2. The commodities in this index are aluminum, antimony, copper, lead, mercury, nickel, silver, platinum, tin, tungsten, zinc, coal, and oil.
3. *Report of the International Conference on Population and Development in Cairo* (New York:
   United Nations, 1994).
4. Paul Ehrlich interviewed on the "Today Show," May 3, 1989.
5. Paul R. Ehrlich, "An Economist in Wonderland," 1981; Lester R. Brown, *State of the World*,
   Worldwatch Institute report (Washington, D.C.: Worldwatch Institute, 1994); Barry Commoner quoted from ABC News, "The Electronic Time Machine," May 18, 1989.

6. Al Gore, *Earth in the Balance* (Boston: Houghton Mifflin, 1992).

7. Donella H. Meadows et al., *The Limits to Growth* (New York: Universe Books, 1972).

8. *Global 2000 Report to the President* (Washington, D.C.: U.S. Government Printing Office, 1980).

9. An assessment of *Global 2000*'s influence is contained in Stephen Moore, "Half-Truths and Consequences: The Legacy of Global 2000," Heritage Foundation *Institution Analysis*, no. 34 (1985).

10. "Are We Running Out of Everything?" *Newsweek*, November 9, 1973.

11. Jerry Taylor, "Sustainable Development," *Regulation* 1 (1994): 37.

12. Ibid., p. 36.

13. For an explanation of prices versus other measures of scarcity, such as "known reserves," see Julian L. Simon, *The Ultimate Resource* (Princeton: Princeton University Press, 1981).

14. A price measures not only the availability of a resource today but also the availability of the resource, and the demand for it, in the future, based on the best information currently accessible to buyers and sellers. There are future markets for most resources, so if speculators believed that wool, or cattle, or oil were going to be in short supply and thus command a higher market price in future years, they could purchase the future rights to these commodities and then resell them at the higher price in the future. If, for example, oil supplies were running out, industry traders would bid up the price for oil futures, which would encourage firms to withhold current supplies from the market, which would raise the price of oil today.

15. Julian L. Simon, *The Ultimate Resource* (Princeton: Princeton University Press, 1980).

16. Quoted in Simon, *Ultimate Resource*, p. 100.

17. The data and analysis in this section were originally published in Stephen Moore, *Doomsday Delayed: America's Surprisingly Bright Natural Resource Future* (Lewisville, Texas: Institute for Policy Innovation, 1992).

18. Paul Ehrlich and Anne Ehrlich, *The End of Affluence* (New York: Ballantine Books, 1976).

19. Meadows et al., *Limits to Growth*.

20. Anita Gordon and David Suzuki, *It's a Matter of Survival* (Cambridge: Harvard University Press, 1990).

21. Ehrlich, "Today Show."

22. G. F. Ray, "The Decline of Primary Producer Power," *National Institute Economic Review* (August 1987): 40.

23. "The Energy Tangle," *Newsweek*, April 16, 1979, p. 22.

24. Quoted in *Time*, October 10, 1977, p. 62.

25. Lawrence Rocks and Richard Runyon, *The Energy Crisis* (New York: Crown Publishers, 1972).

26. John A. Carver, Jr., quoted in ibid., p. 6.

27. "Gas Is at Historic Low, Reckoning for Inflation," *New York Times*, July 25, 1991, Business section.

28. Ibid.

29. Julian L. Simon and Herman Kahn, *The Resourceful Earth: A Response to Global 2000* (New York: Basil Blackwell, 1984).

30. John Tierney, "Betting the Planet," *New York Times Magazine*, December 2, 1990, pp. 52–78.

31. *Time*, March 15, 1982; *Newsweek*, March 7, 1983.

32. See Sheldon Richman, "The U.N.'s Holy War Against Population," *Cato Policy Report* (July–August 1994).

33. For an excellent critique of trend analysis, see Charles Maurice and Charles Smithson, *The Doomsday Machine* (Stanford, Calif.: Hoover Institute Press, 1984), esp. chap. 3.

34. Ibid.

35. Ibid.

36. Joseph L. Bast et al., *Eco-Sanity* (Lanham, Md.: Madison Books, 1994), p. 232.

37. U.S. Office of Technology Assessment, *Technology and the American Economic Transition* (Washington, D.C.: Government Printing Office, 1988).

Chapter 5

THE CAUSES AND PREVENTION OF CANCER
The Role of Environment

Bruce N. Ames and Lois Swirsky Gold

HIGHLIGHTS

- *The idea that there is an epidemic of human cancer caused by synthetic industrial chemicals is false.*

- *If lung cancer (which is primarily due to smoking) is excluded, cancer death rates are decreasing in the United States for all other cancers combined. In addition, there is a steady rise in life expectancy in the developed countries.*

- *Pollution appears to account for less than 1 percent of human cancer, yet public concern and resource allocation for chemical pollution are very high, in good part because of the use of animal cancer tests in cancer risk assessment.*

- *About half of the chemicals tested, whether synthetic or natural, are carcinogenic to rodents at the high doses tested in rodents.*

- *Linear extrapolation from near-toxic doses in rodents to low-level exposure in humans has led to grossly exaggerated mortality forecasts. Such extrapolations cannot be verified by epidemiology. Furthermore, relying on such extrapolations for synthetic chemicals while ignoring the enormous natural background leads to an imbalanced perception of hazard and allocation of resources.*

- *Zero exposure to rodent carcinogens cannot be achieved. Low levels of rodent carcinogens of natural origin are ubiquitous in the environment. It is thus impossible to obtain conditions totally free of exposure to rodent carcinogens or to background radiation.*

- *Risks compete with risks: society must distinguish between significant and trivial risks. Regulating trivial risks or exposure to substances erroneously inferred to cause cancer at low doses can harm health by diverting resources from programs that could be effective in protecting the health of the public.*

- *Epidemiological evidence in humans is sufficient to identify several broad categories of cancer causation for which the evidence is strong and plausible. Since many of these are avoidable, it is possible to reduce incidence rates of many types of cancer.*

- *Tobacco is the most important global cause of cancer and is preventable. Smoking contributes to about one-third of U.S. cancer, about one-quarter of U.S. heart disease, and about 400,000 premature deaths per year in the United States.*

- *The quarter of the population with the lowest dietary intake of fruits and vegetables compared to the quarter with the highest intake has roughly twice the cancer rate for most types of cancer (lung, larynx, oral cavity, esophagus, stomach, colon and rectum, bladder, pancreas, cervix, and ovary).*

- *Decreases in physical activity and increases in smoking, obesity, and recreational sun exposure have contributed importantly to increases in some cancers in the mod-*

ern industrial world, whereas improvements in hygiene have reduced other cancers related to infection.

- *Chronic infections contribute to about one-third of the world's cancer.*

Overview

The idea that synthetic chemicals such as DDT are major contributors to human cancer was inspired, in part, by Rachel Carson's passionate book, *Silent Spring*. This chapter presents the evidence showing that this widely prevailing belief is not true. We also review the latest research on the causes of cancer and explain why much cancer is preventable.

Epidemiological evidence indicates several factors likely to have a major effect on reducing rates of cancer: reduction of smoking, increased consumption of fruits and vegetables, and control of infections. Other factors are avoidance of intense sun exposure, increases in physical activity, and reduction of alcohol consumption and possibly red meat. The risks of many forms of cancer can be reduced, and the potential for further reductions is great. If lung cancer (which is primarily due to smoking) is excluded, cancer death rates are decreasing in the United States for all other cancers combined.

The risks of many forms of cancer can be reduced, and the potential for further reductions is great.

Pollution appears to account for less than 1 percent of human cancer, yet public concern and resource allocation for chemical pollution are very high, in good part because of the use of animal cancer tests in cancer risk assessment. Animal cancer tests, which are done at near-toxic doses, are misinterpreted as meaning that low doses of synthetic chemicals and industrial pollutants are relevant to human cancer. About half of the chemicals tested, whether synthetic or natural, are carcinogenic to rodents at these high doses. A plausible explanation for the high frequency of positive results is that testing at the near-toxic dose frequently can cause chronic cell killing and consequent cell replacement, a risk factor for cancer that can be limited to high doses. Ignoring this effect greatly exaggerates risks. Scientists must determine mechanisms of carcinogenesis for each substance and revise acceptable dose levels as understanding advances.

The vast bulk of chemicals ingested by humans is natural. For example, 99.99 percent of the pesticides we eat are naturally present in plants to ward off insects and other predators. Half of these natural pesticides tested at near-toxic doses are rodent carcinogens. Reducing exposure to the 0.01 percent that are synthetic will not reduce cancer rates. On the contrary, although fruits and vegetables contain a wide variety of naturally occurring chemicals that are rodent

People consume 10,000 times more natural pesticides than they do synthetic pesticides.

carcinogens, inadequate consumption of fruits and vegetables doubles the human cancer risk for most types of cancer. Making these foods more expensive by reducing synthetic pesticide use is likely to increase cancer. Humans also ingest large numbers of natural chemicals from cooking food. Over a thousand chemicals have been reported in roasted coffee; more than half of those tested (nineteen of twenty-six) are rodent carcinogens. There are more rodent carcinogens in a single cup of coffee than potentially carcinogenic pesticide residues in the average American diet in a year, and there are still a thousand chemicals left to test in roasted coffee. This does not mean that coffee is dangerous but rather that animal cancer tests and worst-case risk assessment build in enormous safety factors and should not be considered true risks.

The reason we humans can eat the tremendous variety of natural chemical "rodent carcinogens" is that we, like other animals, are extremely well protected by many general defense enzymes, most of which are inducible (that is, whenever a defense enzyme is in use, more of it is made). These defense enzymes are equally effective against natural and synthetic chemicals. There is no general difference between synthetic and natural chemicals in the ability to cause cancer in high-dose rodent tests.

The idea that there is an epidemic of human cancer caused by synthetic industrial chemicals is false. Linear extrapolation from the near-toxic doses in rodents to low-level exposure in humans has led to grossly exaggerated mortality forecasts. Such extrapolations cannot be verified by epidemiology. Furthermore, relying on such extrapolations for synthetic chemicals while ignoring the enormous natural background leads to an imbalanced perception of hazard and allocation of resources. It is the progress of scientific research and technology that will continue to lengthen human life expectancy; indeed, there has been a steady rise in life expectancy in the developed countries.

Zero exposure to rodent carcinogens cannot be achieved. Low levels of rodent carcinogens of natural origin are ubiquitous in the environment. It is thus impossible to obtain conditions totally free of exposure to rodent carcinogens or to background radiation. Major advances in analytical techniques enable the detection of extremely low concentrations of all substances, whether natural or synthetic, often a million times lower than could be detected thirty years ago.

Risks compete with risks: society must distinguish between significant and trivial risks. Regulating trivial risks or exposure to substances erroneously inferred to cause cancer at low doses can harm health by diverting resources from programs that could be effective in protecting the health of the public. Moreover, wealth creates health: poor people have shorter life expectancy than wealthy people. When money and resources are wasted on trivial problems, society's wealth and hence health is harmed.

Unlike the other chapters in this book, the superscript numbers in the text refer not to consecutively numbered notes but to numbered references at the end of the chapter. Hence, the numbers are not necessarily consecutive in the text.

Trends

Cancer was estimated to cause 23 percent of the person-years of premature loss of life and about 530,000 deaths in the United States in 1993.[1] Four major cancers (lung, colon-rectum, breast, and prostate) account for 55 percent of the deaths. Nevertheless, cancer death rates in the United States are decreasing, after adjusting for age and excluding lung cancer. According to a 1993 update from the National Cancer Institute, the age-adjusted mortality rate for all cancers combined (excluding lung and bronchus) declined from 1950 to 1990 for all individual age groups except 85 and above.[1] The decline ranged from 71 percent in the 0–4-year-old group to 8 percent in the 74–85-year-old group. The update notes that "if lung cancer were eliminated, then the overall cancer death rate would have declined over 14% between 1950 and 1990."[1] One plausible explanation for the only age group that did increase, in the over-85 group (by 6 percent), is that autopsies were less common in the past. Smoking, in addition to causing the bulk of lung cancer, contributes to other malignancies, such as cancers of the mouth, esophagus, pancreas, bladder, leukemia, and possibly colon; if these were taken into account, the decline would be greater (figure 5-1).

If lung cancer is included, overall cancer mortality has decreased more than 25 percent for each age group under 45 and has increased for age groups over 55 years. The decreases in cancer deaths during this period have been primarily from stomach, cervical, uterine, and rectal cancer. The increases have been primarily from lung cancer, which is due to smoking (as is 30 percent of all U.S. cancer deaths) and non-Hodgkin's lymphoma (NHL). Reasons for the increase in NHL are not clear, but smoking may contribute,[2, 3] and HIV is a small but increasing cause.

An analysis by Professor Peto has come to the same conclusion: "The common belief that there is an epidemic of death from cancer in developed countries is a myth, except for the effects of tobacco. In many countries cancer deaths from tobacco are going up, and in some they are at last coming down. But, if we take away the cancer deaths that are attributed to smoking then the cancer death rates that remain are, if anything, declining. This is reassuringly true in Western Europe, Eastern Europe and North America—and, in the 'West,' the death rates from other diseases are falling rapidly. For most non-smokers, the health benefits of modern society outweigh the new hazards. Apart from tobacco (and in places, HIV), the Western world is a remarkably healthy place to live" (figure 5-2).[4]

Although the number of smokers is declining in the United States, overall lung cancer continues to increase because of decades of delay between when a person begins smoking and the onset of the disease. The rate of lung cancer among Ameri-

FIGURE 5-1   Cancer Death Rates by Site, 1930–1990

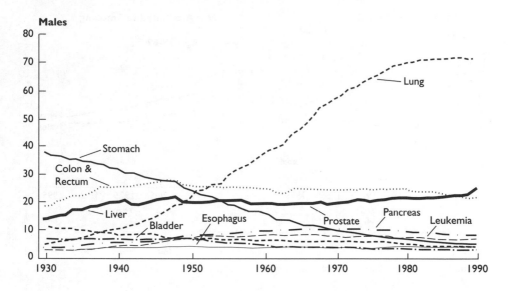

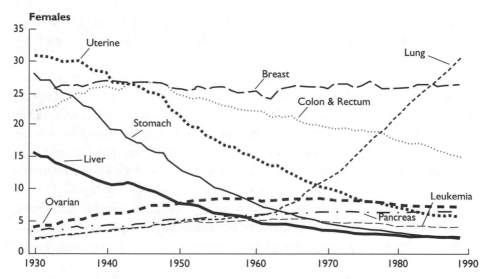

Note: Rates are per 100,000 and are age adjusted to the 1970 U.S. census population.

Source: American Cancer Society, *Cancer Facts and Figures* (Atlanta: ACS, 1994).

FIGURE 5-2   Total Cancer Mortality in the United States, 1955–1990

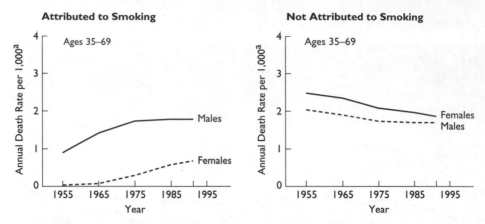

[a] Mean of seven age-specific rates, ages 35–69; annual death rate/1000.

Source: R. Peto, A. D. Lopez, J. Boreham, M. Thun, and C. Heath, Jr., *Mortality from Smoking in Developed Countries, 1950–2000* (Oxford: Oxford University Press, 1994).

can men appears to have peaked, while the rate is still going up for American women, who started smoking more recently than men. Those trends are clear in figure 5-1 for lung cancer and in figure 5-2 for all cancer deaths attributed to smoking.

To interpret changes in mortality rates, one must consider both changes in incidence rates (the number of people newly diagnosed with the cancer) and effects of treatment. Incidence rates have been increasing for some types of cancer. Doll and Peto of Oxford University, two of the world's leading epidemiologists, in their comprehensive study on the causes of cancer point out that incidence rates should not be taken in isolation, because reported incidence rates for a disease might reflect increases in registration of cases and improvements in diagnosis.[5] The reported rise in cancer rates among men born in the 1940s compared to those born in the 1890s[6] may be due to such artifacts. For example, the rapid increase in age-adjusted prostate cancer incidence without any major increases in mortality is almost certainly due largely to increased screening and incidental detection during prostatectomy for benign prostatic hypertrophy.

Major Factors Influencing Cancer Risks

Background Damage

Biochemical studies of carcinogenesis have indicated an important role of metabolic oxidative damage to DNA that is balanced by elaborate defense and repair processes. The rate of cell division, which is influenced by hormones, growth, cell killing, and inflammation, is also key, as this determines the probability of converting DNA lesions to mutations. These mechanisms are likely to underlie many epidemiologic observations and together suggest practical interventions and areas for further research.

Metabolism, like other aspects of life, involves trade-offs. Oxidant by-products of normal metabolism cause extensive damage to DNA, protein, and lipid. We argue that this damage is a major contributor to aging and to degenerative diseases of aging such as cancer, heart disease, cataracts, and brain dysfunction.[7] Antioxidant defenses against this damage include Vitamins C and E and carotenoids. To the extent that the major external risk factors for cancer— smoking, unbalanced diet, and chronic inflammation—are diminished, cancer will appear at a later age, and the proportion of cancer that is caused by normal metabolic processes will increase.

Oxidative damage to DNA, proteins, and other macromolecules accumulates with age and has been postulated to be a major, but not the only, type of metabolic damage leading to aging.[7] By-products of normal metabolism—superoxide, hydrogen peroxide, and hydroxyl radical—are the same oxidative mutagens produced by radiation.[8] Oxidative lesions in DNA accumulate with age, so that by the time a rat is old (two years) it has about a million DNA lesions per cell, about twice that in a young rat.[7] Mutations also accumulate with age. DNA is oxidized in normal metabolism because antioxidant defenses, though numerous, are not perfect.

Metabolic oxidants damage proteins as well as DNA.[9] In two human diseases associated with premature aging, Werner's syndrome and progeria, oxidized proteins accumulate at a much higher rate than normal.[9] Cataracts, which also represent the accumulation of oxidized protein, are a common manifestation of oxidative stresses, such as ultraviolet radiation and smoking, as well as of insufficient antioxidant protection.[7, 10–12]

Diet

Although diet is thought to account for about one-third of cancer risk,[5] the specific factors are only slowly being clarified. We present here a brief overview of the field.

Cancer prevention by calorie or protein restriction. In rodents a calorie-restricted diet markedly decreases tumor incidence and increases lifespan[13–15] but decreases reproduction. Protein restriction appears to have the same effects on rodents as calorie restriction, though it is less well studied.[16] An understanding of mecha-

Although Americans are concerned about hazardous wastes, pollution, and synthetic pesticide residues, epidemiological studies demonstrate that these substances are responsible for a tiny percentage of human cancer at most. Most cancer is due to smoking and dietary imbalance.

nisms for the marked effect of dietary restriction on aging and cancer is becoming clearer and may in good part be due to reduced oxidative damage and reduced cell division rates. Although epidemiological evidence on restriction in humans is sparse, the possible importance of growth restriction in human cancer is supported by epidemiologic studies indicating higher rates of breast and other cancers among taller persons.[17, 18] Japanese women, for example, are now taller and menstruate earlier—and they have increased breast cancer rates. Also, many of the variations in breast cancer rates among countries, and trends over time within countries, are compatible with changes in growth rates and attained adult height.[19]

Cancer prevention by dietary fruits and vegetables. Consumption of adequate fruits and vegetables is associated with a lowered risk of degenerative diseases such as cancer, cardiovascular disease, cataracts, and brain and immune dysfunction.[7] Nearly 200 studies in the epidemiological literature have been reviewed and relate, with great consistency, the lack of adequate consumption of fruits and vegetables to cancer incidence.[20–22] The quarter of the population with the lowest dietary intake of fruits and vegetables compared to the quarter with the highest intake has roughly twice the cancer rate for most types of cancer (lung, larynx, oral cavity, esophagus, stomach, colon and rectum, bladder, pancreas, cervix, and ovary). The protective effect for hormonally related cancers is weaker and less consistent; for breast cancer the protective effect appears to be about 30 percent.[17, 20, 23] Other work also suggests a protective effect of fruit and vegetable consumption on heart disease and other degenerative diseases of aging.[7] Yet only 9 percent of Americans met the intake recommended by the National Cancer Institute and the National Research Council of two servings of fruits and three of vegetables per day.[24, 25]

Laboratory studies suggest that antioxidants such as Vitamins C and E and carotenoids in fruits and vegetables account for a good part of their beneficial effect;[7] however, the effects of dietary intakes of the antioxidants Vitamin C, Vitamin E, and carotenoids are difficult to disentangle by epidemiological studies from other important vitamins and ingredients in fruits and vegetables.[24, 26]

A wide array of compounds in fruits and vegetables in addition to antioxidants may contribute to the reduction of cancer. Folic acid may be particularly important. Low folic acid intake causes chromosome breaks in rodents[27] and humans[28, 29] and increases tumor incidence in some rodent models.[30] Folic acid is required for the synthesis of DNA. Low folate intake has been associated with several neoplasms, including adenomas and cancers of the colon.[31-33] Deficient intake of folic acid is common in U.S. diets and is associated in mothers with neural tube birth defects in their offspring.[34] About 15 percent of the U.S. population[35] and about half of low-income black children[36] or black elderly[37] are at a low folate level where chromosome breaks have been seen.[28] Dietary fiber, obtained only from foods of plant origin, may contribute to lower risk of colon cancer.[38] Plant foods also contain a wide variety of weak estrogens that may act as antiestrogens by competing with estrogenic hormones.[21, 26, 39]

Other aspects of diet. While the benefits of fruits and vegetables in the prevention of cancer are most clearly supported by epidemiologic studies, strong international correlations suggest that animal (but not vegetable) fat and red meat may increase the incidence of cancers of the breast, colon, and prostate.[40] However, large prospective studies of fat intake and breast cancer have consistently shown a weak or no association with the incidence of breast cancer.[17] In contrast, animal fat and red meat have been associated with colon cancer risk in numerous case-control and cohort studies; the association with meat consumption appears more consistent.[41-43] Consumption of animal fat and red meat has been associated with risk of prostate cancer in multiple studies.[42, 44] Hypothesized mechanisms for these associations include the effects of dietary fats on endogenous hormone levels,[45] local effects of bile acids on the colonic mucosa, effects of carcinogens produced in the cooking of meat, and excessive iron intake. Excess iron absorption (particularly heme iron from meat) is a plausible, though unproved, contributor to the production of oxygen radicals.[7] Some of the large geographic differences in colon cancer rates that have been attributed to dietary factors are probably due to differences in physical activity, which is inversely related to colon cancer risk in many studies.[46–48]

Alcoholic beverages cause inflammation and cirrhosis of the liver, and liver cancer.[49] Alcohol is an important cause of oral and esophageal cancer (and is also synergistic with smoking)[49] and possibly contributes to colorectal cancer.[33, 50] Breast cancer is also associated with alcohol consumption.

Cooking food is plausible as a contributor to cancer.[51] A wide variety of chemicals are formed during cooking. Four groups of chemicals that cause tumors in rodents have attracted attention because of mutagenicity, potency, and concentration: nitrosamines, heterocyclic amines, polycyclic hydrocarbons, and furfural and similar furans. Epidemiological studies on cooking are difficult and so far are inadequate to evaluate a carcinogenic effect in humans.[52]

Tobacco

Tobacco is the most important global cause of cancer and is preventable. Smoking contributes to about one-third of U.S. cancer, about one-quarter of U.S. heart disease, and about 400,000 premature deaths per year in the United States.[53] Tobacco is a known cause of cancer of the lung, bladder, mouth, pharynx,

> **Tobacco is the most important global cause of cancer and is preventable. Smoking contributes to about one-third of U.S. cancer, about one-quarter of U.S. heart disease, and about 400,000 premature deaths per year in the United States.**

pancreas, kidney, stomach, larynx, esophagus,[4] and possibly colon.[54–56] It causes even more deaths by diseases other than cancer. Tobacco is causing about 3 million deaths per year worldwide and will, if today's rates of smoking continue, cause about 10 million deaths per year a few decades from now.[53] "Over the whole of the second half of this century (1950–2000) the total number of deaths caused by smoking in developed countries will be about 60 million."[4] The evidence for environmental tobacco smoke as a cause of cancer is much weaker; it has been estimated to cause up to 3,000 additional cases of cancer in the U.S.,[57, 58] though this estimate has been strongly disputed.[59]

The carcinogenic mechanisms of tobacco smoking are not well understood Smoking is a severe oxidative stress, and smoke contains a wide variety of mutagens and rodent carcinogens. The oxidants in cigarette smoke (mainly nitrogen oxides) deplete the body's antioxidants. Thus, smokers must ingest two to three times more ascorbate than nonsmokers to achieve the same level of ascorbate in blood, but they rarely do.[60–62] Inadequate diets (and smoking) of fathers may result not only in damage to their somatic DNA but to the DNA of their sperm. When the level of dietary ascorbate is insufficient to keep seminal fluid ascorbate at an adequate level, then oxidative lesions in sperm DNA are increased two and a half times.[63] An inadequate level of ascorbate is more common among single males, the poor, and smokers.[64] Paternal smoking may possibly increase the risk of birth defects and childhood cancer in offspring.[65]

Chronic Infection, Inflammation, and Cancer

White cells and other phagocytic cells of the immune system combat bacteria, parasites, and virus-infected cells by destroying them with potent mutagenic oxidizing agents. These oxidants protect humans from immediate death from infection but also cause oxidative damage to DNA, mutation, and chronic cell killing with compensatory cell division,[66, 67] thereby contributing to the carcinogenic process. Antioxidants appear to inhibit some of the pathology of chronic inflammation.[7]

Chronic infections contribute to about one-third of the world's cancer. Hepatitis B and C viruses, which infect about 500 million people worldwide, are a major cause of chronic inflammation leading to liver cancer, one of the most common cancers in Asia and Africa.[68–70] Nearly half the world's liver cancer occurs in China.[71] Vaccinating babies at birth is potentially an effective method to reduce liver cancer and is routinely done for hepatitis B in Taiwan.

The mutagenic mold toxin aflatoxin, which is found in moldy peanut and corn products, appears to interact with chronic hepatitis infection in liver cancer development.[72] Biomarker measurements on populations in Africa and China confirm that these populations are chronically exposed to high levels of aflatoxin.[73, 74] In the United States, liver cancer is rare. Although hepatitis B and C viruses infect less than 1 percent of the U.S. population, hepatitis viruses can account for half of liver cancer cases among non-Asians[75] and even higher percentages among Asians.[76]

Another major chronic infection is schistosomiasis, which is widespread in Asia and Egypt. In Asia, the eggs of *Schistosoma japonicum* deposited in the colonic mucosa cause inflammation and subsequent colon cancer.[77] In Egypt, the eggs of *S. haematobium* deposited in the bladder cause inflammation and bladder cancer.[77] *Opisthorchis viverrini*, a liver fluke, infects millions of people in Thailand and Malaysia. The flukes lodge in bile ducts and increase the risk of cholangiocarcinoma.[77] *Chlonorchis sinensis* infections in millions of Chinese increase the risk for biliary tract cancer.[77] *Helicobacter pylori* bacteria, which infect the stomachs of more than one-third of the world's population, are a major cause of stomach cancer, ulcers, and gastritis.[77] In wealthy countries the infection is often asymptomatic, which suggests that inflammation may be at least partially suppressed, possibly by adequate levels of dietary antioxidants.[78]

Chronic inflammation resulting from noninfectious sources can also lead to cancer. For example, asbestos exposure leading to chronic inflammation may be in good part the reason it is a significant risk factor for cancer of the lung.[79, 80]

Human papilloma virus, a major risk factor for cervical cancer, does not appear to work through an inflammatory mechanism.[81] It is spread by sexual contact, an effective way of transmitting viruses.

Hormones

Henderson, Pike, and colleagues have reviewed the extensive literature indicating a large role of sex hormones in cancer causation, possibly contributing to as much as one-third of all cancer cases.[45] Hormones are likely to act by causing cell division. Endometrial cancer appears most exquisitely sensitive to cumulative estrogen exposure, with risks being elevated ten- to twenty-fold by long-term use of exogenous estrogens.[82] Estrogens increase the division of endometrial cells, but progestogens reduce division; thus the addition of progestogens to estrogen therapy after menopause may reduce the risk of endometrial cancer.[45]

Ovarian cancer seems to be related to factors that increase the division of surface epithelial cells; for example, pregnancies substantially reduce the number of ovulations and therefore reduce cell division and the risk of this malignancy.[45] Oral contraceptives, which also block ovulation, decrease risk, by as much as 50 percent with five years of use.[83]

Factors that increase cumulative exposure to estrogens, such as early age at menarche, late menopause, and prolonged estrogen therapy after menopause, increase the risk of breast cancer.[45, 84] Breast cancer cells proliferate in the presence of estrogens, and progestogens also appear to enhance cell division.[45] Moreover, the addition of progestogens to estrogen therapy does not reduce, and may even increase, the risk of breast cancer.[85] Pregnancy has a complex relation with breast cancer; risk is initially increased for one to two decades (probably due to hormonal stimulation), but lifetime incidence is ultimately reduced,[86] possibly due to a permanent differentiation of stem cells resulting in less prolif-eration.[87] Lactation modestly reduces breast cancer incidence.[88, 89]

The evidence that hormones influence the incidence of breast cancer sug-gests ways of reducing incidence. One proposal is to develop a hormonal con-traceptive that mimics the effect of an early menopause; this might reduce breast cancer by half.[90] Exercise may lower breast cancer risk in young women, probably through influencing hormone levels.[91] Alcohol consumption, which has been consistently associated with breast cancer risk in large prospective studies, as well as in most case-control studies,[92] appears to increase endoge-nous estrogen levels;[93] thus, reduced consumption of alcohol may decrease breast cancer risk.

Less Important Factors Influencing Cancer Risks

Occupation

The International Agency for Research on Cancer of the World Health Orga-nization (IARC) evaluates potential cancer risks to humans from a variety of chemical exposures.[94] Half of the sixty chemicals and chemical mixtures they have evaluated as having sufficient evidence of carcinogenicity in humans are occupational exposures, which tend to be concentrated among small groups of people who have been chronically exposed at high levels—for example, work-place exposures such as the rubber industry and coke production and exposure to specific aromatic amines, petrochemicals, metals, and so forth. The issue of how much cancer can be attributed to occupational exposure has been contro-versial, but a few percent seems a reasonable estimate. Doll and Peto have dis-cussed difficulties in making such estimates, including the lack of accurate data on history of exposure and current exposures, as well as confounding factors such as socioeconomic status and smoking.[5] Lung cancer was by far the largest con-tributor to their estimate of the proportion of cancers due to occupation. The preeminence of smoking as a cause of lung cancer confounds the interpretation of rates in terms of particular workplace exposures, such as asbestos, which ap-pears to multiply rather than just add to the effect of smoking. In contrast, as-bestos alone is a known risk factor for mesothelioma. Asbestos was estimated to cause a high proportion of occupational cancers;[5] however, recent estimates for asbestos-related cancer are lower.[95, 96]

Exposures in the workplace can be high compared to other chemical expo-sures to humans, as in food, air, or water. We have argued that increased cell division rates are important in causing mutation and cancer, and therefore extrapolation from the results of high-dose animal cancer tests to low-dose human exposures cannot be done without considering the mechanism of carcinogenesis for the chemical.[97, 98] However, past occupational exposures have often been high, and comparatively little quantitative extrapolation may be required from high-dose rodent tests to high-dose occupational exposures. Since occupational cancer is concentrated among small groups exposed at high levels, there is an opportunity to control or eliminate risks once identified. However, few chemicals are regulated by the U.S. Occupational Safety and Health Administration (OSHA) as potential human carcinogens. For seventy-five rodent carcinogens regulated by OSHA with permissible exposure limits (PELs), we recently ranked potential carcinogenic hazards on an index that compares the permitted dose rate to workers with the carcinogenic dose to rodents.[99] We found that for nine chemicals, the permitted exposures were within a factor of 10 of the rodent carcinogenic dose and for seventeen they were between 10 and 100 times lower. These values are high in comparison to hypothetical risks regulated by other federal agencies. An additional 120 rodent carcinogens had no OSHA PEL, suggesting the need for further regulatory attention.

Sun Exposure

Exposure to the sun is the major cause of skin cancer, with melanoma being of the most importance. Exposure during the early decades of life, particularly when sufficient to cause burns, appears to be the dominant factor.[100] Prevention of skin cancer is feasible if fair-skinned people become aware of this information and take protective measures.

> **Exposure to the sun is the major cause of skin cancer.**

Medical Interventions

Some cancer chemotherapeutic drugs, particularly alkylating agents, cause second malignancies, most commonly leukemias, lymphomas, and sarcomas.[101] Some formerly used drugs, such as phenacetin and diethylstilbesterol, were associated with increased cancer risk.[102] Potent immunosuppressive agents such as cyclosporin also increase the risk of a variety of cancers.[103] Estrogen replacement therapy increases the risk of endometrial and breast cancer. Diagnostic X-rays have also contributed to malignancies.[104] Although these side effects should weigh in therapeutic decisions, the overall contribution of medications and diagnostic procedures to cancer incidence is small.

Pollution

Synthetic pollutants are feared by much of the public as major causes of cancer, but this is a misconception. Even if the worst-case risk estimates for synthetic pollutants that have been made by the EPA were assumed to be true risks, the proportion of cancer that EPA could prevent by regulation would be tiny.[105] Epidemiological studies, moreover, are difficult to conduct because of inadequacies in exposure assessment and failure to account for confounding factors like smoking, diet, and geographic mobility.

Indoor air is generally of greater concern than outside air because 90 percent of people's time is spent indoors, and the concentrations of pollutants tend to be higher than outdoors. The most important carcinogenic air pollutant, however, is likely to be radon, which occurs naturally as a radioactive gas that is generated as a decay product of the radium present in trace quantities in the earth's crust. Radon enters houses primarily in air that is drawn from the underlying soil. Based on epidemiological studies of high exposures to underground miners, radon has been estimated to cause as many as 15,000 lung cancers per year in the United States, mostly among smokers due to the synergistic effect with smoking.[106–108] Epidemiological studies of radon exposures in homes[109, 110] have failed to demonstrate convincingly an excess risk. About 50,000 to 100,000 of the homes in the United States (0.1 percent) are estimated to have annual average radon levels approximately twenty times the national average, and their inhabitants receive annual radiation doses that exceed the current occupational standard for underground miners. Efforts to identify high-radon houses indicate that they occur most frequently in concentrated geographic areas.[111] In high areas, individuals can perform a measurement in their homes for about twenty dollars and, if levels are high reduce them substantially using available contractors, for perhaps fifteen hundred dollars.[107]

Water pollution as a risk factor for cancer appears small. Among potential hazards that have been of concern, the most important are radon (exposure is small compared to air) and arsenate. Natural arsenate is a known human carcinogen at high doses,[112, 113] and further research is needed on the mechanism and dose response in humans. Chlorination of water, an important public health intervention, produces large numbers of chlorine-containing chemicals as byproducts, some of them rodent carcinogens. Evidence that chlorination of water increases human cancer has been judged inadequate.[114]

Hereditary Factors

Inherited factors clearly contribute to some percentage of cancer, particularly childhood cancer and cancer in early adulthood. Overall, cancer increases exponentially with age except for a blip on the curve for childhood cancer, which is thought to be mainly due to inheriting a mutant cancer gene.[115, 116] Heredity is likely to affect susceptibility to all cancers; to what extent is not clear, though it is obvious that skin color plays a large role in sun-associated cancers such as melanoma. With the rapid progress of molecular biology, the genetic factors will soon become understood. Factors other than heredity play the dominant causative role for most major cancers as indicated by the large differences in cancer rates among countries, the observation that migrants adopt cancer rates close to those of their host populations, and the large temporal changes in the rates of many cancers.

Distractions: Animal Cancer Tests and the Rachel Carson Fallacy

The idea that there is an epidemic of human cancer caused by synthetic industrial chemicals is not supported by either toxicology or epidemiology. Though some epidemiologic studies suggest an association between cancer and low levels of industrial pollutants, the studies do not correct for diet, which is a potentially large confounding factor; moreover, the levels of pollutants are low and rarely seem plausible as a causal factor when compared to the background of natural chemicals that are rodent carcinogens.[117]

Rachel Carson had a fundamental misconception: "For the first time in the history of the world, every human being is now subjected to contact with dangerous chemicals, from the moment of conception until death."[118] She was wrong: the vast bulk of the chemicals humans are exposed to are natural, and for every chemical some amount is dangerous.

Animal cancer tests are usually done on synthetic chemicals at near-toxic doses of the chemical. These results have been misinterpreted as meaning that low doses of synthetic chemicals and industrial pollutants are relevant to human cancer. About *half* of the chemicals tested, synthetic or natural, are carcinogenic to rats or mice at these high doses.[97, 119, 120] A plausible explanation for the high proportion of positive results is that testing at near-toxic levels frequently can cause chronic cell killing and consequent cell replacement, which is a risk factor for cancer that can be limited to high doses.[97, 98]

The great bulk of chemicals ingested by humans is natural, by both weight and number. For example, 99.99 percent of the pesticides in the diet are naturally present in plants to ward off insects and other predators.[121] Half of the natural pesticides tested (twenty-nine of fifty-seven) are rodent carcinogens.[117] Reducing exposure to the 0.01 percent that are synthetic chemicals will not appreciably reduce cancer rates. On the contrary, fruits and vegetables are important for reducing cancer; making them more expensive by reducing the use of synthetic pesticides is likely to increase cancer by reducing consumption. People with low incomes eat fewer fruits and vegetables[122] and spend a higher percentage of their income on food.

Humans also ingest large numbers of natural chemicals from cooking food. For example, more than a thousand chemicals have been identified in roasted coffee; more than half of those tested (nineteen of twenty-six) are rodent carcinogens (table 5-1).[117] There are more natural rodent carcinogens by weight in a single cup of coffee than potentially carcinogenic synthetic pesticide residues in the average U.S. diet in a year, and there are still a thousand known chemicals in roasted coffee that have not been tested. This does not necessarily mean that coffee is dangerous, but that animal cancer tests and worst-case risk assessments build in enormous safety factors and should not be considered true risks.

Because of their unusual lipophilicity and long environmental persistence, there has been particular concern for a small group of polychlorinated synthetic chemicals such as DDT and PCBs. There is no convincing epidemiological evidence,[123] nor is there much toxicological plausibility,[117] that the levels usually found in the environment are likely to be a significant contributor to cancer. TCDD, which is produced naturally by burning when chloride ion is present, as in forest fires and as an industrial by-product, is an unusually potent rodent carcinogen but seems unlikely to be a significant human carcinogen at the levels to which the general population is exposed.

The reason we humans can eat the tremendous variety of natural "rodent carcinogens" in our food is that, like other animals, we are extremely well protected by many general defense enzymes, most of which are inducible; that is whenever a defense enzyme is in use, more of it is made.[124] Defense enzymes are effective against both natural and synthetic chemicals, such as potentially mutagenic reactive chemicals. One does not expect, nor does one find, a general difference between synthetic and natural chemicals in ability to cause cancer in high-dose rodent tests.[97, 117]

We have ranked possible carcinogenic hazards from known rodent carcinogens, using an index that relates human exposure to carcinogenic potency in ro-

TABLE 5-1   Carcinogenicity Status of Natural Chemicals in Roasted Coffee

Positive	Not Positive	Yet to Test
Acetaldehyde	Acrolein	~1000 chemicals
Benzaldehyde	Biphenyl	
Benzene	Eugenol	
Benzofuran	Nicotinic acid	
Benzo(*a*)pyrene	Phenol	
Caffeic acid	Piperidine	
Catechol	[Uncertain: Caffeine]	
1,2,5,6-Dibenzanthracene		
Ethanol		
Ethylbenzene		
Formaldehyde		
Furan		
Furfural		
Hydrogen peroxide		
Hydroquinone		
Limonene		
MeIQ		
Styrene		
Toluene		

Source: L. Gold et al., "Rodent Carcinogens: Setting Priorities," *Science*, October 9, 1992, p. 262.

dents (HERP) (table 5-2). [117] Our ranking does not estimate risks, which current science does not have the ability to do. Rather, possible hazards of synthetic chemicals are put into perspective against the background of naturally occurring rodent carcinogens in typical portions of common foods. The residues of synthetic pesticides or environmental pollutants rank low in comparison to the background, despite the fact that such a comparison gives a minimal view of hypothetical background hazards because so few chemicals in the natural world have been tested for carcinogenicity in rodents. Our results indicate that many ordinary foods would not pass the regulatory criteria used for synthetic chemicals. However, these results do not necessarily indicate that coffee consumption, for example, is a significant risk factor for human cancer even though the HERP of caffeic acid in a cup of coffee is about a thousand-fold greater than the HERP equivalent of the one-in-a-million worst-case risk used by the EPA (Table 5–2).

TABLE 5-2   A Ranking of Possible Carcinogenic Hazards from Natural and Synthetic Chemicals

Possible Hazard: HERP (%)	Daily Human Exposure	Human Dose of Rodent Carcinogen
140	EDB: workers, daily intake (high exposure)	EDB, 150 mg (before 1977)
17	Clofibrate (avg daily dose)	Clofibrate, 2 g
16	Phenobarbital, 1 sleeping pill	Phenobarbital, 60 mg
[14]	Isoniazid pill (prophylactic dose)	Isoniazid, 300 mg
6.2	Comfrey-pepsin tablets, 9 daily	**Comfrey root, 2.7 g**
[5.6]	Metronidazole (therapeutic dose)	Metronidazole, 2 g
4.7	**Wine (250 ml)**	**Ethyl alcohol, 30 ml**
4.0	Formaldehyde: workers' avg daily intake	Formaldehyde, 6.1 mg
2.8	**Beer (12 oz; 354 ml)**	**Ethyl alcohol, 18 ml**
1.4	Mobile home air (14 hour/day)	Formaldehyde, 2.2 mg
1.3	Comfrey-pepsin tablets, 9 daily	**Symphytine, 1.8 mg**
0.4	Conventional home air (14 hr/day)	Formaldehyde, 598 µg
[0.3]	Phenacetin pill (avg dose)	Phenacetin, 300 mg
0.3	**Lettuce, 1/8 head (125 g)**	**Caffeic acid, 66.3 mg**
0.2	**Natural root beer (12 oz; 354 ml)**	**Safrole, 6.6 mg (banned)**
0.1	**Apple, 1 whole (230 g)**	**Caffeic acid, 24.4 mg**
0.1	**1 Mushroom (15 g)**	**Mix of hydrazines, etc.**
0.1	**Basil (1 g of dried leaf)**	**Estragole, 3.8 mg**
0.07	**Mango, 1 whole (245 g; pitted)**	**d-Limonene, 9.8 mg**
0.07	**Pear, 1 whole (200 g)**	**Caffeic acid, 14.6 mg**
0.07	**Brown mustard (5 g)**	**Allyl isothiocyanate, 4.6 mg**
0.06	Diet cola (12 oz; 354 ml)	Saccharin, 95 mg
0.06	**Parsnip, 1/4 (40 g)**	**8-Methoxypsoralen, 1.28 mg**
0.04	**Orange juice (6 oz; 177 ml)**	**d-Limonene, 5.49 mg**
0.04	**Coffee, 1 cup (from 4 g)**	**Caffeic acid, 7.2 mg**
0.03	**Plum, 1 whole (50 g)**	**Caffeic acid, 6.9 mg**
0.03	**Safrole: U.S. avg from spices**	**Safrole, 1.2 mg**
0.03	**Peanut butter (32 g; 1 sandwich)**	**Aflatoxin, 64 ng**
0.03	**Comfrey herb tea (1.5 g)**	**Symphytine, 38 µg**
0.03	**Celery, 1 stalk (50 g)**	**Caffeic acid, 5.4 mg**

TABLE 5-2   Continued

0.03	Carrot, 1 whole (100 g)	Caffeic acid, 5.16 mg
0.03	Pepper, black: U.S. avg (446 mg)	d-Limonene, 3.57 mg
0.02	Potato, 1 (225 g; peeled)	Caffeic acid, 3.56 mg
0.008	Swimming pool, 1 hour (for child)	Chloroform, 250 µg
0.008	Beer, before 1979 (12 oz; 354 ml)	Dimethylnitrosamine, 1 µg
0.006	Bacon, cooked (100 g)	Diethylnitrosamine, 0.1 µg
0.006	Well water, 1 liter contaminated (worst in Silicon Valley, CA)	Trichloroethylene, 2.8 mg
0.005	Coffee, 1 cup (from 4 g)	Furfural, 630 µg
0.004	Bacon, pan fried (100 g)	N-nitrosopyrrolidine, 1.7 µg
0.003	Nutmeg: U.S. avg (27.4 mg)	d-Limonene, 466 µg
0.003	1 mushroom (15 g)	Glutamyl p-hydrazino-benzoate, 630 µg
0.003	Conventional home air (14 hr/day)	Benzene, 155 µg
0.003	Sake (250 ml)	Urethane, 43 µg
0.003	Bacon, cooked (100 g)	Dimethylnitrosamine, 300 ng
0.002	White bread, 2 slices (45 g)	Furfural, 333 µg
0.002	Apple juice (6 oz; 177 ml)	UDMH, 5.89 µg (from Alar, 1988)
0.002	Coffee, 1 cup (from 4 g)	Hydroquinone, 100 µg
0.002	Coffee, 1 cup (from 4 g)	Catechol, 400 µg
0.002	DDT: daily dietary avg	DDT, 13.8 µg (before 1972 ban)
0.001	Celery, 1 stalk (50 g)	8-Methoxypsoralen, 30.5 µg
0.001	Tapwater, 1 liter	Chloroform, 83 µg (U.S. avg)
0.001	Heated sesame oil (15 g)	Sesamol, 1.13 mg
0.0008	DDE: daily dietary avg	DDE, 6.91 µg (before 1972 ban)
0.0006	Well water, 1 liter contaminated (Woburn, MA)	Trichloroethylene, 267 µg
0.0005	1 mushroom (15 g)	p-Hydrazinobenzoate, 165 µg
0.0005	Hamburger, pan fried (3 oz; 85 g)	PhIP, 1.28 µg
0.0005	Jasmine tea, 1 cup (2 g)	Benzyl acetate, 460 µg
0.0005	Salmon, pan fried (3 oz; 85 g)	PhIP, 1.18 µg
0.0004	EDB: Daily dietary avg	EDB, 420 ng (from grain; before 1984 ban)
0.0004	Beer (12 oz; 354 ml)	Furfural, 54.9 µg
0.0003	Well water, 1 liter contaminated (Woburn, MA)	Tetrachloroethylene, 21 µg

TABLE 5-2   Continued

0.0003	Carbaryl: daily dietary avg	Carbaryl, 2.6 µg (1990)[a]
0.0002	Apple, 1 whole (230 g)	UDMH, 598 ng (from Alar, 1988)
0.0002	**Parsley, fresh (1 g)**	**8-Methoxypsoralen, 3.6 µg**
0.0002	Toxaphene: daily dietary avg	Toxaphene, 595 ng (1990)[a]
0.00008	**Hamburger, pan fried (3 oz; 85 g)**	**MeIQx, 111 ng**
0.00008	DDE/DDT: daily dietary avg	DDE, 659 ng (1990)[a]
0.00003	**Whole wheat toast, 2 slices (45 g)**	**Urethane, 540 ng**
0.00002	Dicofol: daily dietary avg	Dicofol, 544 ng (1990)[a]
0.00002	**Cocoa (4 g)**	**α-Methylbenzyl alcohol, 5.2 µg**
0.00001	**Lager beer (12 oz; 354 ml)**	**Urethane, 159 ng**
0.000008	**Hamburger, pan fried (3 oz; 85 g)**	**IQ, 23.4 ng**
0.000001	Lindane: daily dietary avg	Lindane, 32 ng (1990)[a]
0.0000004	PCNB: daily dietary avg	PCNB (Quintozene) 19.2 ng (1990)[a]
0.0000001	**Hamburger, pan fried (3 oz; 85 g)**	**MeIQ, 1.28 ng**
0.0000001	Chlorobenzilate: daily dietary avg	Chlorobenzilate, 6.4 ng (1989)[a]
<0.00000001	Chlorothalonil: daily dietary avg	Chlorothalonil, <6.4 ng (1990)[a]
0.000000008	Folpet: daily dietary avg	Folpet, 12.8 ng (1990)[a]
0.000000007	**Coffee, 1 cup (from 4 g)**	**MeIQ, 0.064 ng**
0.000000006	Captan: daily dietary avg	Captan, 11.5 ng (1990)[a]

Source: L. Gold et al., "Rodent Carcinogenes: Setting Priorities," *Science*, October 9, 1992, p. 117.
Notes: Daily human exposure: Reasonable daily intakes are used to facilitate comparisons. The calculations assume a daily dose for a lifetime; where drugs are normally taken only for a short period, we have bracketed the HERP *(human exposure rodent potency index)*. *Possible hazard:* HERP. The human dose of rodent carcinogen is divided by 70 kg to give a mg/kg of human exposure, and this dose is given as the percentage of the daily dose of the chemical tested in animals that will reduce by half the proportion of animals without tumor. These values used in the HERP calculation are averages calculated by taking the harmonic mean of these doses in the positive tests in that species from the Carcinogenic Potency Database. Average values have been calculated separately for rats and mice, and the more sensitive species is used for calculating possible hazard.

Natural chemicals are in boldface type.

[a] Estimate is based on average daily dietary intake for 60–65-year-old females, the only adult group reported for 1990. Because of the agricultural usage of these chemicals and the prominence of fruits and vegetables in the diet of older Americans, the residues are generally slightly higher than for other adult age groups.

Adequate risk assessment from animal cancer tests requires more information about many aspects of toxicology, such as effects on cell division, induction of defense and repair systems, and species differences.

Linear extrapolation from the near-toxic dose in rodents to low-level exposure in humans for synthetic chemicals, while ignoring the enormous natural background, has led to exaggerated cancer risk estimates and an imbalance in the perception of hazard and the allocation of resources. Although some epidemiologic studies find an association between cancer and low levels of industrial pollutants, the studies do not correct for diet, a potentially large confounding factor, and the levels of pollutants are low and rarely seem plausible as a causal factor.[117] The idea that there is an epidemic of human cancer caused by synthetic industrial chemicals is not supported by either toxicology or epidemiology.

If the costs were minor the issue of putting hypothetical risks into perspective would not be so important, but the costs are huge.[125, 126] Costs escalate as cleanliness approaches perfection. The idea of trade-offs is not adequately dealt with in most attempts to deal with pollutants; instead it is assumed that upper-

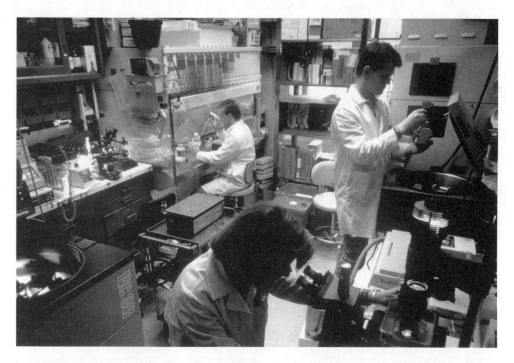

Biomedical research will shed light on how living cells become cancerous. It is this research, rather than maximum tolerated dosage tests on rats, that will lead to breakthroughs in the war against cancer.

bound risk assessment to one in a million protects the public. The Office of Management and Budget Report[127] and also the Harvard Center for Risk Analysis report[128] that compared costs for risk reduction among government agencies concluded that the money spent to save a life by EPA is often orders of magnitude higher than that spent by many other government agencies. EPA risk estimates are based on "risk assessment" (default, worst-case, linear extrapolations to one-in-a-million risk), unlike most other government agencies, so the actual discrepancy between EPA and many other agencies is even greater. Many scholars have pointed out that expensive regulations intended to save lives[129] may actually lead to increased deaths, in part by diverting resources from important health risks and in part because higher incomes are associated with lower mortality risks.[130, 131] Worst-case assumptions in risk assessment is a policy decision, not a scientific one, and confuses attempts to allocate money effectively for risk abatement. Regulating trivial risks impedes effective risk management.[132]

Discussion

Epidemiological evidence in humans is sufficient to identify several broad categories of cancer causation for which the evidence is strong and plausible. Since many of these are avoidable, it is possible to reduce incidence rates of many types of cancer. In a monumental 1981 review of avoidable risks of cancer in the United States,[5] Doll and Peto attributed 30 percent of cancer deaths to tobacco and 35 percent to dietary factors, although the plausible contribution of diet ranged from 10 to 70 percent. Other factors were judged to contribute far less. Since that time the contribution of smoking appears to have increased somewhat (35 percent seems more likely), although the prevalence of smoking in adults has decreased, because the relative risk due to smoking has greatly increased for almost all cancers as well as cardiovascular disease.[53] This is probably due to both a declining risk of cancer death in nonsmokers and to the fact that the lifetime impact of smoking since adolescence is being experienced only now. Available data on diet and cancer have increased many-fold since 1981 and generally support the earlier estimate; a slightly narrower estimated range of 20 to 40 percent seems most plausible.[133] In general, new data have most strongly emphasized the inadequate consumption of protective factors rather than an excessive intake of harmful factors. The estimate for diet is revised slightly downward largely because the international contrasts in colon cancer rates are probably due to differences in physical activity as well as diet. The Doll and Peto estimate for the dietary contribution to breast cancer of 50 percent is still plausible, even though this may not be avoidable in a practical sense if rapid growth rates are the most important underlying nutritional factor. The estimate for alcoholic beverages can be increased slightly from 3 percent (plus or minus 1 percent) to 5 percent (plus or minus 1 percent) as many new studies have supported associations with breast and colon cancer. Data subsequent to 1981 have not provided a basis to alter the earlier estimates for other causes appreciably.

One approach to estimating the population impact of adopting major lifestyle factors associated with low cancer risk is to compare cancer incidence and mortality rates of the general population to those of Seventh-Day Adventists, who generally do not smoke, drink heavily, or eat much meat but do eat a diet rich in fruits and vegetables.[134, 135] Substantially lower mortality rates of lung, bladder, and colon cancers are experienced in this group; overall cancer mortality is about half that of the general U.S. population. Although this comparison has limitations—better use of medical services may contribute to reduced mortality, and imperfect compliance with recommendations may underestimate the impact of lifestyle—the results strongly suggest that a large portion of cancer deaths can be avoided using knowledge at hand. Incidence rates rather than mortality rates provide a similar picture, although the differences are somewhat less. For breast cancer the healthy behavior of Seventh-Day Adventists was not sufficient to have a major impact on risk.

> **Decreases in physical activity, and increases in smoking, obesity, and recreational sun exposure, have contributed importantly to increases in some cancers in the modern industrial world, whereas improvements in hygiene have reduced other cancers related to infection. There is no good reason to believe that synthetic chemicals underlie the major changes in incidence of some cancers.**

Decreases in physical activity, and increases in smoking, obesity, and recreational sun exposure, have contributed importantly to increases in some cancers in the modern industrial world, whereas improvements in hygiene have reduced other cancers related to infection. There is no good reason to believe that synthetic chemicals underlie the major changes in incidence of some cancers. In the United States and other industrial countries life expectancy is steadily increasing and will increase even faster as smoking declines.

REFERENCES

This work was supported by National Institute of Environmental Health Sciences Center Grant ESO1896 and National Cancer Institute Outstanding Investigator Grant CA39910 to B.N.A.;

and by the Director, Office of Energy Research, Office of Health and Environmental Research, of the U.S. Department of Energy under Contract DE-AC03-76SF00098 to L.S.G. Parts of this chapter have been adapted from reference 136, and we thank W. C. Willett for his help.

1. B. A. Miller, L. A. G. Ries, B. F. Hankey, C. L. Kosary, A. Harras, S. S. Devesa, and B. K. Edwards, *SEER Cancer Statistics Review: 1973–1990*, National Cancer Institute, NIH Pub. No. 93-2789 (1993).

2. L. M. Brown, G. D. Everett, R. Gibson, L. F. Burmeister, L. M. Schuman, and A. Blair, "Smoking and risk of non-Hodgkin's lymphoma and multiple myeloma," *Cancer Causes Control* 3 (1992): 49–55.

3. M. S. Linet, J. K. McLaughlin, A. W. Hsing, S. Wacholder, H. T. Co Chien, L. M. Schuman, E. Bjelke, and W. J. Blot, "Is cigarette smoking a risk factor for non-Hodgkin's lymphoma or multiple myeloma? Results from the Lutheran Brotherhood Cohort Study," *Leukemia Res.* 16 (1992): 621–624.

4. R. Peto, A. D. Lopez, J. Boreham, M. Thun, and C. Heath, Jr., *Mortality from Smoking in Developed Countries, 1950–2000* (Oxford: Oxford University Press, 1994).

5. R. Doll and R. Peto, "The causes of cancer. Quantitative estimates of avoidable risks of cancer in the United States today," *J. Natl. Cancer Inst.* 66 (1981): 1191–1308.

6. D. L. Davis, G. E. Dinse, and D. G. Hoel, "Decreasing cardiovascular disease and increasing cancer among whites in the United States from 1973 through 1987," *JAMA* 271 (1994): 431–437.

7. B. N. Ames, M. K. Shigenaga, and T. M. Hagen, "Oxidants, antioxidants, and the degenerative diseases of aging," *Proc. Natl. Acad. Sci. USA* 90 (1993): 7915–7922.

8. C. Von Sonntag, *The Chemical Basis of Radiation Biology* (London: Taylor & Francis, 1987).

9. E. R. Stadtman, "Protein oxidation and aging," *Science* 257, no. 5074 (1992): 1220–1224.

10. S. E. Hankinson, W. C. Willett, G. A. Colditz, J. M. Seddon, B. Rosner, F. E. Speizer, and M. J. Stampfer, "A prospective study of cigarette smoking and risk of cataract surgery in women," *JAMA* 268, no. 8 (1992): 994–998.

11. S. E. Hankinson, M. J. Stampfer, J. M. Seddon, G. A. Colditz, B. Rosner, F. E. Speizer, and W. C. Willett, "Nutrient intake and cataract extraction in women: A prospective study," *Br. Med. J.* 305 (1992): 335–339.

12. P. F. Jacques, S. C. Hartz, L. T. J. Chylack, R. B. McGandy, and J. A. Sadowski, "Nutritional status in persons with and without senile cataract: Blood vitamin and mineral levels," *Am. J. Clin. Nutr.* 48, no. 1 (1988): 152–158.

13. F. J. C. Roe, P. N. Lee, G. Conybeare, G. Tobin, D. Kelly, D. Prentice, and B. Matter, "Risks of premature death and cancer predicted by body weight in early adult life," *Hum. Exp. Toxicol.* 10 (1991): 285–288.

14. F. J. C. Roe, "Non-genotoxic carcinogenesis: Implications for testing extrapolation to man," *Mutagenesis* 4 (1989): 407–411.

15. R. K. Boutwell and M. W. Pariza, "Historical perspectives: Calories and energy expenditure in carcinogenesis," *Am. J. Clin. Nutr.* 45 (Suppl.) (1987): 151–156.

16. L. D. Youngman, J.-Y. K. Park, and B. N. Ames, "Protein oxidation associated with aging is reduced by dietary restriction of protein or calories," *Proc. Natl. Acad. Sci. USA* 89 (1992): 9112–9116.

17. D. J. Hunter and W. C. Willett, "Diet, body size, and breast cancer," *Epidemiol. Rev.* 15, no. 1 (1993): 110–132.

18. C. A. Swanson, D. Y. Jones, A. Schatzkin, L. A. Brinton, and R. G. Ziegler, "Breast cancer risk assessed by anthropometry in the NHANES I epidemiological follow-up study," *Cancer Res.* 48 (1988): 5363–5367.

19. W. C. Willett and M. J. Stampfer, "Dietary fat and cancer: Another view," *Cancer Causes Control* 1 (1990): 103.

20. G. Block, B. Patterson, and A. Subar, "Fruit, vegetables and cancer prevention: A review of the epidemiologic evidence," *Nutr. Cancer* 18 (1992): 1–29.

21. K. A. Steinmetz and J. D. Potter, "Vegetables, fruit, and cancer. I. Epidemiology," *Cancer Causes Control* 2, no. 5 (1991): 325–357.

22. M. J. Hill, A. Giacosa, and C. P. J. Caygill, eds., *Epidemiology of Diet and Cancer* (West Sussex, England: Ellis Horwood Limited, 1994).

23. G. R. Howe, T. Hirohata, and T. G. Hislop, "Dietary factors and risk of breast cancer: Combined analysis of 12 case-control studies," *J. Natl. Cancer Inst.* 82 (1990): 561–569.

24. G. Block, "The data support a role for antioxidants in reducing cancer risk," *Nutr. Rev.* 50 (1992): 207–213.

25. B. H. Patterson, G. Block, W. F. Rosenberger, D. Pee, and L. L. Kahle, "Fruit and vegetables in the American diet: Data from the NHANES II survey," *Am. J. Public Health* 80 (1990): 1443–1449.

26. K. A. Steinmetz and J. D. Potter, "Vegetables, fruit, and cancer. II. Mechanisms," *Cancer Causes Control* 2, no. 6 (1991): 427–442.

27. J. T. MacGregor, R. Schlegel, C. M. Wehr, P. Alperin, and B. N. Ames, "Cytogenetic damage induced by folate deficiency in mice is enhanced by caffeine," *Proc. Natl. Acad. Sci. USA* 87 (1990): 9962–9965.

28. B. C. Blount, "Detection of DNA damage caused by folate deficiency and chronic inflammation" (1994). Ph.D. thesis, University of California, Berkeley.

29. R. B. Everson, C. M. Wehr, G. L. Erexson, and J. T. MacGregor, "Association of marginal folate depletion with increased human chromosomal damage *in vivo*: Demonstration by analysis of micronucleated erythrocytes," *J. Natl. Cancer Inst.* 80 (1988): 525–529.

30. A. Bendich and C. E. Butterworth Jr., eds., *Micronutrients in Health and in Disease Prevention* (New York: Marcel Dekker, 1991).

31. S. A. Glynn and D. Albanes, "Folate and cancer: A review of the literature," *Nutr. Cancer* 22 (1994): 101–119.

32. E. Giovannucci, M. J. Stampfer, G. A. Colditz, E. B. Rimm, D. Trichopoulos, B. A. Rosner, F. E. Speizer, and W. C. Willett, "Folate, methionine, and alcohol intake and risk of colorectal adenoma," *J. Natl. Cancer Inst.* 85 (1993): 875–884.

33. J. L. Freudenheim, S. Graham, J. R. Marshall, B. P. Haughey, S. Cholewinski, and G. Wilkin-

son, "Folate intake and carcinogenesis of the colon and rectum," *Int. J. Epidemiol.* 20, no. 2 (1991): 368–374.

34. D. Rush, "Periconceptional folate and neural tube defect," *Am. J. Clin. Nutr.* 59 (1994): 511S–516S.

35. F. R. Senti and S. M. Pilch, "Analysis of folate data from the second National Health and Nutrition Examination Survey (NHANES II)," *J. Nutr.* 115 (1985): 1398–1402.

36. L. B. Bailey, P. A. Wagner, G. J. Christakis, C. G. Davis, H. Appledorf, P. E. Araujo, E. Dorsey, and J. S. Dinning, "Folacin and iron status and hematological findings in black and Spanish-American adolescents from urban low-income households," *Am. J. Clin. Nutr.* 35 (1982): 1023–1032.

37. L. B. Bailey, P. A. Wagner, G. J. Christakis, P. E. Araujo, H. Appledorf, C. G. Davis, J. Masteryanni, and J. S. Dinning, "Folacin and iron status and hematological findings in predominately black elderly persons from urban low-income households," *Am. J. Clin. Nutr.* 32 (1979): 2346–2353.

38. B. Trock, E. Lanza, and P. Greenwald, "Dietary fiber, vegetables, and colon cancer: Critical review and meta-analyses of the epidemiologic evidence," *J. Natl. Cancer Inst.* 82, no. 8 (1990): 650–661.

39. S. H. Safe, "Dietary and environmental estrogens and antiestrogens and their possible role in human disease," *Environ. Sci. Pollution Res.* 1, (1994): 29–33.

40. B. Armstrong and R. Doll, "Environmental factors and cancer incidence and mortality in different countries, with special reference to dietary practices," *Int. J. Cancer* 15, no. 4 (1975): 617–631.

41. W. C. Willett, M. J. Stampfer, G. A. Colditz, B. A. Rosner, and F. E. Speizer, "Relation of meat, fat, and fiber intake to the risk of colon cancer in a prospective study among women," *N. Engl. J. Med.* 323, no. 24 (1990): 1664–1672.

42. E. Giovannucci, E. B. Rimm, J. J. Stampfer, G. A. Colditz, A. Ascherio, and W. C. Willett, "Intake of fat, meat, and fiber in relation to risk of colon cancer in men," *Cancer Res.* 54, no. 9 (1994): 2390–2397.

43. R. A. Goldbohm, P. A. van der Brandt, P. van 't Veer, H. A. M. Brants, E. Dorant, F. Sturmans, and R. J. J. Hermus, "A prospective cohort study on the relation between meat consumption and the risk of colon cancer," *Cancer Res.* 54 (1994): 718–723.

44. L. Le Marchand, L. N. Kolonel, L. R. Wilkens, B. C. Myers, and T. Hirohata, "Animal fat consumption and prostate cancer: A prospective study in Hawaii," *Epidemiology* 5 (1994): 276–282.

45. B. E. Henderson, R. K. Ross, and M. C. Pike, "Toward the primary prevention of cancer," *Science* 254 (1991): 1131–1138.

46. M. Gerhardsson, B. Floderus, and S. E. Norell, "Physical activity and colon cancer risk," *Int. J. Epidemiol.* 17, no. 4 (1988): 743–746.

47. M. L. Slattery, M. C. Schumacher, K. R. Smith, D. W. West, and N. Abd-Eghany, "Physical activity, diet, and risk of colon cancer in Utah," *Am. J. Epidemiol.* 128 (1988): 989–999.

48. M. J. Thun, E. E. Calle, M. M. Namboodiri, W. D. Flanders, R. J. Coates, T. Byers, P. Bof-

fetta, L. Garfinkel, and C. W. J. Heath, "Risk factors for fatal colon cancer in a large prospective study," *J. Natl. Cancer Inst.* 84, no. 19 (1992): 1491–1500.

49. International Agency for Research on Cancer, *Alcohol Drinking*, IARC Monograph (Lyon, France: International Agency for Research on Cancer, 1988).

50. E. Giovannucci, E. B. Rimm, A. Ascherio, M. J. Stampfer, G. A. Colditz, and W. C. Willett, "Alcohol, methyl-deficient diets and risk of colon cancer in men" (submitted for publication).

51. T. Sugimura, S. Sato, H. Ohgaki, S. Takayama, M. Nagao, and K. Wakabayaski, *Genetic Toxicology of the Diet* (New York: Alan R. Liss, 1986).

52. International Agency for Research on Cancer, *Some Naturally Occurring Substances: Food Items and Constituents, Heterocyclic Aromatic Amines and Mycotoxins.* IARC Monographs (Lyon, France: International Agency for Research on Cancer, 1993).

53. R. Peto, A. D. Lopez, J. Boreham, M. Thun, and C. Heath, Jr., "Mortality from tobacco in developed countries: Indirect estimation from national vital statistics," *Lancet* 339 (1992): 1268–1278.

54. E. Giovannucci, E. B. Rimm, M. J. Stampfer, G. A. Colditz, A. Ascherio, J. Kearney, and W. C. Willett, "A prospective study of cigarette smoking and risk of colorectal adenoma and colorectal cancer in U.S. men," *J. Natl. Cancer Inst.* 86 (1994): 183–191.

55. E. Giovannucci, G. A. Colditz, M. J. Stampfer, D. Hunter, B. A. Rosner, W. C. Willett, and F. E. Speizer, "A prospective study of cigarette smoking and risk of colorectal adenoma and colorectal cancer in U.S. women," *J. Natl. Cancer Inst.* 86 (1994): 192–199.

56. J. E. Fielding, "Preventing colon cancer: Yet another reason not to smoke," *J. Natl. Cancer Inst.* 86 (1994): 162–164.

57. U.S. Environmental Protection Agency, *Respiratory Health Effects of Passive Smoking: Lung Cancer and Other Disorders*, Office of Health and Environmental Assessment, Office of Research and Development, Washington, D.C. (1992).

58. E. T. H. Fontham, P. Correa, P. Reynolds, A. Wu-Williams, P. A. Buffler, R. S. Greenberg, V. W. Chen, T. Alterman, P. Boyd, D. F. Austin, and J. Liff, "Environmental tobacco smoke and lung cancer in nonsmoking women," *JAMA* 271 (1994): 1752–1759.

59. G. Huber, R. Brockie, and V. Mahajan, "Smoke and mirrors: The EPA's flawed study of environmental tobacco smoke and lung cancer," *Regulation* 16, no. 3 (1993): 46.

60. G. Schectman, J. C. Byrd, and R. Hoffman, "Ascorbic acid requirements for smokers: Analysis of a population survey," *Am. J. Clin. Nutr.* 53, no. 6 (1991): 1466–1470.

61. G. G. Duthie, J. R. Arthur, and W. P. T. James, "Effects of smoking and vitamin E on blood antioxidant status," *Am. J. Clin. Nutr.* 53 (1991): 1061S–1063S.

62. M. H. Bui, A. Sauty, F. Collet, and P. Leuenberger, "Dietary vitamin C intake and concentrations in the body fluids and cells of male smokers and nonsmokers," *J. Nutr.* 122 (1991): 312–316.

63. C. G. Fraga, P. A. Motchnik, M. K. Shigenaga, H. J. Helbock, R. A. Jacob, and B. N. Ames, "Ascorbic acid protects against endogenous oxidative damage in human sperm," *Proc. Natl. Acad. Sci. USA* 88 (1991): 11003–11006.

64. B. Patterson and G. Block, "Fruit and vegetable consumption: national survey data," in A. Bendich and C. E. J. Butterworth, eds., *Micronutrients in Health and in Disease Prevention* (New York: Marcel Dekker, 1991), pp. 409–436.

65. B. N. Ames, P. A. Motchnik, C. G. Fraga, M. K. Shigenaga, and T. M. Hagen, "Antioxidant Prevention of Birth Defects and Cancer," in D. R. Mattison and A. Olshan, eds., *Male-Mediated Developmental Toxicity* (New York: Plenum Publishing Corporation, 1994), pp. 243–259.

66. E. Shacter, E. J. Beecham, J. M. Covey, K. W. Kohn, and M. Potter, "Activated neutrophils induce prolonged DNA damage in neighboring cells [published erratum appears in Carcinogenesis 1989 Mar;10(3):628]," *Carcinogenesis* 9, no. 12 (1988): 2297–2304.

67. K. Yamashina, B. E. Miller, and G. H. Heppner, "Macrophage-mediated induction of drug-resistant variants in a mouse mammary tumor cell line," *Cancer Res.* 46 (1986): 2396–2401.

68. R. P. Beasley, "Hepatitis B virus," *Cancer* 61 (1987): 1942–1956.

69. E. Tabor and K. Kobayashi, "Hepatitis C virus, a causative infectious agent of non-A, non-B hepatitis: Prevalence and structure. Summary of a conference on hepatitis C virus as a cause of hepatocellular carcinoma," *J. Natl. Cancer Inst.* 84 (1992): 86–90.

70. M.-W. Yu, S.-L. You, A.-S. Chang, S.-N. Lu, Y.-F. Liaw, and C.-J. Chen, "Association between hepatitis C virus antibodies and hepatocellular carcinoma in Taiwan," *Cancer Res.* 51 (1991): 5621–5625.

71. D. M. Parkin, J. Suernsward, and C. S. Muir, "Estimates of the worldwide frequency of twelve major cancers," *Bull. WHO* 62 (1984): 163–182.

72. G.-S. Qian, R. K. Ross, M. C. Yu, J.-M. Yuan, Y.-T. Gao, B. E. Henderson, G. N. Wogan, and J. D. Groopman, "A follow-up study of urinary markers of aflatoxin exposure and liver cancer risk in Shanghai, People's Republic of China," *Cancer Epidemiol. Biomarkers Prev.* 3 (1994): 3–10.

73. J. D. Groopman, J. Zhu, P. R. Donahue, A. Pikul, L.-S. Zhang, J. S. Chen, and G. N. Wogan, "Molecular dosimetry of urinary aflatoxin DNA adducts in people living in Guangxi Autonomous Region, People's Republic of China," *Cancer Res.* 52 (1992): 45–51.

74. W. A. Pons, "High pressure liquid chromotagraphy determinations of aflatoxins in corn," *J. Assoc. Off. Anal. Chem.* 62 (1979): 584–586.

75. M. C. Yu, M. J. Tong, S. Govindarajan, and B. E. Henderson, "Nonviral risk factors for hepatocellular carcinoma in a low-risk population, the non-Asians of Los Angeles County, California," *J. Natl. Cancer Inst.* 83 (1991): 1820–1826.

76. F.-S. Yeh, M. C. Yu, C.-C. Mo, S. Luo, M. J. Tong, and B. E. Henderson, "Hepatitis B virus, aflatoxins, and hepatocellular carcinoma in southern Guangxi, China," *Cancer Res.* 49 (1989): 2506–2509.

77. International Agency for Research on Cancer, *Schistosomes, Liver Flukes and* Helicobacter Pylori, IARC Monograph (Lyon, France: International Agency for Research on Cancer, 1994).

78. C. Howson, T. Hiyama, and E. Wynder, "The decline in gastric cancer: Epidemiology of an unplanned triumph," *Epidemiol. Rev.* 8 (1986): 1–27.

79. L. G. Korkina, A. D. Durnev, T. B. Suslova, Z. P. Cheremisina, N. O. Daugel-Dauge, and

I. B. Afanas'ev, "Oxygen radical-mediated mutagenic effect of asbestos on human lymphocytes: Suppression by oxygen radical scavengers," *Mutat. Res.* 265 (1992): 245–253.

80. J. P. Marsh and B. T. Mossman, "Role of asbestos and active oxygen species in activation and expression of ornithine decarboxylase in hamster tracheal epithelial cells," *Cancer Res.* 51 (1991): 167–173.

81. D. R. Lowy, R. Kirnbauer, and J. T. Schiller, "Genital human papillomavirus infection," *Proc. Natl. Acad. Sci. USA* 91, no. 7 (1994): 2436–2440.

82. H. Jick, A. M. Walker, R. N. Watkins, D. C. D'Ewart, J. R. Hunter, A. Danford, S. Madsen, B. J. Dinan, and K. J. Rothman, "Replacement estrogens and breast cancer," *Am. J. Epidemiol.* 112, no. 5 (1980): 586–594.

83. S. E. Hankinson, G. A. Colditz, D. J. Hunter, T. L. Spencer, B. Rosner, and M. J. Stampfer, "A quantitative assessment of oral contraceptive use and risk of ovarian cancer," *Obstet. Gynecol.* 80, no. 4 (1992): 708–714.

84. J. R. Harris, M. E. Lippman, U. Veronesi, and W. C. Willett, "Breast cancer," *N. Engl. J. Med.* 327, no. 5 (1992): 319–328.

85. G. A. Colditz, M. J. Stampfer, W. C. Willett, D. J. Hunter, J. E. Manson, C. H. Hennekens, B. A. Rosner, and F. E. Speizer, "Type of postmenopausal hormone use and risk of breast cancer: 12-year follow-up from the Nurses' Health Study," *Cancer Causes Control* 3, no. 5 (1992): 433–439.

86. B. Rosner, G. Colditz, and W. C. Willett, "Reproductive risk factors in a prospective study of breast cancer: The Nurses' Health Study," *Am. J. Epidemiol.* 139 (1994): 819–835.

87. J. Russo, G. Calaf, N. Sohi, Q. Tahin, P. L. Zhang, M. E. Alvarado, S. Estrada, and I. H. Russo, "Critical steps in breast carcinogenesis," *Ann. N.Y. Acad. Sci.* 698, no. 1 (1993): 1–20.

88. P. A. Newcomb, B. E. Storer, M. P. Longnecker, R. Mittendorf, E. R. Greenberg, R. W. Clapp, K. P. Burke, W. C. Willett, and B. MacMahon, "Lactation and a reduced risk of premenopausal breast cancer," *N. Engl. J. Med.* 330, no. 2 (1994): 81–87.

89. T. Byers, S. Graham, T. Rzepka, and J. Marshall, "Lactation and breast cancer. Evidence for a negative association in premenopausal women," *Am. J. Epidemiol.* 121 (1985): 664–674.

90. B. Henderson, R. Ross, and M. Pike, "Hormonal chemoprevention of cancer in women," *Science* 259 (1993): 633–638.

91. L. Bernstein, B. E. Henderson, R. Hanisch, J. Sullivan-Halley, and R. K. Ross, "Physical exercise and reduced risk of breast cancer in young women," *J. Natl. Cancer Inst.* 86 (1994): 1403–1408.

92. M. P. Longnecker, "Alcoholic beverage consumption in relation to risk of breast cancer: Meta-analysis and review," *Cancer Causes Control* 5, no. 1 (1994): 73–82.

93. J. F. Dorgan, M. E. Reichman, J. T. Judd, C. Brown, C. Longcope, A. Schatzkin, W. S. Campbell, C. Franz, L. Kahle, and P. R. Taylor, "The relation of reported alcohol ingestion to plasma levels of estrogens and androgens in premenopausal women (Maryland, United States)," *Cancer Causes Control* 5, no. 1 (1994): 53–60.

94. International Agency for Research on Cancer, *Some Industrial Chemicals*, IARC Monographs (Lyon, France: International Agency for Research on Cancer, 1994).

95. R. R. Connelly, R. Spirtas, M. H. Myers, C. L. Percy and J. F. Fraumeni, Jr., "Demographic patterns for mesothelioma in the United States," *J. Natl. Cancer Inst.* 78 (1987): 1053–1060.

96. T. Reynolds, "Asbestos-linked cancer rates up less than predicted," *J. Natl. Cancer Inst.* 84 (1992): 560–562.

97. B. N. Ames and L. S. Gold, "Chemical carcinogenesis: Too many rodent carcinogens," *Proc. Natl. Acad. Sci. USA* 87 (1990): 7772–7776.

98. B. N. Ames, M. K. Shigenaga, and L. S. Gold, "DNA lesions, inducible DNA repair, and cell division: Three key factors in mutagenesis and carcinogenesis," *Environ. Health Perspect.* 93 (1993): 35–44.

99. L. S. Gold, G. B. Garfinkel, and T. H. Stone, "Setting priorities among possible carcinogenic hazards in the workplace," in C. M. Smith, D. C. Christiani, and K. T. Kelsey, eds., *Chemical Risk Assessment and Occupational Health: Current Applications, Limitations, and Future Prospects* (Westport, CT: Auburn House, 1994).

100. International Agency for Research on Cancer, *Solar and Ultraviolet Radiation*, IARC Monograph (Lyon, France: International Agency for Research on Cancer, 1992).

101. M. Ellis and M. Lisher, "Second malignancies following treatment in non-Hodgkin's lymphoma," *Leuk. Lymphoma* 9 (1993): 337–342.

102. International Agency for Research on Cancer, *Overall Evaluations of Carcinogenicity: An Updating of* IARC Monographs *Volumes 1 to 42, Suppl. 7*, IARC Monographs (Lyon, France: International Agency for Research on Cancer, 1987).

103. B. Ryffel, "The carcinogenicity of cyclosporin," *Toxicology* 73 (1992): 1–22.

104. S. Preston-Martin, D. C. Thomas, M. C. Yu, and B. E. Henderson, "Diagnostic radiography as a risk factor for chronic myeloid and monocytic leukaemia (CML)," *Br. J. Cancer* 59 (1989): 639–644.

105. M. Gough, "How much cancer can EPA regulate anyway?" *Risk Anal.* 10 (1990): 1–6.

106. G. Pershagen, G. Akerblom, O. Axelson, B. Clavensjo, L. Damber, G. Desai, A. Enflo, F. Lagarde, H. Mellander, M. Swartengren, and G. A. Swedjemark, "Residential radon exposure and lung cancer in Sweden," *N. Engl. J. Med.* 330 (1994): 159–164.

107. A. V. Nero, "A national strategy for indoor radon," *Issues Sci. Tech.* 9, no. 1 (1992): 33–40.

108. J. H. Lubin, J. D. Boice Jr., C. Elding, R. W. Hornint, G. Howe, E. Kunz, R. A. Kusiak, H. I. Morrison, E. P. Radford, J. M. Samet, M. Tirmarche, A Woodward, Y. S. Xiang, and D. A. Pierce, *Radon and Lung Cancer Risk: A Joint Analysis of 11 Underground Miner Studies*, U.S. Department of Health and Human Services, NIH Publication No. 94-3644 (1994).

109. E. G. Létourneau, D. Krewski, N. W. Choi, M. J. Goddard, R. G. McGregor, J. M. Zielinski, and J. Du, "Case-control study of residential radon and lung cancer in Winnipeg, Manitoba, Canada," *Am. J. Epidemiol.* 140, no. 4 (1994): 310–322.

110. J. H. Lubin, "Invited commentary: Lung cancer and exposure to residential radon," *Am. J. Epidemiol.* 140, no. 4 (1994): 323–332.

111. A. Nero, "Developing a methodology for identifying high-radon areas," *Center for Building Science News (Lawrence Berkeley Laboratory)* 1, no. 3 (1994): 4–5.

112. A. H. Smith, R. C. Hopenhayn, M. N. Bates, H. M. Goeden, P. I. Hertz, H. M. Duggan,

R. Wood, M. J. Kosnett, and M. T. Smith, "Cancer risks from arsenic in drinking water," *Environ. Health Perspect*. 97, no. 259 (1992): 259–267.

113. M. N. Bates, A. H. Smith, and R. C. Hopenhayn, "Arsenic ingestion and internal cancers: A review," *Am. J. Epidemiol*. 135, no. 5 (1992): 462–476.

114. International Agency for Research on Cancer, *Chlorinated Drinking-Water: Chlorination By-products; Some Other Halogenated Compounds; Cobalt and Cobalt Compounds*, IARC Monograph (Lyon, France: International Agency for Research on Cancer, 1991).

115. A. Knudsen, "Hereditary cancers: Clues to mechanisms of carcinogenesis," *Br. J. Cancer* 59 (1989): 661–666.

116. B. Ponder, "Inherited predisposition to cancer," *Trends in Genetics* 6 (1990): 213–218.

117. L. S. Gold, T. H. Slone, B. R. Stern, N. B. Manley, and B. N. Ames, "Rodent carcinogens: Setting priorities," *Science* 258 (1992): 261–265.

118. R. Carson, *Silent Spring* (Boston: Houghton Mifflin, 1962).

119. L. S. Gold, T. H. Slone, and L. Bernstein, "Summary of carcinogenic potency and positivity for 492 rodent carcinogens in the Carcinogenic Potency Database," *Environ. Health Perspect*. 79 (1989): 259–272.

120. L. S. Gold, N. B. Manley, T. H. Slone, G. B. Garfinkel, L. Rohrbach, and B. N. Ames, "The fifth plot of the Carcinogenic Potency Database: Results of animal bioassays published in the general literature through 1988 and by the National Toxicology Program through 1989," *Environ. Health Perspect*. 100 (1993): 65–135.

121. B. N. Ames, M. Profet, and L. S. Gold, "Dietary pesticides (99.99% all natural)," *Proc. Natl. Acad. Sci. USA* 87 (1990): 7777–7781.

122. B. H. Patterson and G. Block, "Food choices and the cancer guidelines," *Am. J. Public Health* 78 (1988): 282–286.

123. T. Key and G. Reeves, "Organochlorines in the environment and breast cancer," *Br. Med. J*. 308 (1994): 1520–1521.

124. B. N. Ames, M. Profet, and L. S. Gold, "Nature's chemicals and synthetic chemicals: Comparative toxicology," *Proc. Natl. Acad. Sci. USA* 87 (1990): 7782–7786.

125. R. Crandall, *Why Is the Cost of Environmental Regulation So High?* (St. Louis, MO: Center for the Study of American Business, 1992).

126. B. Bartlett, "The High Cost of Turning Green," *Wall Street Journal*, September 14, 1994.

127. Office of Management and Budget, "Office of Management and Budget Regulatory Program of the United States Government, April 1, 1991–March 31, 1992" (1993).

128. T. O. Tengs, M. E. Adams, J. S. Pliskin, D. G. Safran, J. E. Siegel, M. C. Weinstein, and J. D. Graham, "Five-hundred life-saving interventions and their cost-effectiveness," *Risk Anal*. (in press).

129. R. L. Keeney, "Mortality risks induced by economic expenditures," *Risk Anal*. 10 (1990): 147–159.

130. A. Wildavsky, *Searching for Safety* (New Brunswick, N.J.: Transaction Press, 1988).

131. W. K. Viscusi, *Fatal Trade-offs* (Oxford: Oxford University Press, 1992).

132. S. Breyer, *Breaking the Vicious Cycle: Toward Effective Risk Regulation* (Cambridge: Harvard University Press, 1993).

133. W. C. Willett, presentation at the President's Cancer Panel (in press).
134. R. L. Phillips, L. Garfinkel, J. W. Kuzma, W. L. Beeson, T. Lotz, and B. Brin, "Mortality among California Seventh-day Adventists for selected cancer sites," *J. Natl. Cancer Inst.* 65 (1980): 1097–1107.
135. P. K. Mills, W. L. Beeson, R. L. Phillips, and G. E. Fraser, "Cancer incidence among California Seventh-day Adventists," *Am. J. Clin. Nutr.* 59 (Suppl.) (1994): 1136S–1142S.
136. B. N. Ames, L. S. Gold, and W. C. Willett, "The causes and prevention of cancer," *Proc. Natl. Acad. Sci. USA* (in press).

Chapter 6

FORESTS
Conflicting Signals

Roger A. Sedjo

HIGHLIGHTS

- *Though nearly 75 percent of the total industrial wood production comes from Northern Hemisphere industrial countries, the temperate forestlands of this region are expanding.*

- *The world's current industrial wood consumption requirements could be produced on only 5 percent of the world's total current forestland.*

- *About one-third of the earth's terrestrial area remains wilderness.*

- *Tropical forests cover a land area that is almost exactly the size of South America.*

- *Tropical and temperate forests together cover a land area the size of the Western Hemisphere.*

- *The rate of conversion of the tropical forests increased from 0.6 percent in 1980 to about 0.8 percent in 1990.*

- *Fully two-thirds of the deforestation in the United States occurred in the sixty years prior to 1910 and most of the other third before 1850.*

- *Although the United States has been the world's number one timber producer since World War II, U.S. forests have experienced an increase in volume in the past fifty years and have maintained roughly the same area over the past seventy-five years.*

- *The total area of forest and other wooded land in Europe increased by 2 million hectares annually between 1980 and 1990.*

- *The total forest area of the temperate region's industrialized countries increased between 1980 and 1990.*

- *During 1993, an estimated 4 million trees were planted in the United States each day.*

- *The forest biomass in the northern Rockies has increased by 30 percent or more since the middle of the eighteenth century.*

- *In recent years, private forestlands have accounted for 85 percent of total tree planting and seeding in the United States.*

- *The expansion of American forests has been made possible by improved tree-growing technology, the advent of tree plantations, improved control over wildfire, and the reversion of many agricultural lands, especially in the South and East, to forest.*

- *Deforestation in the Americas was probably greater before the Columbian encounter than it was for several centuries after.*

- *The recent proliferation of plantation forests in both the temperate and the tropical world has helped to deflect harvesting pressure from natural forests.*

- *Almost all of the timber harvested in the United States, Europe, and Nordic countries comes from second-growth/plantation forests.*

- *Commercial logging is not a major cause of deforestation; expanding agriculture is. In temperate countries, which provide over three-fourths of the world's industrial wood, reforestation is the rule, while in tropical countries forestland conversion to agriculture remains common.*

- *The developed countries in the temperate regions appear to have largely completed forestland conversion to agriculture and have achieved relative land use stability. By contrast, the developing countries in the tropics are still in a land conversion mode. This suggests that land conversion stability correlates strongly with successful economic development.*

For tens of thousands of years humans met most of their basic needs by foraging the land. Game, forest foods, clean water, fuelwood, as well as poles and construction materials were available for only the cost of extraction. Gradually hunting evolved into herding, and humans eventually invented agriculture.

The expansion of human populations led to the need for more agricultural lands. In Europe, India, and China, as well as in pre-Columbian America and elsewhere, lands were gradually modified and converted to accommodate cropping and pasture. Although timber and other products continued to be drawn from the forest, the major force behind the reduction of forests was the expansion of agriculture.

It was not until the seventeenth century that the large-scale expansion of cropland and pasture became a worldwide phenomenon. Even since then, deforestation pressures have been uneven. One or two centuries ago land clearing pressures were greatest in the temperate areas, principally North America and Europe. Although the tropics experienced deforestation in some places, the impact was far smaller. Today the situation is reversed. In many areas of the tropics, governments and individuals still see the forest as an impediment to economic development. They view forest clearing as a social good that provides the society with increased cropland. Just as forestlands were cleared along the eastern seaboard two centuries ago in America to provide land for agriculture, lands today are being cleared in the frontier areas of tropical South America and Asia.

Commercial logging apparently is not a major cause of deforestation, however, since temperate forests, which produce most of the world's timber, are essentially stable. The past few decades have seen the rapid expansion of tree planting and the establishment of forest plantations. Although agriculture replaced foraging for food in much of the world millennia ago, the transition from wood foraging to growing wood is only now taking place. Just as humanity moved from hunting and gathering wild resources to herding and farming, forestry today is in transition from simply drawing on nature's bounty to planting, managing, and harvesting trees utilizing the agricultural model for wood production.

Today much of the world's industrial wood comes from plantation forests established throughout the globe, from Spain to New Zealand, and Chile to South Africa. Almost all of the timber harvested in the United States comes from second-growth forests (those that have regrown after an earlier clearing) or plantation forests. The undisturbed native forests of the United States are almost wholly unavailable for timber harvests by virtue of their being part of the park, preservation, and wilderness system or in other public management. The U.S. Forest Service has almost entirely ceased timber harvesting. Similarly, most of the timber harvests in Europe come from planted or second-growth forests. Many of these lands had earlier been cleared and farmed before being returned to forest as plantations.

The use of a cropping mode to produce wood has had and will continue to have profound effects on the remaining natural forests. Just as agriculture reduced food foraging pressures on natural habitats by increasing crop productivity, so too plantation forests, with their high level of productivity, are reducing the pressures on the remaining natural forests.

> **Just as agriculture reduced food foraging pressures on natural habitats by increasing crop productivity, so too plantation forests, with their high level of productivity, are reducing the pressures on the remaining natural forests.**

This trend is likely to strengthen since the economics of plantation production are favorable in many locations.[1] The world's current industrial wood consumption requirements could be produced on about 200 million hectares of good forestland, an area only about 5 percent of the world's total current forestland. Finally, large areas of the world's forests are protected either by the establishment of reserves or through their own inaccessibility.

How are these forests to be viewed in any broad global assessment?

The Beginnings

Forests have been affected by human activities for tens of thousands of years. Early humans used fire to drive game, remove cover, and remove obstacles to travel. Evidence suggests, for example, that fire-prone vegetation became more common in Australia with the arrival of humans about 40,000 years ago. The shaping of the forest-shrub-grass mixture in Australia goes far back into human prehistory.

The experience of Britain provides an interesting example. By 3000 B.C. Britain was almost entirely covered by virgin forests, which had recolonized Britain with continental tree species after the ending of the last ice age, some 10,000 years ago, which had scraped away any previous forest. Gradually Bronze Age people in Britain cleared small areas of forest. The clearing accelerated with the arrival of the Celts in about 400 B.C., and the rate of deforestation accelerated over the 400 years of Roman occupation, ending in the fifth century A.D. Much of this early clearing in Britain, and indeed throughout Europe, was probably of the temporary slash-and-burn type common in tropical forests today.

Similarly, the forestlands surrounding the Mediterranean were reduced as farming expanded to feed a growing population. In addition, the Mediterranean forests contracted as they were used to provide fuel and materials for shipbuilding and other construction. Over the centuries, the forests of India and China also gradually shrank. There were minimal differences in the scale of deforestation and landscape changes between feudal Europe and imperial China.[2]

Over thousands of years, not only did the forests contract, but the underlying ecosystems adapted to an environment that included the continuous and disturbing presence of humans. This was true in Europe and in varying degrees in the forest/grasslands interface of Africa and throughout the native societies of what would become the New World. One of the greatest obstacles to regeneration in the Mediterranean forests was the introduction of goats, whose browsing behavior prevented forest regrowth.[3]

Medieval Europe

The conversion of large forest areas into agricultural lands in central Europe was completed during the twelfth and thirteenth centuries. Ore smelting and glassmaking, whose development there began in the fourteenth century, relied on wood for energy. By the end of the sixteenth century, large portions of the Alps had been deforested to provide wood for blast furnaces, with consequences that included Alpine torrents and violent flooding.[4]

The scarcity of ship timber at the end of the eighteenth century caused some European governments to promote tree planting, and during the nineteenth century tree plantations became more common throughout Europe.[5] Europe's

FIGURE 6-1   Expansion of Forest Area in France Since the Late Eighteenth Century

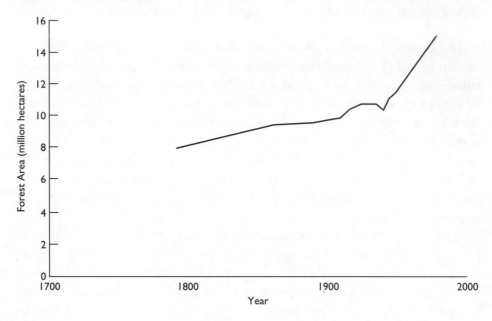

Source: J. G. Laarman and R. A. Sedjo, *Global Forests: Issues for Six Billion People* (New York: McGraw-Hill, 1992), p.64.

forests were probably at their smallest extent more than a century ago.[6] European forest regrowth probably began about the middle of the nineteenth century. Figure 6-1 shows the expansion of the forest area in France since 1800.

Forests in the New World

Although European explorers viewed the forests of the New World as essentially undisturbed by human activity, in fact, "in large part, the forest of the Americas, from Canada to Argentine, were ... highly disturbed or modified by Amerindian use by 1492."[7] The pre-Columbian disturbances included shifting cultivation, forest culling, and the use of fire. That there were large populations of indigenous peoples in the tropics suggests that major land use adaptations were required, and there is evidence of large-scale irrigation and substantial land clearing. In addition to human impacts, the forests were buffeted and modified by disease, insect infestation, storms, wind, and natural fire.

The settlement of the New World by Europeans began slowly. By 1600, the end of the first one hundred years, only 175,000 Spaniards had colonized the

Americas, and British colonization had not yet begun. The forests of eastern North America were impressive and gave rise to the view of a pristine wilderness; in reality, however, "many of the primeval forests that were supposedly encountered by the Europeans and that remain today, including forests with higher biodiversity were not 'pristine' or 'virgin' but were the product of extensive use and modification by the Amerindians."[8] The frequent references of Europeans in their writings to "fields," "meadows," "openings," and so forth leave little doubt of the land-clearing activities of native peoples.[9]

The destruction of large portions of the Amerindian populations, often from diseases brought by the Europeans to which the local populations were highly susceptible, changed the character of the land occupancy in the Americas.[10] This reduction of Amerindian populations by disease may ultimately have created the "wilderness" that awaited the arrival of European settlers.

Thus, European settlement did not disturb wild, virgin, and pristine forests but simply continued to modify forests that had experienced a minor respite after an extended period of intensive change by Amerindian cultures. Two commentators say that "deforestation in the Americas was probably greater before the Columbian encounter than it was for several centuries thereafter."[11]

U.S. Forests

Forests covered 1,044 million acres, or about 40 percent of the land area of the United States at the time of European colonization.[12] At first the impacts of Europeans on the native forests were modest. Early farming was small scale and low intensity, with few permanent effects on forest vegetation. Land clearing was slow, and often deforestation was not permanent. It was common for farmers to abandon depleted fields after a decade or two and move on to clear new forest. The abandoned fields would then regenerate forests naturally. One estimate is that by 1800 the amount of forestland conversion in the United States was "modest."[13]

New England, which was 90 to 95 percent forested when the first colonists arrived, experienced large declines in forest area. By 1850 the forests of Vermont, Rhode Island, Connecticut, and Massachusetts covered only 35 percent of the land area. They were reduced to 50 percent of the land area in New Hampshire and 74 percent of Maine's land area.[14] The forests of the mid-Atlantic states and the southern coastal areas experienced significant forest reductions as well. Nevertheless, by 1850 "only" about 114 million acres of forest had been cleared in the 200 years of European settlement, and most of this was after 1800.[15] This figure is not surprising since the U.S. population, which was only 5.8 million in 1800, had mushroomed to 23.2 million by 1850, and large-scale land clearing began in the eastern United States only after the mid-1800s.

In many respects, the North American forests of the precolonial period were probably not so different from those of today. Pine dominated the southeastern states, and wildfires, created by both natural forces and native peoples, were frequent. The pine forest of the South gradually gave way to the deciduous forest of the North, which in turn gave way to spruce and fir forests to the North and in eastern Canada.

Although old stands were undoubtedly common in colonial America, natural disturbances, including insect infestation, disease, and fire, as well as the actions of the native peoples, probably limited their occurrence. For example, infestation, such as the spruce budworm, was common and tended to operate in long-term cycles that did large amounts of damage and then receded for decades. Older stands tend to be more susceptible to both infestation and disease, thereby limiting the age of the stands. Also, weakened, diseased, and infested stands were vulnerable to wildfire, as were stands damaged by hurricane, windstorms, and other natural forces. All of these factors tended to reduce the longevity of forest timber stands.

In the arid West, natural forces tended to place age limits on many of the forests. Fire was common in the dry areas, which experienced periodic drought. Infestation too was common. In addition, periodic prairie fires, both natural and caused by the native peoples, tended to limit the extent of forest expansion. In fact, recent studies indicate that the forest biomass in the northern Rocky Mountains is 30 percent greater now than in the mid-eighteenth century, just before Europeans arrived. The dramatic reduction of wildfire is believed to be a major factor in the buildup of forest biomass over the past several decades.[16]

Until the mid-1800s most forest clearing was done to open agricultural lands, and wood still provided 90 percent of the nation's energy output. After 1850, however, forest clearing accelerated to meet the growing industrial demand for wood. This period of large-scale forest clearing in America lasted a relatively short period: from about 1850 to 1910. Almost 50 million acres were cleared in the decade 1870–1879 alone and perhaps 200 million acres over the sixty-year period.[17] Between 1870 and 1910, the huge forests of the Lake states were cut to supply the material requirements of rapid industrialization. Railroads too had huge appetite for wood. Forests were felled to supply crossties and bridges for the rapidly expanding rail system. Railroads consumed nearly 25 percent of the wood used in the 1800s.[18]

Three hundred years after the arrival of Europeans, about 300 million acres of forests had been cleared in the United States, fully two-thirds of it in the sixty years prior to 1910 and the other third before 1850. Forests were reduced from 1,044 million acres, or about 40 percent of the total land area, in 1630, to 760 million acres, or about 30 percent of the land area, in 1907. After 1920, U.S. forest area began a modest rebound as agricultural lands were abandoned.

> **Three hundred years after the arrival of Europeans, about 300 million acres of forests had been cleared in the United States, fully two-thirds of it in the sixty years prior to 1910.**

By one measure, U.S. forests have expanded unambiguously since 1920. Six national timber inventories have been undertaken since 1952, with the most recent in 1991, and each showed a greater forest volume than the previous inventory. U.S. forest biomass has increased despite timber harvests that have continued to provide the country with large volumes of wood over the past seventy-five years. The United States is still the world's major industrial wood producer, supplying roughly 25 percent of the world's total.

The ability of American forests to expand physically and provide large volumes of timber is the result of a combination of factors: improved tree-growing technologies, which lead to better forest management; the advent of tree plantations; and improved control of wildfire in the forest.[19] Nevertheless, forest volume cannot rise indefinitely.[20] Net growth is a sign of a young forest. At some point natural mortality will equal growth, even in the absence of any human involvement.

Tropical Forests in the Americas and Elsewhere

The image of pristine New World tropical forests undisturbed by humans is almost surely wrong. The Amerindian populations of Central and South America were substantial; some 54 million lived in the Western Hemisphere in 1492, over 50 million of them south of current U.S. boundaries.[21] These populations considerably modified the natural environment. Terracing, irrigation, agroforestry, and so forth were common, even pervasive. An estimated 76 percent of the population of the Americas south of the present-day United States was eliminated between 1492 and 1650.[22] The overall effect was the drastic reduction of agriculture and the consequent afforestation of many tropical lowlands. Turner and Butzer argue that "the scale of deforestation, or forest modification,

Traditional slash-and-burn farmers cleared this land in Indonesia. Clearing land for such low-yield farming and for firewood—not commercial logging—is the main cause for tropical deforestation.

in the American tropics has only recently begun to rival that undertaken prior to the Columbian encounter."[23]

In the tropics, as in the United States, the nineteenth century saw a decrease in the area of forests largely due to pressures for increased agricultural lands. Logging in the tropics did not constitute a serious threat to the forest since the logging was typically selective, felling only a few trees in any area, the rate of logging was slow, and regeneration typically occurred. In southern Brazil, the native peoples used fire to clear land for slash-and-burn cultivation. This practice was imitated and expanded by the Europeans and gradually gave way to more permanent conversion to pastures and croplands in the twentieth century.[24] In Central America forests were cleared to grow coffee after 1830, bananas after the 1890s, cotton after the 1940s, and cattle beginning in the late 1950s.[25] In the Philippines, forest was cleared to grow sugar cane and rice, as well as for ranching.[26] In parts of India, forestland was cleared for cotton and tea production.[27] Land in Burma was cleared to increase rice production,[28] and in Thailand, both rice and rubber cultivation contributed to forest area reductions.[29]

Commercial Forestry

Commercial forestry provides wood for processing into various commodities and products but not for fuel. Industrial wood obtained from commercial harvests is grouped into two categories: solidwood and fiber. Solidwood refers to products such as lumber and wood panels. Fiber products initially referred to products such as wood pulp (an intermediate product) and paper products. In recent years the distinction has been clouded by the development of a number of composite solidwood products manufactured from wood chips, such as fiber board and wafer board. Global timber production and consumption has increased only about 1 percent annually over the past two decades.[30]

Fully 73.6 percent of total industrial wood production in 1991 came from temperate forests of the industrial countries of the Northern Hemisphere (table 6-1). Without exception, the temperate countries that are supplying the vast majority of the world's industrial wood are the same countries in which the forest estate, both forestland area and forest stocks, has remained stable in recent years. Another 9.3 percent of this production comes from the plantation regions

TABLE 6-1   Global Industrial Wood Production, 1991 (million cubic meters)

	Production	Percentage	Cumulative Percentage
United States	409.9	25.6	25.6
Former Soviet Union	274.3	17.1	42.7
Europe, excluding Nordic countries	195.6	12.5	55.2
Canada	171.2	11.0	66.2
Nordic countries	88.4	5.6	71.8
Japan	27.9	1.8	73.6
China	90.0	5.7	79.3
Argentina, Chile, Brazil, South Africa	101.7	6.5	85.8
Malaysia, Indonesia, Philippines	76.9	4.9	90.7
New Zealand and Australia	42.9	2.8	93.5
Other	104.2	6.5	100.0

Source: FAO, Forest Products, 1980–1991 (Rome: FAO, 1993).

TABLE 6-2   Countries Providing a Large Portion of Their Industrial Wood from Plantation Sources

| | Total Area (thousand hectares) | | Share of Plantations (%) | |
	Natural Forests	Plantations	Total Forest Area	Total Industrial Wood Production
New Zealand	6,270	1,240	19	93
Brazil	396,000	6,500	2	60
Chile	6,300	1,400	22	95
Argentina	36,000	800	2	60
Zimbabwe	28,800	117	0.4	50
Zambia	12,900	60	0.4	50

Source: Devendra Pandey, *Assessment of Tropical Forest Plantation Resources* (Uppsala: Swedish University of Agricultural Sciences, Department of Forest Survey, October 1992).

of Oceania, South America, and South Africa. Although some of those countries are experiencing deforestation, most of their domestic industrial wood production is provided by the plantation forests (table 6-2).

In all of the data presented, the countries provide a high fraction of their industrial wood from their plantation resources. Plantations have inherent advantages in terms of location, accessibility, wood type, and wood quality. Natural forests no longer serve as a major source of industrial wood.

Although many environmentalists claim that commercial logging is a major cause of deforestation, most knowledgeable observers recognize that commercial timber harvests generally do not permanently convert land to another use. Rather, forestlands that are commercially harvested typically remain as forestlands. Reforestation occurs either through tree planting or, less commonly today, natural regeneration. In recent years between 2.5 and 3.0 million acres (about 1.0 to 1.2 million hectares) have been planted annually in the United States (figure 6-2). This amounts to the planting of 4 million to 6 million seedlings *each day*. Additionally, an estimated 4 million hectares of forest plantations are being established worldwide, much of it for commercial purposes.

Temperate forests provide about 75 percent of total world production of forest products. Meanwhile, the world's major temperate timber producers are not experiencing either reduced land area in forests or a reduction of their forest stocks. This means that the harvests are less than the forest regrowth. For example, for the six forest inventories taken in the United States since 1950, net forest growth always exceeded harvests (figure 6-3). Thus, although the United

FIGURE 6-2   Tree Planting in the United States, 1930–1933

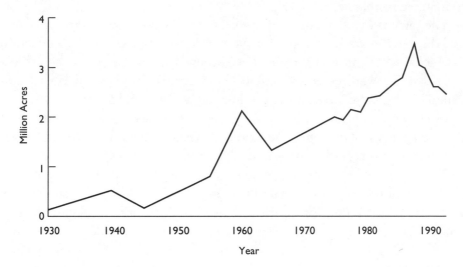

Source: U.S. Department of Agriculture, Forest Service, State and Private Forestry, Cooperative Forestry, *Tree Planting in the United States–1993* (Washington, D.C.: Government Printing Office, 1993).

FIGURE 6-3   U.S. Timber Growth and Removals, 1920–1991

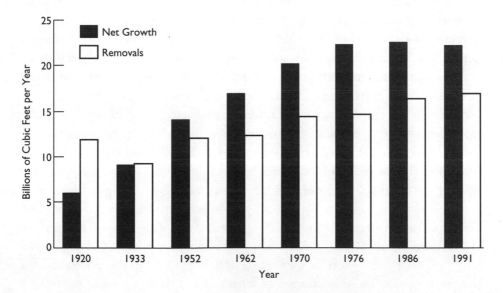

Source: Unpublished data provided by Douglas MacCleery of the U.S. Forest Service.

States is the world's number one timber producer, U.S. forests have continued to increase in volume.

The same is true for Russia and Canada, the wood producers ranked second and third in the world, as well as for Western Europe, the collective production of which is significantly larger than that of Canada. Thus, although almost three-fourths of total world commercial timber production occurs in the northern temperate forests, these forests are expanding in both area and forest volume.

By contrast, only about 20 percent of the world's harvests occurs in the tropics, and even less of the total production is tropical timbers from natural forests, since much of this production is from plantations.

The regions that are producing most of the world's commercial timber are also the regions whose forests are stable or expanding (the temperate region), while the region producing a relatively small fraction of total industrial wood production (the tropics) is experiencing significant deforestation.

The driving force behind tropical deforestation is clearly not commercial harvests. Rather, tropical deforestation is driven primarily by desire for agricultural land.

Forest Renewal and Protection

Although much of the focus of public attention has been on deforestation, in fact, many regions of the world are experiencing substantial reforestation. This resurgence is the result of stabilized or increasing net forest area over vast regions of temperate forests, increased cropping of industrial timber through forest plantations, increased interest in agroforestry, and increased reservation of forest areas for national parks, ecological reserves, and other protected status.[31]

Forests have an amazing capacity for renewal. This should not be surprising since they have been ravaged by nature for millennia by fire, insects, disease, wind storms, periodic ice ages, and other climatic changes. The process of natural reforestation only awaits the opportunity. Reforestation began in parts of the eastern United States as early as the mid-1800s, not as the outgrowth of specific policy but as the result of benign neglect due to the abandonment of the agricultural fields that had been chopped out of the native forest. Some farms began reverting to forests as early as the mid-nineteenth century.[32] By the 1990s Vermont was 75 percent forested; New Hampshire 86 percent; Connecticut, Massachusetts, and Rhode Island almost 60 percent; and Maine about 90 percent.[33] Similarly, the abandonment of depleted farmland in the South and the Lake states led to the reforestation of large areas of those regions.[34] Little of it required human intervention.

Eco-tourists walk in the rain forest on the Caribbean island of St. Lucia. Such tourism is providing local people with an incentive to preserve the rain forests.

Areas where forests occur naturally and precipitation is plentiful, such as the eastern United States, have little problem with natural forest regeneration. Under most circumstances abandoned fields spontaneously begin to regenerate, often in pine. Within a decade or two, the forest is firmly reestablished. In New England these regenerated forests were commonly called "old field pines," reflecting the ability of pines to establish on "old fields." Over time, the pines, which require abundant sun to grow, are gradually replaced in the forest understory by more shade-tolerant conifer species or by a hardwood forest, such as is common in much of New England today.

The former agricultural fields of the South also often tended to regenerate in pines, although hardwood regeneration is common on certain sites under various circumstances. Similarly, many of the former pinelands of the Lake states that were deforested and converted to agriculture are now reforested in a variety of tree species, some quite different from the earlier pine.

Reforestation is not unique to the United States. Major forest regrowth began first in Europe, and forests there are still undergoing the greatest degree of net reforestation in the world. Today most of Europe is rather heavily forested. (figure 6-4).

Even tropical forests, often characterized as fragile and difficult to regenerate, usually renew themselves if they are not impeded by alternative land uses. For example, much of the tropical forest of Central America was believed to be pristine, but now we know that this forest has overgrown early native civilizations that had severely disturbed it. Similarly, the banks of the Panama Canal, which were almost wholly defoliated during the construction of the canal early in this century, are now covered with lush tropical forests. This regeneration is due entirely to natural processes.

FIGURE 6-4   Estimated Changes in the Area of Europe's Forest and Other Wooded Land, 1980–1990

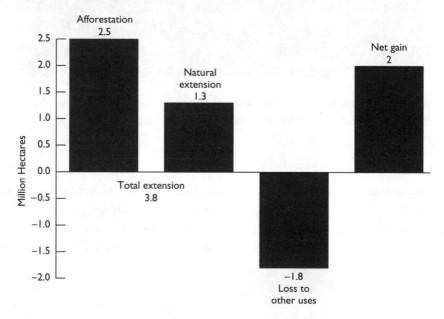

Source: A. V. Korotkov and T. J. Peck, "Forest Rescues of the Industrialized Countries: An ECE/FAO Assessment," *Unisylva* 44, no. 174 (1993): 24.

Protected Areas

The increase in the number of protected areas is one response to concerns over the losses of wilderness, wild and undisturbed areas, including forest. Recently the areas under protected status have expanded rapidly. Figure 6-5 shows the cumulative area under protected status since 1900. Although it took approximately sixty years, until 1960, for the first 1 million square kilometers to achieve protected status, the second 1 million acres took only about ten years. In the most recent fifteen years reported, 1975 to roughly the end of the 1980s, almost 3 million additional square kilometers have been placed in protected status.

Table 6-3 presents an estimate of the extent of wild areas in the world by major region. The study it comes from identified global land areas with a minimum size of 400,000 hectares having no roads, buildings, transport infrastructure, powerlines, pipelines, mines, dams, canals, aqueducts, reservoirs, or oil wells.[35] It revealed nearly 5 billion hectares of "wilderness" lands, or about one-third of the global terrestrial area. Some 60 percent of the wilderness is tundra, deserts,

FIGURE 6-5   Cumulative World Area Under Protected Status Since 1900

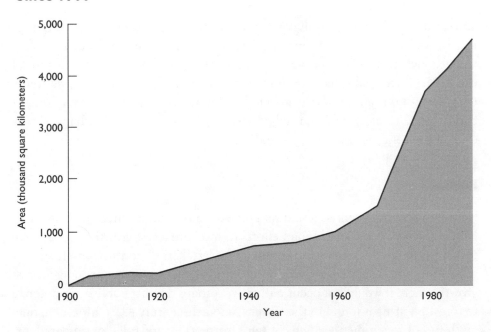

Source: Walter V. Reid and Kenton R. Miller, *Keeping Options Alive: The Scientific Basis for Conserving Biodiversity* (Washington, D.C.: World Resources Institute, 1993), p. 71.

TABLE 6-3   Extent of Wild Area in the World, by Major Regions

	"Wild Area"[a] (thousand km^2)	As Proportion of Total Land (%)
Antarctica	13, 210	100
Africa	8,230	27
Soviet Union	7,520	34
North America	6,850	37
Asia (excluding Soviet Union)	3,780	14
Latin America	3,750	21
Australia and Oceania	2,370	28
Greenland	2,170	99
Europe (excluding Soviet Union and Greenland)	140	3
World	48,020	32

Source: J. M. McCloskey and H. Spalding, "A Reconnaissance-Level Inventory of the Amount of Wilderness Remaining in the World," *Ambio* 18, no. 4 (1989): 222, 226.

[a] "Wild Area": Land units with a minimum size of 400,000 hectares having no roads, buildings, transportation infrastructure, power lines, pipelines, mines, dams, canals, aqueducts, reservoirs, or oil wells.

and similar lands. Forests of all types, but mainly cool coniferous forest and tropical moist forests, constitute another 30 percent, or 1.5 billion hectares.

Although this estimate is clearly crude, it does suggest that large portions of the earth's terrestrial surface are unoccupied by humans and remain essentially in a wild condition, minimally affected by human structures and direct disturbances. Additionally, a substantial portion of the earth's forest (perhaps 40 percent of the total) is found in this wild area.

Plantations

Reforestation can occur as a natural process via natural regeneration, or it can be the result of conscious human efforts to promote forest growth through tree planting (figure 6-6). Planting is usually undertaken to control which species grow on a site, to accelerate the reforestation process, or to encourage forest growth where it would not occur naturally. Perhaps the first published reference to forest planting is found in the *Guanzi*, a fourth-century B.C. Chinese manual on the art of government, which contains instructions and rules for timber planting, management, and protection to ensure a continuous supply of wood.[36] For-

FIGURE 6-6   Expansion of World Land Areas Under Regular Cropping, 1860–1978

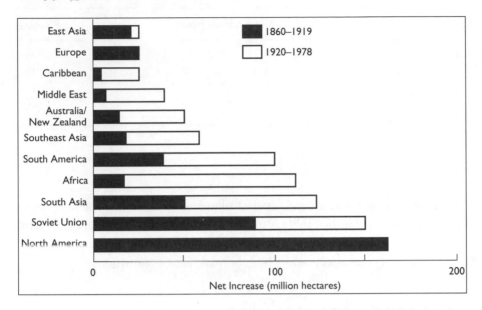

Source: John F. Richards "World Environmental History and Economic Development," in W. C. Clark and R. E. Munn, eds., *Sustainable Development of the Biosphere* (Cambridge: Cambridge University Press, 1986).

est plantations were established in the fourteenth century in Italy in the Florentine mountains and in the fifteenth century in England and Scotland. By the seventeenth century there were calls to plant trees in Germany, France, and Britain. During the nineteenth century large areas of Europe were planted, largely in conifers, both for the timber values and to protect the valleys from the catastrophic flooding and avalanches that had resulted from earlier logging and livestock grazing on steep mountainsides.[37]

The latter part of the twentieth century has seen a burgeoning of forest regeneration activity in much of the world. One estimate is that plantations had attained an area of 120 to 140 million hectares by the mid-1980s.[38] In the tropical countries of Asia and South America especially, plantations are being established at an accelerating rate. In 1992 a study by the Food and Agriculture Organization (FAO) revealed that there were about 30.7 million hectares of plantation forests in the tropics and that the 1990 net increase was about 1.82 million hectares, a sharp increase over the 11.876 million hectares of plantations in the tropical region reported by FAO for 1975.[39] Tables 6-4 and 6-5 provide, respectively, estimates of plantation areas and annual planting rates in the

TABLE 6-4   Net Forest Plantation Areas in the Tropics, 1990 (thousands of hectares)

Region	Net Establishment	Net Plantings[a]
Africa	2,100	90
Asia and the Pacific	22,600	1,470
Latin America and Caribbean	6,000	260

Source: Ibid.

[a] Net makes adjustment for mortality and other losses.

tropics and the size and annual level of planting of the ten tropical countries that do the most planting. Table 6-6 presents an estimate of the area of plantations in some nontropical developing countries. China in particular has both a high number of plantations and a high annual rate of planting.

Plantations in the Temperate Regions

Although plantation forests are increasingly important in the tropics, even more plantations are being established in the temperate countries. Table 6-7 contains

TABLE 6-5   Tropical Plantations, Top Ten Countries, 1992 (thousands of hectares)

Countries	Total	Annual
1. India	18,900	1,700
2. Indonesia	8,750	100
3. Brazil	7,000	200
4. Vietnam	2,100	125
5. Thailand	775	40
6. Venezuela	362	30
7. Cuba	350	30
8. Bangladesh	335	10
9. Myanmar	334.8	40
10. Madagascar	310	5

Source: Pandey, *Assessment*, p. 15.

TABLE 6-6   Forest Plantations in Nontropical Developing and Developed Countries, 1990 (thousands of hectares)

Nontropical Developing Countries			Developed Countries		
Countries	*Total*	*Annual*	*Countries*	*Total*	*Annual*
China	36,000	1,250	United States	31,850	1,000
Republic of Korea	2,000	50	Undivided Soviet Union	23,800	1,000
Chile	1,450	80	Japan	10,670	50
Republic of South Africa	1,333	30	Canada	5,023	400
Argentina	800	25	New Zealand	1,240	10
Morocco	526	30	Australia	965	30
Uruguay	208	2			

estimates of the extent of industrial plantation forests in 1985, including the United States, the Soviet Union, Western Europe, and Japan. Roughly two-thirds of the world plantation forests are located in the temperate forest area.

In 1993, tree planting in the United States occurred on almost exactly 1 million hectares.[40] Although this is down from its 1988 peak of 1.37 million hectares (3.39 million acres) under the Conservation Reserve Program, the amount of tree planting is still impressive. The number of plantings in the United States was estimated for 1993 as 4 million trees per day each day of the year. This, it might be noted, is below the highest levels experienced, a few years earlier, of 6 million trees planted per day.

Taken in its totality, the portion of the total forestlands in plantations is continuing to increase, although the United States has maintained roughly the same area of forest over the past seventy-five years.

Globally, Pandey estimates about 4 million hectares of trees are planted annually. This compares with Mather's 1990 estimate of about 10 million hectares annually in the temperate region. Mather's number appears somewhat high, although planting rates can vary. For example, New Zealand is currently planting about 100,000 hectares annually, well above the 10,000 reported for 1990 in the Pandey study.[41] This increase is due to the completion of institutional reforms in New Zealand's land tenure system.

Plantation forests represent an extension of the agricultural cropping model to forestry. No longer do humans forage from natural forests for their wood ma-

TABLE 6-7   Global Area of Forest Plantations, c. 1985

	Plantation Area (thousands of hectares)
Industrialized countries	
Soviet Union	21,900
Western Europe	13,000
United States	12,100
Japan	9,600
Canada	1,500
New Zealand	1,100
Australia	800
Subtotal	60,000
Developing countries	
China	12,700
Brazil	6,100
India	3,100
Indonesia	2,600
Republic of Korea	2,000
Chile	1,200
Argentina	800
Others	7,400
Subtotal	35,900

Source: Sandra Postal and Lori Heiss, *Reforesting the World*, Worldwatch Paper 83 (Washington, D.C.: Worldwatch, April 1988), p. 28.

terials. As in agriculture, where cropping replaces the gathering of food, forest plantations replace wood that would otherwise be harvested from natural forests.

Global Forests: The Current Situation

The precise amount of deforestation both worldwide and in the tropics is difficult to determine, as is the precise amount of current forested area. Nevertheless, systematic global estimates were developed in the early 1980s and updated

Tree plantations are expanding rapidly worldwide.

in the early 1990s. These, despite some limitations, provide what most believe are fairly accurate estimates of the global areas that are forested and deforested.

Tropical Forests

The evidence indicates that the rate of conversion of the tropical forests has continued to increase. Tropical forests were estimated at covering 1,910 million hectares in 1980.[42] By 1990, tropical forest cover was reduced to 1,756 million hectares. Thus, the average annual area of tropical deforestation amounted to 15.4 million hectares, or an 0.8 percent annual rate of deforestation over the decade (table 6-8). This compares with the 1980 estimate of tropical deforestation at 11.3 million hectares annually, or a 0.6 percent annual rate.

Tropical forests still cover over 13 percent of the global land area, or an area almost exactly that of South America. By ecological zone, 76 percent of the tropical rain forest zone is still covered with forest: 46 percent of the moist deciduous area, 30 percent of the dry deciduous, and 19 percent of the dry and very dry zones taken together.

TABLE 6-8   Tropical Deforestation Rates, 1990 (millions of hectares)

Geographic Subregion/ Region	Number of Countries	Land Area	Forest Cover		Annual Deforestation, 1981–1990	
			1980	1990	Million Hectares	% per Annum
Africa	40	2,236.1	568.6	527.6	4.1	0.7
West Sahelian Africa	6	528.0	43.7	40.8	0.3	0.7
East Sahelian Africa	9	489.7	71.4	65.5	0.6	0.9
West Africa	8	203.8	61.5	55.6	0.6	1.0
Central Africa	6	398.3	215.5	204.1	1.1	0.5
Tropical Southern Africa	10	558.1	159.3	145.9	1.3	0.9
Insular Africa	1	58.2	17.1	15.8	0.1	0.8
Asia and Pacific	17	892.1	349.6	310.6	3.9	1.2
South Asia	6	412.2	69.4	63.9	0.6	0.8
Continental Southeast Asia	5	190.2	88.4	75.2	1.3	1.6
Insular Southeast Asia	5	244.4	154.7	135.4	1.9	1.3
Pacific	1	45.3	37.1	36.0	0.1	0.3
Latin America and Caribbean	33	1,650.1	992.2	918.1	7.4	0.8
Central America and Mexico	7	239.6	79.2	68.1	1.1	1.5
Caribbean	19	69.0	48.3	47.1	0.1	0.3
Tropical South America	7	1,341.6	864.6	802.9	6.2	0.7
Total	90	4,778.3	1,910.4	1,756.3	15.4	0.8

Source: FAO, *Forest Resources Assessment 1990: Tropical Countries,* FAO Forestry Paper 112 (Rome: FAO, 1993), p. 156.

Temperate Forests

Temperate forests have not been as systematically examined as tropical forests. However, based on the 1990 assessment by the Economic Commission on Europe (ECE) and the Food and Agriculture Organization (FAO), Korotkov and Peck determined that the area of forest in Europe increased, as did that of the former Soviet Union.[43] The assessment concludes that the United States experienced a modest decline in forest area during the last decade, while Canada appears to be roughly in balance.[44]

Japan reestablished its depleted forests in the early postwar period with a vigorous program of reforestation. The Japanese forest area showed only very small decline over the past decade. Finally, in the past three decades China, basically a temperate climate country, has undertaken a massive and largely successful program of reforestation, as has South Korea.

Thus, based on the ECE/FAO 1990 assessment, Korotkov and Peck conclude that for temperate-region countries as a whole, there was a slight increase in the total forest area between 1980 and 1990.[45] In the United States, the period of rapid deforestation was between 1850 and 1910; in the tropical world the period of rapid deforestation appears to be occurring now.

Acid Rain and Forest Dieback

Other reports bolster the ECE/FAO assessment on European forest regrowth. A recent study by the European Forestry Institute found that forests in Europe have been increasing substantially in total land area and forest volume since 1950.[46]

Perhaps the most puzzling aspect of these findings was the large increase, about 35 percent, in the rate of forest growth that occurred during the 1980s. This surge took place broadly throughout Europe at the same time as the highly popularized fears over forest dieback, allegedly caused by acid rain, were peaking. Some environmentalists claimed that the sulfur dioxide released into the atmosphere as a result of burning coal to produce electricity was damaging forests when it was deposited with rainfall. The new findings contradict the notion that acid rain pollution had severe effects on European forests. Careful surveys of tree conditions in Europe did not find widespread damage as a result of acid rain. Some in the scientific community now regard claims about the extent of damage to forests caused by acid rain, the "dieback theory," as having been refuted by the evidence.[47]

The experience in the United States has been similar to that of Europe with respect to the effects of acid rain on forests. The $600 million long-term National Acid Precipitation Assessment Program found little evidence that acid rain was causing significant damage to forests in the United States.

The effects of air pollution on forests are complex. Clearly, intense exposure to sulfur dioxide can cause local damage, but the dire effects of acid rain have been widely exaggerated.

Total Global Forests

The most recent ECE/FAO estimates estimate the current forested area as 2.06 million hectares in the industrial countries and 1.756 million in the tropics, for a total of 3.82 million hectares. This number apparently excludes the temperate forests of northern China and Korea and so underestimates the world's total. A reasonable estimate of total world forested area currently is just a bit below 4 billion hectares.

Markets, Ownership, and Forests

Worldwide, about 77 percent of the global forest area is publicly owned and 23 percent private,[48] although some estimate that 90 percent of the world's forests are publicly owned.[49] Thus, whatever forests have suffered and what the future holds for them lies, to a large extent, in the hands of government.

This ownership pattern represents a combination of accident and design. In the United States, for example, the lands to the west of the eastern seaboard were initially public. For decades, U.S. policy was for these lands to be privatized via the Homestead Act, transfers to railroads, and so forth. Nevertheless, toward the end of the nineteenth century, as public concerns about the condition of the forests increased, large areas of unprivatized forestlands were placed into public reserves; most later became part of the national forest system.

Similarly, large areas of forestlands in South America, Asia, Africa, Russia, and Canada are state lands. Some countries promote privatization as a vehicle to frontier development, as in Brazil. Russia is considering long-term concession arrangements with private firms, but most countries are not moving toward privatization.

In Canada, for example, most of the forestlands, especially in the west, are Crown lands controlled by provincial governments. Most Canadian timber production comes from Crown lands under long-term (typically, twenty years) concession agreements. Although flexibility exists in the agreements, penalties are assessed for nonperformance, including failure to meet the harvest goals, as well as for harvests in excess of the targets.

Similar systems exist elsewhere. The Indonesian government awards twenty-year concessions for harvesting state forestlands to private companies. In the tropics the harvests are almost always selective rather than clear-cuts, and the concessionaire is responsible for reforestation of the harvested area. Typically the reforestation is in native tropical species, although plantations are also being established in some places.

In the United States the Forest Service normally holds competitive auctions for timber harvest rights. Harvest must be completed usually within three to five years, and payment is made to the Forest Service on the basis of the volumes and species harvested. Long-term management responsibility, however, resides with the Forest Service. In addition to timber, the Forest Service has responsibility for the production of multiple outputs, including wildlife, recreation, and water quality. In recent years the role of the public lands in industrial wood production has declined dramatically, while that of the private lands has expanded production, in part to fill the void left by public harvest declines.

Private forestlands in the United States and elsewhere have tended to focus on the production of timber and industrial wood. In recent years private forestlands have accounted for about 85 percent of the total tree planting and seeding in the United States. Many of the management innovations, which have increased tree-growing productivity substantially, were developed and adopted by the private sector. Similarly, private forestlands are major wood suppliers in eastern Canada, the Nordic countries, and throughout Europe, as well as plantation regions of South America, New Zealand, and elsewhere.

Factors Affecting Deforestation

The situation in the temperate regions is quite different from that of the tropical regions. The temperate forests have been essentially stable or expanding for the past several decades to several hundred years. By contrast, the tropical forests are still being reduced.

Deforestation remains largely the result of the development and expansion of agriculture. Land conversion from forests to agriculture appears to have largely ceased in the temperate world; in fact, reversion back to forests is more common. In the tropical world, however, the process of land conversion to agriculture continues.

Commercial logging is not a major cause of deforestation; expanding agriculture is. In temperate countries, which provide over three-fourths of the world's industrial wood, reforestation is the rule; in tropical countries, land conversion to agriculture remains common. The developed countries in the temperate regions appear to have largely completed forestland conversion to agriculture and have achieved relative land use stability. By contrast, the developing countries in the tropics are still in a land conversion mode. This suggests that land conversion stability correlates strongly with successful economic development.

Countries that have achieved economic development almost always had an early period in which forestlands were rapidly converted to productive agricultural lands. Thus, conversion is not necessarily undesirable. Forested areas provided the agricultural land that allowed the development of Europe and North America.

Temperate forests are stable because developed countries have little incentive to expand their agricultural land base; they have already dramatically increased crop yields on the acres under cultivation.[50] By contrast, the developing countries' crop yields are still quite low, so there is pressure to expand the areas under cultivation.[51]

Additionally, industrialized countries have well-developed institutions of land tenure, property rights, enforcement capabilities, judicial systems, and so forth. Well-defined, secure property rights and markets provide clearer and less ambiguous signals. In the developing countries, deforestation has been exacerbated by the absence of clear property rights to the land and forest.[52] In areas where large tropical forests exist, landownership is often unclear and weakly held. Enforcement capacities are limited, judicial systems weak, and so forth. Weak or unclear land tenure encourages slash-and-burn agricultural systems in tropical forests.

Insecure tenure additionally limits the incentive to make long-term commitments to the land, so peasant farmers quickly clear the land, make modest plantings, and abandon the site after only a few years. Since they do not own the land, they have no incentive to engage in long-term activities, such as forestry, whose financial returns would exceed that of shorter-term agricultural activities. Therefore, the incentive structure is often tilted against forestry, even when the underlying situation is favorable.

In addition, governments often provide incentives for land conversion. In the nineteenth-century United States, the Homestead Act provided "free" title to public lands that were "improved" and "developed" and then occupied for seven years. Land clearing was viewed as de facto evidence of improvement. Similar programs are often undertaken today in the developing world to encourage the development of "frontier" lands. In the late 1970s, 2 million hectares of tropical forest were cleared annually for cattle ranching in Latin America.[53] In many countries converting tropical forests into pasture was heavily subsidized. Governments also indirectly subsidized land clearing through extensive road construction, which provides low-cost access to potential farmers.

Further, deforestation can occur when traditional tenure arrangements break down, when the government lacks the power to enforce property rights and/or manage and protect public lands, or when the government contributes to destabilization through perverse policies. The government of Nepal's nationalization of the forests generated destructive instability by alienating the local villagers. Infuriated, they began illegally to log the forests that they had previously protected when they were communal property.[54]

In summary, economic development promotes forest stability through well-defined and recognized property rights, the enforcement of property rights, the absence of government subsidies to encourage land clearing, and high levels and growth rates of agricultural productivity.

In summary, economic development promotes forest stability through well-defined and recognized property rights, the enforcement of property rights, the absence of government subsidies to encourage land clearing, and high levels and growth rates of agricultural productivity.

Conclusion

Historically, humans everywhere have interacted with and "disturbed" forests. As humans progressed and became agriculturists rather than simple hunter-gatherers, they found it necessary to modify natural ecosystems. Land management tended toward replacing complex multispecies systems with ecologically simpler and more productive farming systems. In purely economic terms, the actual stock of forestlands exceeded the stock desired by humanity, while the actual stock of farmland was less than desired. Thus there were incentives for adjusting those stocks by converting forestlands into more highly desired agricultural lands.

New growth in a clear-cut area in the Flathead National Forest, Montana. Properly managed forests can naturally regenerate.

This adjustment does not imply that the conversion trend will continue until all forests are gone, however. People have long recognized the value of forests, as a source of construction materials and fuelwood, of other forest outputs (e.g., game and recreation), and of environmental services (e.g., watershed protection and flood control). The land conversion process, moreover, is not one way, flowing only from forests to cleared lands. Lands once cleared may be converted back into forests, as either plantations or naturally regenerated forests.

Forests are not and have never been unchangeable. Being biological systems, they have an amazing resiliency and ability to adapt to fluid conditions, whether these changes are the result of nature or humans. Thus, although the forests of the temperate world have experienced many anthropogenic disturbances over the millennia, they are in remarkably good condition in many respects. In vast areas of the globe, much of the natural forest is intact and minimally affected by human disturbances. In addition, plantation forests are growing in importance and increasingly deflecting timber harvesting pressures away from natural forests.

The first sentence in Marion Clawson's *Forests for Whom and for What?* begins: "Forests serve the American people in many ways and have the potential to serve more people in better and more generous ways."[55] This same insight can be expanded to include all of humanity and all the world's forests.

NOTES

1. Roger A. Sedjo, *The Economics of Plantation Forests: Resources for the Future* (Washington, D.C.: Washington Resources for the Future, 1983).

2. N. Menzies, "Sources of Demand and Cycles of Logging in Premodern China," in J. Darcavel and R. P. Tucker, eds., *Changing Pacific Forests: Historical Perspectives on the Forest Economy of the Pacific Basin* (Durham, N.C.: Duke University Press, 1992); B. L. Turner II and Karl L. Butzer, "The Columbian Encounter and Land-Use Change," *Environment* 34, no. 8 (1992): 16–20, 37–44.

3. J. V. Thirgood, "Man's Impact on the Forests of Europe," *Journal of World Forest Resource Management* 4, no. 2 (1989): 127–167.

4. Peter Santorius and Hans Henle, *Forestry and Economic Development* (New York: Praeger, 1968).

5. E. L. Johan, "The Impact of Industry on the Landscape and Environment of Austria Prior to the First World War," *Forest Conservation History* 34 (1990): 122–129.

6. Thirgood, "Man's Impact."

7. Turner and Butzer, "Columbian Encounter," p. 37.

8. Ibid., p. 42.

9. M. Williams, "Forests," in B. L. Turner et al., *The Earth as Transformed by Human Action* (Cambridge: Cambridge University Press with Clark University, 1990).

10. W. M. Denevan, *The Population of the Americas in 1492*, 2d ed. (Madison: University of Wisconsin Press, 1992).

11. Turner and Butzer, "Columbian Encounter," p. 42.

12. Douglas W. MacCleery and Brad Smith, personal communication, unpublished forest service data, August 19, 1994.

13. Williams, "Forests"; M. Clawson, *Forests for Whom and for What?* (Baltimore: Johns Hopkins Press, 1975).

14. John W. Barrett, "The Northeast Region," in his *Regional Silviculture of the United States* (New York: Wiley, 1988).

15. Michael Williams, "Industrial Impacts of the Forests of the United States, 1860-1920," *Journal of Forest History* 31, no. 3 (1987): 108–121.

16. M. Williams, *Americans and Their Forests, A Historical Geography* (New York: Cambridge University Press, 1989).

17. Ibid.

18. Douglas W. MacCleery, "What on Earth Have We Done to Our Forests?" *USDA/Forest Service*, January 10, 1994.

19. Roger A. Sedjo, "Forest Resources," in *America's Renewable Resources*, ed. Kenneth Frederick and Roger Sedjo (Washington, D.C.: Resources for the Future, 1991).

20. Clawson, *Forests for Whom?*

21. Deneven, *Population of the Americas*.

22. Ibid.

23. Turner and Butzer, "Columbian Encounter," p. 42.

24. Williams, "Forests."

25. Ibid.

26. Dennis M. Roth, "Philippine Forests and Forestry: 1565–1920," in R. P. Tucker and J. F. Richards, *Global Deforestation and the Nineteenth-Century World Economy* (Durham, N.C.: Duke University Press, 1983).

27. John F. Richards and M. B. McAlpin, "Cotton Cultivation and Land Clearing in the Bombay Deccan and the Karnatak: 1818–1920," in Tucker and Richards, *Global Deforestation*; R. P. Tucker "The British Empire and India's Forest Resources: The Timberlands of Assam and Kumaon, 1914–1950," in R. P. Tucker and J. F. Richards, *World Deforestation in the Twentieth Century* (Durham, N.C.: Duke University Press, 1988), pp. 91–111.

28. Michael Adas, "Colonization, Commercial Agriculture, and the Destruction of the Deltaic Rainforests of British Burma in the Late Nineteenth Century," in Tucker and Richards, *Global Deforestation*.

29. David Feeny, "Agricultural Expansion and Forest Depletion in Thailand, 1900-1975," in Tucker and Richards, *World Deforestation*, pp. 112–146.

30. Roger A. Sedjo and Kenneth S. Lyon, *The Long-Term Adequacy of World Timber Supply* (Washington, D.C.: Resources for the Future, 1991).

31. J. G. Laarman and R. A. Sedjo, *Global Forests: Issues for Six Billion People* (New York: McGraw-Hill, 1992).

32. Roland M. Harper, "Changes in the Forest Area of New England in Three Centuries," *Journal of Forestry* 16 (1918): 442–452.

33. Barrett, "Northeast Region"; MacCleery "What on Earth."

34. Fraser J. Hart, "Loss and Abandonment of Cleared Farm Land in the Eastern United States," *Annuals* (Washington, D.C.: Association of American Geographers, 1968). Also see Williams, *Americans*.

35. J. M. McCloskey and H. Spalding, "A Reconnaissance-level Inventory of the Amount of Wilderness Remaining in the World," *Ambio* 18, no. 4 (1989): 221–227.

36. Menzies, "Sources of Demand."

37. Thirgood, *Man's Impact*.

38. J. Evans, "Plantation Forestry in the Tropics: Trends and Prospects," *International Tree Crops Journal* 4 (1986).

39. FAO, *Forest Resources Assessment 1980: Tropical Countries*, FAO Forestry Paper 30 (Rome: FAO, 1983).

40. U.S. Department of Agriculture, Forest Service, State and Private Forestry, Cooperative Forestry, *Tree Planting in the United States—1993* (Washington, D.C.: Government Printing Office, 1993).

41. Devandra Pandey, "Assessment of Tropical Forest Plantation Resources" (Uppsala: Swedish University of Agricultural Sciences, Department of Forest Survey, October 1992); A. S. Mather, *Global Forest Resources* (London: Belhaven Press, 1990); J. P. McLauren, *Radiata Pine Growers' Manual*, FRI Bulletin 4 (Rutorua: New Zealand Forest Research Institute, 1993).

42. FAO, *Forest Resources Assessment 1990: Tropical Countries*, FAO Forestry Paper 112 (Rome: FAO, 1993).

43. A. V. Korotov and T. J. Peck, "Forest Resources of the Industrialized Countries: An ECE/FAO Assessment," *Unisylva* 44 (1993).

44. Ibid.; T. G. Horner, W. R. Clark, and S. L. Gray, "Determining Canada's Forest Area and Wood Volume Balance, 1977–86" (paper presented at the Conference on Canada's Timber Resources, June 3–6, 1990, Victoria British Columbia); ECE/FAO, *Outlook for the Forest Products Sector of the USSR*, ECE/TIM/48 (New York: United Nations, 1989).

45. Korotov and Peck, "Forest Resources."

46. Kullerro Kuusela, *Forest Resources in Europe* (Cambridge: Cambridge University Press, 1994).

47. O. Kandler, "The Air Pollution/Forest Decline Connection: The 'Waldsterben' Theory Refuted," *Unisylva* 44, no. 174 (1993).

48. FAO, *World Forest Inventory 1963* (Rome: FAO, 1963).

49. P. J. Steward, "The Dubious Case for State Control," *Ceres* 18, no. 2 (1988): 14–19.

50. Yujiro Hayami and Vernon W. Ruttan, *Agriculture Development: An International Perspective* (Baltimore: Johns Hopkins Press, 1971).

51. Douglas Southgate and Morris Whitaker, *Economic Progress and the Environment* (New York: Oxford University Press, 1994).

52. L. Fortmann and John W. Bruce, *Whose Trees?* (Boulder: Proprietary Dimensions of Forestry, Westview Press, 1988).

53. World Conservation Monitoring Centre Staff, *Global Biodiversity: Statue of the Earth's Living Resources* (London: Chapman Hall, 1992).

54. J. E. M. Arnold and J. G. Campbell, "Collective Management of Hill Forests in Nepal," in *Common Property Resource Management* (Washington, D.C.: National Academy Press, 1986), pp. 425–454.

55. Clawson, *Forests for Whom?*

Chapter 7

CONSERVING BIODIVERSITY
Resources for Our Future

Stephen R. Edwards

HIGHLIGHTS

- An estimated 8 to 10 million different species live on the planet. Of these, only about 1.5 million have been named.

- Documented animal extinctions peaked in the 1930s, and the number of extinctions has been declining since then.

- Thirty-nine percent of the extinctions resulted from the introduction of nonnative species, 36 percent from habitat destruction, and 23 percent from hunting (for sport and products for sale) and deliberate elimination of a species. The remaining 2 percent became extinct for a variety of causes, including pollution.

- More than 75 percent of the land on every continent except Europe is available for wildlife. It is this "undeveloped" land in developing countries that is of greatest importance to conserving biodiversity over the long term.

- Well over a billion hectares of the earth's surface have been set aside as protected areas, representing over 10 percent of the land area.

- Since 1973, only twenty-three species have been removed from the U.S. Endangered Species list: eight were listed in error, a court invalidated the listing of one, seven recovered, and seven became extinct. In other words, there is an equal probability that a species listed on the Endangered Species Act will recover or become extinct.

- Governments and conservation organizations need to begin to promote harvesting and using wild species rather than stopping such uses.

- Trade in wildlife products is of the same order of magnitude as trade in forestry and fish products. In 1986, international trade in primates, ivory, orchids, reptile skins, fur-bearers, and tropical fish was worth $5 billion.

- To enhance the probability of sustaining the wild harvest or use, the rural people living in developing countries with wildlife resources must have sufficient incentives to conserve the resources, free from excessive government interference. Environmentalists need to recognize that rural people can be allies in conservation, and not treat them as environmental criminals.

- Listing species to achieve legal protection is not working. More land and marine areas are being designated as protected; however, it is difficult to assess how effective protected areas are at conserving biodiversity.

It was hot, about 45°C (114°F) in the shade. A group of about thirty rural people from the vicinity of Bufoulabe, Mali, had gathered to discuss with me and colleagues how they use the wildlife and what assistance they would like to manage these resources better. In the course of the meeting, a farmer stated that he would like all birds killed because they ate his millet. Another allowed that his livelihood was dependent on those birds, which he harvested and sold. The farmer was equally adamant that his family depended on the cash he earned from his crops. As the debate unfolded, sometimes with great emotion, an elderly gentlemen stood. He said, "It is not God's will to kill all of the birds." However, recognizing the conflict between the other two, he went on to suggest, "It would be fair for the man who harvested the birds to share some of his income with the farmer in compensation for his losses."

I tell this story because it illustrates many of the issues inherent in the debate over alternative approaches to conservation of biodiversity. It underscores the conflict between rural peoples' dependence on agriculture and their use of wild species; illustrates the conflicts conservationists have with preservationists when they try to help rural people manage wild species in rural community development; documents the innate commitment of humans to conserve the wild species with which we live; and provides a solution to preserving biodiversity that recognizes the needs of both stakeholders.

This story is not unique. Around the world—from Pakistan to Panama, Niger to Chile, Nicaragua to Zimbabwe—I have witnessed similar exchanges in dozens of rural villages over the past five years. What is even more remarkable is that in many of these villages, the people are doing something about conserving their wild resources on their own, often in spite of government policies.

My aim is to provide a reference point and baseline of information on biodiversity conservation that can be used to monitor and evaluate the field in the future. The key questions are: What is at stake? How can we solve the problem? What are we doing wrong? And what lies ahead in the future? The picture I present is one of opportunity, not despair. (See box 7-1.)

What Is at Stake?

We—the human species—have been dependent on other species since the beginning of our time, estimated at about 500,000 years. We eat other species, wear them, and use them to warm our houses, heat our water, and cook our food. We make and color our fabrics using other species; build our houses; make the furniture we sit on and the utensils that make our lives easier; and prevent and cure illnesses. As such, our development (both social and, in more recent times, economic) has been inextricably linked to our capacity to derive benefits from uses of wild species, which appears to be limited only by our creativity. As conservationists gain a better understanding of how biological systems work, they in-

---

BOX 7-1   Definitions of Terms

Biodiversity: "The sum total of all the plants, animals, fungi and microorganisms in the world, or a particular area; all of their individual variation; and all of the interactions between them."[1]

Conservation versus preservation[2]

Conservation: The saving of living natural resources *for use.*

Preservation: The saving of living natural resources *from use.*

Genetic diversity: The variation among individuals of the same species.

Species: The different kinds of organisms that comprise the basic units of biodiversity. Different species do not generally interbreed with each other under natural conditions, but when it does happen, the offspring are normally sterile.

Taxa: Scientists who describe species give them two-part names. The first is the genus, the second its specific name. Species believed to have close evolutionary relationships are given the same genus name. A genus may have many species. Genera are grouped in higher taxonomic categories called Families, then Orders, and so on. Each level in this hierarchy is referred to as a taxon.

[1] P. H. Raven, "Defining Biodiversity," *Nature Conservancy* (January–February 1994): 11–15.
[2] Based on definitions provided by J. Passmore, *Man's Responsibility for Nature* (New York: Scribner's, 1974).

---

creasingly recognize the need to conserve not only species or habitats but also the processes and variability that comprise the essential fabric of life on this planet.

Nevertheless, our understanding of biodiversity comes from a very limited knowledge about individual species and their ecological relationships. The composition of species that form different habitats varies according to nonliving conditions (e.g., availability of water, mineral content and pH of the soil, amount of sunlight). Within an area, different individuals of the same species vary, called *genetic diversity*, as do the number of species that comprise different habitats and the number of individuals of each species.

With very few exceptions, we do not know how all of these different pieces work together. We do not know how many species there are. It is estimated that there are between 8 million and 10 million different species on the planet; of these, only about 1.5 million have been named.[1] About 90 percent of the named species occur on land, and most are found in temperate regions of North America, Europe, Russia, and Australia—because that is where the greatest number of scientists are located who study them.

In no case do individual species exist in a vacuum. All species, including humans, are dependent on others. Some species are eaten by others; some are necessary for the reproduction of others; some serve as "housing" or nesting material for others. We know that all of these interrelationships between different plant and animal species are means of capturing and conveying the energy from the sun, which drives all of life's processes on the planet.

Therefore, at one level, what is at stake is our own survival. Humans are no less dependent on the effectiveness of these processes than other species. This is not an abstraction. In Niger, West Africa, over 80 percent of the domestic energy consumption by rural people to heat their huts, boil water, and cook comes from wild-harvested wood.[2] In Africa, Asia, and Latin America, virtually all of the rural poor depend on harvests of some wild resources every day of their lives. Therefore, conservation of nature's energy conversion system is also critical to world stability. Sustained economic and social development throughout the world depends on how we manage these renewable wild resources.

Are Species Extinctions a Threat to Our Survival?

At some level, I believe that all humans intuitively understand that we need the natural environment. Species like sequoias, tigers, elephants, aardvarks, alligators, snails, and butterflies capture our imagination. They are the subjects of stories, art, poetry, music, movies, and television. They are an integral part of our cultural and spiritual heritage no matter what part of the world we live in. Nearly everyone can visualize an elephant lumbering across the African savanna with its trunk swinging side to side and the sun reflecting off of its tusks—even if they have neither been to Africa nor seen an elephant in the wild. Because of such personal feelings, it is easier to grasp the importance of protecting an elephant, a whale, an eagle, a redwood, or even an ant than it is to grapple with the abstract concept of conserving biodiversity.

We are also sensitized, through the media and by fund-raising campaigns, to the plight of individual species. Webb notes that a standard approach in fund raising for conservation "is to make people 'like' a particular animal or aspect of the environment, such that they feel guilty about its neglect . . . , become sensitive about anything [that] threatens . . . its survival, and become more conscious about looking after it in the present."[3] Arne Kalland, of the Nordic Institute of Asian Studies, notes that species protection campaigns often are accompanied by public education programs that "totemize" groups of species into composites, such as elephant, whale, seal, cat, primate, and kangaroo.[4] The fact that each composite includes several species is lost along with the fact that not all species within each composite face the same threats. Nonetheless, the public identifies with the need to "save the seal" or "save the whale" and contributes accordingly.

We also read dire predictions of massive extinctions by recognized authorities, and their statements grab our attention:

- "Since more than nine-tenths of the original tropical rain forests will be removed in most areas within the next 30 years or so, it is expected that half of the organisms in these areas will vanish with it. By the end of the next century perhaps two-thirds of all species on Earth can be expected to vanish."[5]

- "The rate of extinction of bird and mammal species between 1600 and 1975 has been estimated to be between five and fifty times higher than it was through most of the eons of our evolutionary past. . . . In the last decades of the twentieth century, that rate is expected to rise to some forty to four hundred times 'normal.' "[6]

- "As many as 25 to 30 percent of the species on Earth will go extinct by the year 2,000."[7]

These scholars and many others are deeply concerned about the impact that massive extinctions will have on the survival of the planet. At the same time, we know that species have gone extinct in the past; habitats are converted all the time to agriculture and shopping malls; roads are built through forests; forests are harvested for timber. Our lives, at least in industrialized countries, are better today than they were fifty or a hundred years ago. And there has been no catastrophe. So what is the big deal?

From a conservationist's standpoint, the answer is simple. The human population will reach 10 billion by the year 2050, and demands on wild resources for commercial and subsistence purposes will likely reach levels never before known on the planet. At the same time, at the most basic level, we only know about 15 percent of the species we live with on the planet and even less about what they do and how they interact. Under such circumstances, even the most optimistic person would agree that common sense calls for caution.

Measuring Extinctions

Scientists use two ways to estimate extinctions. The first is straightforward: we count them. The second is by extrapolating losses of species by measuring the reduction in different kinds of habitat.

With regard to counting known extinctions, the International Union for the Conservation of Nature and Natural Resources (IUCN) has reported 593 extinctions of animal species (table 7-1). Heywood and Stuart reported 384 vascular plants have gone extinct.[8] However, all documentation to designate a species as extinct is based on negative evidence: the species has not been seen in the wild for over fifty years. But this does not necessarily mean that the species is lost, nor do the numbers include all species that could have gone extinct. We just do not know, in part because sufficient time has not passed to meet the criteria for designating a species as extinct, in part because we cannot monitor all species, and in part because we do not know all of the species.

Documented extinctions are often used to examine trends in extinction rates. The common way of illustrating the pattern of these extinctions is shown in fig-

TABLE 7-1 Documented Extinctions of Animal Species, by Region, 1600–1994

	Mammals	Birds	Reptiles	Amphibians	Fishes	Invertebrates	Total
Africa	7	34	12	1	1	50	105
Antarctica	1	0	0	0	0	0	1
Asia	8	12	0	1	2	25	48
Europe	3	5	0	0	0	24	32
North and Central America	37	27	6	2	30	162	264
Oceania	25	36	2	0	2	104	169
South America	2	0	0	0	1	3	6
Unknown	0	1	0	0	0	0	1
Total	83	115	20	4	36	368	626

Source: B. Groombridge, ed., *1994 IUCN Red List of Threatened Animals* (Gland, Switzerland: IUCN, 1993).

FIGURE 7-1 **Frequency of Documented Extinctions of Animals, 1600–1949**

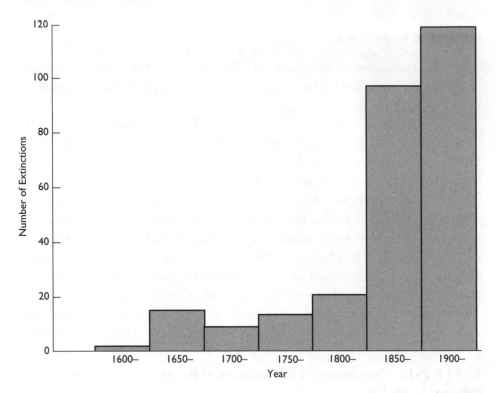

Source: Ibid.

ure 7-1.[9] The data through 1949 give the impression that we are experiencing a very rapid increase in extinctions, but this may not be the case. When data from 1950 to 1990 are included, there appears to be a decline in the number. When the data are viewed in ten-year increments, as in figure 7-2, it is clear that documented extinctions peaked in the 1930s and that the number of extinctions has been declining since then.

The reduction in the number of extinctions over the past sixty years may be because we do not have sufficient data to designate the species as extinct. Brian Groombridge, who edited the most recent edition of the *IUCN Red List of Threatened Animals*, states that "around 75 percent of recorded extinctions . . . have occurred on islands; almost all bird and mollusc extinctions have been recorded on islands." He goes on to say, "Very few extinctions have been recorded in continental tropical forest habitat, where mass extinction events have been predicted to be underway."[10] Island species are more vulnerable than

FIGURE 7-2   Frequency of Animal Extinctions, 1900–1990

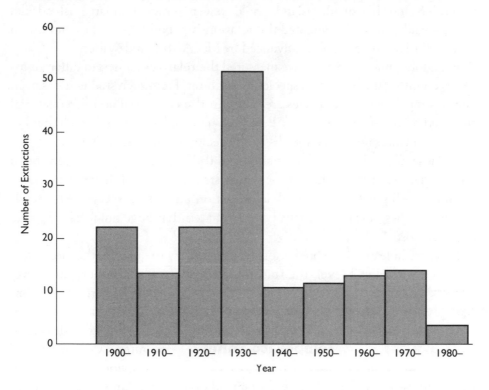

Source: Ibid.

species on large continents. For one thing, island species are trapped on their islands and cannot escape to another part of their range as continental species might. Additionally, islands tend to have more species that are found solely on particular islands, and they have fewer total numbers of species, so they may be better known to scientists and therefore there is a greater probability that their extinction would be documented.

The extrapolation method to estimate extinctions is more complex. It is based on the premise that the number of species found in a particular habitat is proportional to the area of that habitat on a logarithmic scale.[11] This means that, on average, you can expect to find twice as many species in 10 hectares than in 1 hectare of a particular habitat. The opposite is also true: that is, if a particular habitat is reduced from 10 hectares to 1 hectare, you would expect to find only half as many species remaining in the habitat.

The correlation between area and numbers of species is commonly referred to as the "species-area relationship" and provides a model that can be tested by collecting data on the number of species found in discrete habitats. Intuitively we understand that smaller islands "hold" fewer species than larger islands do. Several studies have documented this relationship, building on the equilibrium theory of island biogeography advanced by MacArthur and Wilson.[12]

For a long time biologists have measured the relative richness of different areas by counting the number of species in a habitat. From such studies it has been demonstrated that more species are found in the equatorial and subequatorial tropics than in temperate zones.[13] Based on such studies, it is estimated that between two-thirds and three-fourths of all species are located in the tropics.

Because of the greater species richness in the tropics and the apparent rapid rates of deforestation in these areas, scientists have focused their attention on the relationship between tropical deforestation and loss of species. As a result of these studies, several scientists have estimated that large numbers of species are being lost.[14]

How credible are these predictions? When we try to quantify the rate of forest loss at the global level, the statistics are confusing. Whitmore and Sayer review the various problems associated with using published statistics on deforestation rates.[15] One problem is the varying definitions of

More species are found in the equatorial and subequatorial tropics than in temperate zones.

what a forest is. Statistics have been lumped on different kinds of forests (e.g., tropical closed forests, dry forests, moist forests), and they do not reflect different kinds of forest usage. Complete deforestation (clear-cutting) will have a much different impact on the species composition than the harvest of selected trees, for example.

Irrespective of the validity of the data on deforestation rates, several scientists have also begun to question whether the species-area relationship is a viable model for predicting extinctions for other reasons. Simberloff points out that

> using the species-area relationship . . . [to predict species loss from deforestation] . . . entails the assumption that a single uniform-habitat forest with species spread uniformly throughout it is reduced to another single uniform-habitat forest with species spread uniformly throughout it. However, all forests are patchwork quilts of various habitats, so that species are not uniformly distributed.[16]

Simberloff is saying that even within what we perceive as a single habitat, there is considerable variation. Some areas within a forest may be especially species rich and others less so. We know this to be true just by taking a walk in the woods. Bracket fungi do not grow on every tree, berry bushes do not occur everywhere, and different birds and insects are heard in different parts of the forest. We also know that the forests of the eastern United States have been reduced significantly over the past 200 years. According to Simberloff, only between 1 and 2 percent of the original forest remains in fragmented patches.[17] Yet only three species of birds are known to have gone extinct: the Carolina parakeet, the ivory-billed woodpecker, and the passenger pigeon. Obviously there are dozens of species of birds associated with these forests that have not gone extinct, albeit there are probably fewer individuals of each species remaining today.

Vernon Heywood and Simon Stuart, who head IUCN's Plant and Species Conservation Programmes, conclude a review of species extinctions in tropical forests by noting that "there is currently very little evidence of extinctions at the rates predicted by some theoretical models."[18] They point out that predicting extinctions is fraught with many inherent difficulties:

- We do not know how many species there are.

- We do not know how many species occur in rain forests or how they are distributed.

- We do not know how valid it is to extrapolate bird and mammal extinction rates to other taxonomic groups.

- We do not know how valid it is to extrapolate data on species diversity in the New World tropics to other tropical areas in the world.

- We do not know the size of the populations of most tropical species.

- We know very little about the capacity of primary forest species to adapt to secondary forest.

- We cannot confirm whether an extinction has occurred.

- We do not fully understand the dynamics of an inherent extinction process.

- We do not know what effect positive conservation actions may have on the extinction process.

Causes of Extinctions

Yet although the record of documented extinctions may not be complete, it is possible to assess the overt causes that led to the loss of these species. Based on an analysis conducted by the World Conservation Monitoring Centre, 39 percent of the extinctions resulted from the introduction of nonnative species, 36 percent by habitat destruction, and 23 percent from hunting (for sport and products for sale) and deliberate elimination of a species (figure 7-3).[19] The remaining 2 percent went extinct for a variety of causes, including pollution.

Population biologists recognize that extinction is a process rather than a here today–gone tomorrow phenomenon. Species do not disappear from one day to the next. Their loss is gradual and most often attributable to a multiplicity of causes. During this process, genetic erosion, or loss of genetic variability in species, may be very important to the long-term maintenance of habitats.

That Heywood and Stuart challenge the efficacy of theoretical models used to predict extinctions does not mean they are unconcerned. They underscore the urgency of conservation interventions in relation to species that are "committed to extinction" if something is not done.[20] They believe that at some point in the process of genetic erosion and/or loss of habitat, populations of species may become committed to extinction. Based on work by Simberloff and Reid and Miller, they point out that in a given area, "when species-area relationships reach an equilibrium, particular species are destined eventually to decline and become extinct locally or totally."[21] In this argument, reduction of the area of a particular habitat changes the equilibrium between the species in that area and

FIGURE 7-3 Causes of Documented Animal Extinctions

Source: World Resources Institute, *World Resources 1994–95* (New York: Oxford University Press, 1994).

could lead to suites of species' "committing" themselves to extinction. This leads to the conclusion that any perturbation of a habitat, such as the introduction of an exotic species, conversion to agriculture, or overexploitation, will have a consequence.

The crucial point is that human activity, not the species, must be managed.

How Can Biodiversity Be Conserved?

The rural poor who live closest with wild resources are essential to successful conservation. *The World Conservation Strategy*, published in 1980, provided the first comprehensive, integrated strategy to conserve wild species and habitats. Central to this strategy is the integration of conservation objectives into the economic development process. It recognizes that "hundreds of millions of rural people in developing countries" are malnourished and destitute and "compelled to destroy the resources necessary to free them from starvation and poverty."[22]

Ten years later a successor volume, *Caring for the Earth: A Strategy for Sustainable Living*, was published. This updated strategy recognizes that within the limits of the earth, people have "the right . . . to the resources needed for a decent standard of living, and hence a right to derive economic and other benefits from wild species." At the same time people have the responsibility to ensure "that their uses of renewable natural resources are sustainable." *Caring for the Earth* establishes target goals for achieving sustainability in using wild species by the end of this decade.[23]

The trend to "humanize" conservation is also underscored in *Our Common Future* and more recently in *Agenda 21*.[24] Both recognize the central role of people throughout the world in conserving natural resources on which they depend. The same philosophy is embodied in the Convention on Biological Diversity which was launched at the Earth Summit and calls on parties to manage biological resources for sustainable use.

From another perspective, rural people are important because the majority of habitat, and associated wild species, are outside protected areas, and it is in these areas that these people live with the wild species we want to conserve. Rowan Martin, deputy director for research for the Department of Parks and Wild Life Management in Zimbabwe, has summarized the land distribution by crops and settlements, protected areas, and "wildlife potential" by continent (figure 7-4). Based on his figures, more than 75 percent of the land on every continent, except Europe, is available for wildlife.[25] It is this "undeveloped" land in developing countries that is of greatest importance to conserving biodiversity over the long term.

FIGURE 7-4   Type of Land Use, by Region

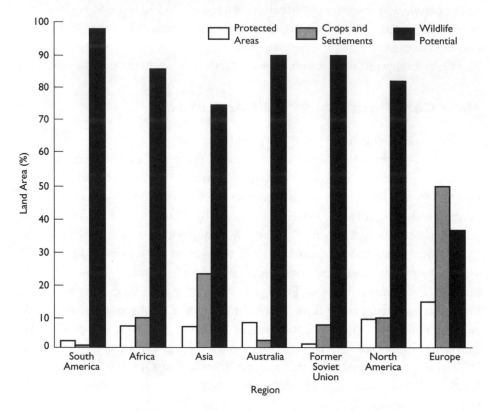

Source: R. B. Martin, "Should Wildlife Pay Its Way?" (Keith Roby Address, Perth, Australia, December 8, 1993).

The sustainable use of wildlife is an approach to conserving biodiversity that views people as the solution rather than the problem. The concept does not discriminate between consumptive and nonconsumptive uses of wild resources. It strives to integrate the social and biological sciences in conservation actions. It also recognizes that wild species and natural habitats may be the predominant resources available to rural poor people. This interdependence of humans on wild species and habitats provides the basis for building incentive systems that promote conservation.

Recognizing that wild resources have value to rural people—for food, medicine, religious purposes, and trade—is essential to this human-centered approach to conservation. It holds that people have an incentive to maintain a species or habitats if they derive some benefit from their use. Only by enhancing the value of wild resources can habitats compete with other forms of land use.

Richard Sandbrook, founder and head of the International Institute for Environment and Development (IIED), illustrates the relationships of three principal factors—ecological, economic, and social and cultural—that operate on wild resources (figure 7-5). Although each factor could be argued as the "most important," he stresses that sustainability of uses of wild resources derives from a balanced, holistic consideration of all three factors.[26]

The Department of Parks and Wild Life Management of Zimbabwe has been on the front line since 1975 in trying to balance the needs of the rural poor with efforts to conserve that country's wild resources for future generations. As a re-

FIGURE 7-5   **Factors Contributing to Sustainable Use of Wild Resources**

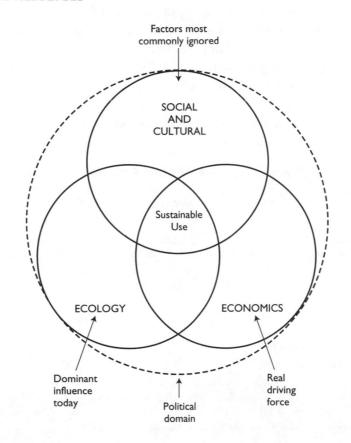

Source: Ibid.

sult, much of what we know about sustainable use comes from the pioneer work of public and private researchers in that country.

The cornerstone of Zimbabwe's approach is its "Parks and Wild Life Act of 1975, under which landholders were granted full rights to use wildlife without government interference (with the exception of Specially Protected Species.)"[27] This law effectively gave responsibility for managing wild resources to land-holders. In 1982 the act was amended to grant district councils authority over wildlife use on traditional communal lands. By 1986 a collaborative program involving the Department of Parks and Wild Life Management, University of Zim-babwe (Centre for Applied Social Sciences) donors, and nongovernment organizations (World Wide Fund for Nature and Zimbabwe Trust) was established to help rural villages learn how to manage wild species for their own benefit. This program is called CAMPFIRE (Communal Area Management Programme for Indigenous Resources).

The government promotes multiple uses of wildlife. Considerable effort has gone into documenting and explaining the values of different uses of wildlife to the landholders and the leaders in communal areas. Table 7-2 summarizes the different types of wildlife and land management that have been promoted and theoretical net returns from each per 1,000 square kilometers, assuming a full complement of large mammal species.[28] Under these conditions, with maximum use being made of an area it would produce a net profit of U.S.$84 per hectare compared to a net profit of U.S.$3 per hectare from cattle ranching.[29]

TABLE 7-2   Different Uses of Wildlife Promoted in Zimbabwe

Type of Management	Gross Return (U.S.$/ha)	Assumed Profit Return (%)	Net Return (U.S.$/ha)
Mass tourism	100	100	50
Exclusive tourism	50	100	25
International safari hunting	7.5	200	5
Sale of live animals	5	100	2.5
Meat, hides, and products	2.5	66	1
Subsistence hunting	1	100	0.5
Cattle ranching	15	20	3

Source: R. B. Martin, "Should Wildlife Pay Its Way?" (Keith Roby address, Perth, Australia, December 1993); Department of National Parks and Wild Life Management, Zimbabwe.

Although these figures are theoretical, there is compelling evidence that the economic incentives are in fact working in favor of conserving biodiversity. Today more than 75 percent of the privately owned ranches in Zimbabwe derive some income from wildlife; more than 500 farmers derive all or part of their income from wildlife management. For example, in order to make money from tourism for game viewing or safari hunting, or by selling meat to restaurants, landowners must invest in managing the wildlife populations. It is this approach that traditional conservationists find hard to grasp: profitable businesses must nurture the wildlife to make money from it. The area allocated to wildlife on private lands is increasing at the rate of 6 percent per year according to a recent survey of landholders.[30] The first two district councils were granted formal authority by government to use wild species in 1988. By 1993, twenty-two district councils had joined the program. Overall, the amount of land dedicated to wildlife use in Zimbabwe has increased significantly over the past fifty years. Today national parks account for less than 30 percent, while privately owned and communal lands account for nearly 50 percent.[31]

This situation is not unique to Zimbabwe. Throughout southern Africa, between 10 and 20 percent of the commercial farmers were involved in game ranching by 1991.[32] In Kenya, David Hopcroft has shown that 50 to 100 percent more meat can be produced per acre from game ranching because more game species can be kept per unit of area than cattle. He found that in 10 acres he could keep eight gazelle but only one cow.[33] Additionally, his net profit in 1994 from a multiple of uses (including tourism, sale of meat to restaurants, training, and research) of fifteen species of wildlife is ten times greater than what he would have gotten from traditional cattle ranching.[34] In 1994, Thompson's gazelle produced 3.65 times more lean meat than range cattle in a given area.

I recently visited the village of Masoka in northern Zimbabwe which is participating in the CAMPFIRE program. When I approached this village of some 120 households, the first thing I noticed was an electric fence that, I was told, stretched 19 kilometers around the village. This fence separates the villagers from the wild animals that are both revered and feared—feared because they kill people and destroy crops. Those animals that we see on our televisions and consider cute are all too real to them. But it is more than that. The fence also means that the village children can go to school instead of spending their days in the fields protecting the crops from elephants and buffalo. Villagers can sleep at night without worrying about what might happen to their children. Life is better.

During my visit, I met with the village chief and members of the village Wildlife Committee to discuss their views of the CAMPFIRE program. Our meeting was held in the office of the principal of the village school, a point of extreme pride. With a background of laughing children, the committee members proudly told me of their accomplishments. Last year each household in the village received about U.S.$450 from sales to a safari operator for the animals that his clients had shot in the vicinity of the village. That may not seem like much; however, compared to the U.S.$650 average annual per capita gross national product of the country, it is significant income in anyone's terms. Most of this money was being used for upkeep on the electric fence, to buy school supplies, and to build a clinic. To the villagers of Masoka, wildlife is an asset, and they want it to stay—even if it is dangerous.

Use of wildlife by private landowners or rural people under CAMPFIRE is not a panacea. Not all people in Zimbabwe agree to this approach. Further, not all villages view wildlife in the same way that the people of Masoka do. Not too far from Masoka, three people were killed by elephants in the last year. Most of the rural people in this district were outspoken in their desire to have the animals shot. One reason they feel this way is that there are far more people in that district than in Masoka, and the people had received only Z.$1 each from wildlife uses in the previous year.

Despite some bureaucratic problems, CAMPFIRE's main achievement is developing the self-reliance of the rural people. It provides a structure to identify problems and to solve them. Villages make decisions responsive to the desires of their residents. The districts participating in the program have formed their own trade association to promote their views and needs with the government. The government is considering further decentralization of authority, from the district council level directly to the villages.

Principles of Sustainable Use

Grahame Webb, an Australian ecologist who has specialized in crocodile management for the past twenty years, notes that "to sustain . . . means to keep things going." Although it is not possible to guarantee in advance that a particular activity is sustainable, nevertheless, it is feasible to determine the probability that it can be sustained. Therefore, sustainable use "can be approached objectively, rationally and unemotionally," according to Webb.[35]

Webb offers a simple model defining the crucial elements of the sustainable use of wildlife (figure 7-6). In order to strengthen the sustainability of any wildlife use, money must be spent on monitoring wildlife to measure changes in the population or habitat and on corrective actions. This is no different from what people already do when they raise chickens or cattle. To use wildlife sustainably, the cost of producing it must be factored in—something we do not even

FIGURE 7-6   Critical Elements of Wildlife Use Programs

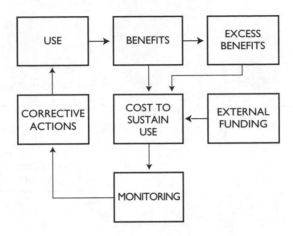

Source: G. W. Webb, "Criteria for the Sustainable Use of Wildlife," presented in the Workshop on Sustainable Use of Renewable Wild Resources, Nineteenth Session of the IUCN General Assembly, Buenos Aires, Argentina, January 1994.

think about when we purchase a dozen eggs or a pound of hamburger in the grocery store. However, these costs have not normally been explicitly incorporated into the sale price of products derived from wild species, as they are in Zimbabwe. Traditionally, the costs of conservation measures are borne by governments, through departmental budgets, or by donor agencies.

Adaptive Management

Webb calls monitoring and taking corrective action *adaptive management* (figure 7-7). Richard Bell, who has spent over twenty years studying wildlife in several countries in southern and eastern Africa, describes adaptive management as simply common sense. In 1986, Bell recognized that wildlife management depends on ecology, sociology, and economics, each of which is "complex and poorly understood. Therefore, conservationists and wildlife managers characteristically operate in situations where the outcome of their actions is uncertain."[36] Understanding the factors that contribute to the uncertainties are as much human centered as they are biological is crucial.

Conservation is really about managing human impacts on wild ecosystems, whether for aesthetic reasons or income.[37] Limiting the number of tourists walking through a park, as in Yosemite National Park, is as much a management de-

FIGURE 7-7 Adaptive Management

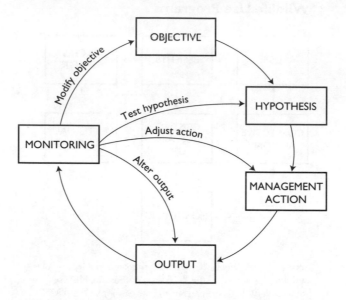

Source: Adapted from R. B. Martin, "Alternative Approaches to Sustainable Use: What Does and Doesn't Work," presented in Symposium 1: Conservation Benefits from Using Wildlife, Conservation through Sustainable Use of Wildlife Conference, University of Queensland, Australia, February 8–11, 1994.

cision to conserve the resource as adjusting a harvest quota for a fishery because census or harvest information indicates a decline in the target population.

Villagers in Nicaragua were keen to manage green iguana lizards to increase their harvest for subsistence and sale. The lizards, and their eggs, are a favorite food in the region. Each female produces a lot of eggs, but the majority of the young hatchlings do not survive in the wild. A survey of the lizard population near the village showed that the species had been overharvested. After reviewing the survey data, the villagers decided to establish a "reserve" in which no hunting would take place. They collected a large number of hatchlings and eggs and placed them in an enclosed area, where they were protected from predators. They fed the animals while they were in captivity. After a few months, the captive animals were released into the reserve. Hunting was restricted to an area adjacent to the reserve. This elegant management plan promoted the conservation of the species while accommodating the needs of the villagers. Ultimately, the sustainability of this activity will be based on the ability of the villagers to market the lizards. Unfortunately, since formal assistance was begun in 1992, no effort has been made to help the villagers develop markets. Neither the government nor private agencies that are supporting the village have the technical skills necessary to take this crucial step to enhance the sustainability of the management activity.

Human Factors

The role of humans in natural resource management has been studied by Albertson et al. and by Faulkner and Albertson.[38] Their model emphasizes a cyclic process in rural community development, with each cycle providing a stage for developing greater capacity and more sophisticated activities (figure 7-8). The spectrum of natural resources used changes with each successive cycle.

Outsiders, like development assistance agencies, can only facilitate decisions in rural villages. The villagers themselves must be committed to the development and conservation objectives and design their own activities to achieve those objectives. This may appear to be a trivial point, but I cannot count the number of villages I have visited in which the leaders thanked me for asking them what they wanted to do rather than telling them what we would do for them.

Villagers in the Petén region of Guatemala heavily depend on wild-harvested game meat for food and for sale to local restaurants catering to tourists visiting the local Mayan ruins. A study of the hunting efficiency (that is, the time needed for a hunter to harvest the game) showed that the activity was not sustainable. Rather than immediately addressing wildlife management issues, the first year's activities were devoted to establishing a working relationship with the villagers and gaining their trust. The key activity that cemented the relationship was a play depicting the history of the village in which the villagers played the various parts. Only after these steps were taken was it possible to deal with wildlife conservation and management issues.

There are few rural communities that have not been the subject of one development project or another over the years. Invariably such development schemes are the brainchildren of someone sitting at a desk in an air-conditioned office, far removed from the realities that the villagers are trying to cope with. Often the villagers' own government can be described at best as anathema. There is no trust. At the same time, villagers see donor agencies as sources of money that will help, no matter how silly the donors' programs are. It is no surprise that most development schemes stop once the funding dries up. This is

FIGURE 7-8 Sustainable Development Wheel

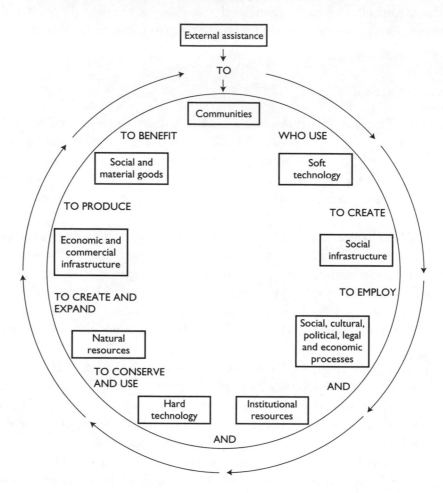

Source: Modified from A. O. Faulkner and M. Albertson, "Tandem Use of Hard and Soft Technology: An Evolving Model for Third-World Village Development," *International Journal on Applied Engineering Education* 2, no. 2 (1986).

why a great deal of effort must go into establishing a collaborative relationship with villagers before any sustainable development activity can proceed.

Dependence on Wild Harvests

The Convention on International Trade in Endangered Species of Wild Fauna and Flora (CITES) organization maintains data on international trade in animal and plant species listed under the convention. Fitzgerald, using CITES trade data, calculated that trade in primates, ivory, orchids, reptile skins, furbearers, and tropical fish was worth U.S.$5 billion in 1986.[39] Exports from developing countries of wood products were estimated to be worth U.S.$6 billion, and fisheries exports were valued at U.S.$5.5 billion in 1982.[40] Although these figures are dated, they do indicate that trade in wildlife products is of the same order of magnitude as trade in forestry and fish products. When all forms of wildlife harvests, including subsistence and domestic trade, are considered, it is conceivable that trade in wildlife products far exceeds the forestry and fisheries sectors.

There is no comprehensive source of data to document the degree to which wild species are harvested for subsistence or domestic trade. Most of our knowledge of wild species harvests by rural people is anecdotal or based on unpublished consultants' reports. Makombe provides an overview of the degree to which wild resources are harvested and used in Africa.[41] One study reported that 83 percent of 714 ethnic groups in Africa and Madagascar were dependent on wild resources.[42] Fifty percent relied on fishing, hunting, and gathering of wild plant products. In Niger, 99 percent of the domestic energy (and 80 percent of primary energy) requirements are met from wild-harvested wood.[43] Anstey estimated that 75 percent of the meat production in Liberia was derived from wild harvests in 1991.[44]

In more developed communities, especially where there is greater investment in agricultural production, species that are not essential for subsistence may be harvested for sale or trade with others in the village or with neighboring villages to obtain cash to buy essentials. Rural peoples' dependence on wild-harvested products increases during droughts, civil conflicts, when crops are lost to pests, and after natural disasters. Where there is high reliance on agricultural products to obtain cash to purchase essentials, the impact on wild harvests is exacerbated.

Most often, in developing countries, rural people do not distinguish between harvests for personal consumption and harvests for sale. The fact that the sale may result in the animal or plant's being exported to another country is of no consequence to them. Rarely do they know whether a particular wildlife product has an international market, let alone the value. While they may know something about national markets, again because of their isolation, the local value is established by middlemen who transport the wildlife products to national markets and the points of export. (Several examples of rural-based harvests for domestic and international trade are included in box 7-2.) If this trade is to be sustainable, it is essential that an equitable share of the value now going to the middlemen and exporters be used to manage the resources.

BOX 7-2   Examples of Rural Harvests for Commercial Purposes

GAYA DISTRICT, NIGER

The Roneraie palm (*Borassus aethiopium*) is used by people for a number of purposes: male flowers provide fodder for livestock; an embryonic structure high in carbohydrates, locally known as miritchi, is cultivated and sold; fruits are eaten and sold; leaf stems are used for making furniture; leaf fibers are used for making fish traps; and the trunk is split and used for house construction.[1]

ECUADOR

William T. Vickers spent years working in an Ecuadorian village in the Amazon Basin to document the hunting patterns of the villagers. Based on a sample of 863 hunting days spanning ten years, he determined that the village harvested forty-eight species of game. Four species (woolly monkey, white-lipped peccary, collared peccary, and Salvin's curassow) constituted 52 percent of the game species that were harvested; eight (with the addition of blue-throated piping guan, agouti, Spix's guan, and howler monkey) constituted 81 percent of the total. He concluded that the hunters compensated when a "favored" species became more difficult to find by taking alternate species.[2]

PETÉN, GUATEMALA

Rural people harvest pimienta gorda (used in food preservation and to provide the scent in Old Spice cologne), chicle (a natural latex used in chewing gum), and xate (pronounced *chate*; small palms used in the flower trade). Reining and Heinzman documented how the people harvest these products, all destined for international trade, at different times of the year.[3]

EAST AND WEST KALIMANTAN, INDONESIA

Rural farmers augment their income by collecting reptiles for the exotic leather trade. One modern tannery in East Kalimantan processes 350,000 skins a year. Species include the reticulated python, veranus lizard, and whip snake.

(continued)

KALMYKIA, RUSSIA

The saiga antelope (*Saiga tatarica tatarica*) has been harvested by local people for the meat, hides, and horn for decades. The meat is consumed locally and has provided an important source of protein. The horn is used in the Asian medicine market. Two years ago the horn sold for about U.S.$400 per kilogram in Asia.

NORTH AMERICA

Sport hunting in the United States and Canada generates about U.S.$70 billion per year in income.[4]

[1] A. Tiega, "Wildlife Use in Niger: Historical Trends and Future Perspectives," in *Proceedings of the First International Wildlife Management Congress, San Jose, Costa Rica* (forthcoming); D. D. Marthe, "Etude socio-economique pour l'elaboration et la mise en oeuvre d'un programme d'appui à la gestion locale de la roneraie" (IUCN-Niger, 1993).

[2] W. T. Vickers, "Hunting Yields and Game Composition over Ten Years in an Amazon Indian Territory," in J. G. Robinson and K. H. Redford, eds., *Neotropical Wildlife Use and Conservation* (Chicago: University of Chicago Press, 1992).

[3] C. Reining and R. Heinzman, "Nontimber Forest Products in Petén, Guatemala: Why Extractive Reserves Are Critical for Both Conservation and Development," in M. Plotkin and L. Famolare, eds., *Sustainable Harvest and Marketing of Rain Forest Products* (Washington, D.C.: Island Press, 1992).

[4] V. Geist, "Wildlife Conservation as Wealth," *Nature* 368(1994):491–492.

What Are We Doing Wrong?

Incomplete Data on the Status of Species

We know a lot more about large animals, especially those that are "charismatic," than we do about smaller, more obscure species. This bias affects species listings, fund raising, research, and conservation actions. Lists of threatened species prepared by the IUCN are important because they are often used to justify the designation of endangered species on international treaties, in the United States, and other countries. The IUCN has defined several categories of threat to wild species (box 7-3). Based on these categories, nearly 6,000 animal species were considered to be subject to varying degrees of threat (table 7-3) in 1993.[45] The 1990 list reported about 4,500 species as threatened.

There is considerable variance in the percentage of described species that have been assessed in each of the groups considered. Birds are the only group in which 100 percent of the species have been assessed. In this case, the number considered threatened was reduced from 1,026 in 1990 to 970 in 1993, after the assessment was

TABLE 7-3 Numbers of Threatened Animal Species, by Category of Threat

	Endangered	Vulnerable	Rare	Indeterminate	Subtotal	Insufficient Information	Total
Vertebrates							
Mammalia	177	199	89	68	533	208	741
Aves	188	241	257	176	862	108	970
Reptilia	47	88	79	43	257	59	316
Amphibia	32	32	55	14	133	36	169
Cephalaspidomorphi	0	1	2	0	3	2	5
Elsamobranchii	0	0	0	1	1	2	3
Actinopterygii	158	224	244	303	929	41	970
Sarcopterygii	0	1	0	0	1	0	1
Subtotal	602	786	726	605	2,719	456	3,175
Invertebrates							
Anthozoa	1	1	0	0	2	1	3
Turbellaria	2	0	1	0	3	0	3
Enopla	2	0	4	0	6	0	6
Gastropoda	264	412	138	253	1,067	29	1,096
Bivalvia	45	10	1	59	115	4	119
Polychaeta	0	0	0	0	0	1	1
Hirudinea	0	0	0	1	1	0	1
Oligochaeta	2	145	1	0	148	0	148
Merostomata	0	0	0	0	0	4	4

Arachnida	3	2	4	8	17	1	18
Crustacea	10	6	28	82	126	32	158
Insecta	252	119	241	538	1,150	34	1,184
Onychophera	1	7	4	0	12	0	12
Echinoidea	0	0	0	0	0	1	1
Subtotal	582	702	422	941	2,647	107	2,754
Total	1,184	1,488	1,148	1,546	5,366	563	5,929

Source: B. Groombridge, ed., *1994 IUCN Red List of Threatened Animals* (Gland, Switzerland: IUCN, 1993), table 1.
Note: See box 7-3 for definitions of categories of threat.

BOX 7-3   IUCN Categories of Threat

Extinct: Species not definitely located in the wild during the past fifty years (criterion as used by the Convention on International Trade in Endangered Species of Wild Fauna and Flora).

Endangered: Taxa in danger of extinction and whose survival is unlikely if the causal factors continue operating. Included are taxa whose numbers have been reduced to a critical level or whose habitats have been so drastically reduced that they are deemed to be in immediate danger of extinction. Also included are taxa that may be extinct but have definitely been seen in the wild in the past fifty years.

Vulnerable: Taxa believed likely to move into the endangered category in the near future if the causal factors continue operating. Included are taxa of which most or all of the populations are decreasing because of overexploitation, extensive destruction of habitat, or other environmental disturbances; taxa with populations that have been seriously depleted and whose ultimate security has not yet been ensured; and taxa with populations that are still abundant but are under threat from severe adverse factors throughout their range.

Rare: Taxa with small world populations that are not at present endangered or vulnerable but are at risk. These taxa are usually localized within restricted geographic areas or habitats or are thinly scattered over a more extensive range.

Indeterminate: Taxa known to be endangered, vulnerable, or rare but there is not enough information to say which of the three categories is the appropriate one.

Insufficiently known: Taxa that are suspected but not known to belong to any of the above categories, because of lack of information.

Source: B. Groombridge, ed., *1994 IUCN Red List of Threatened Animals* (Gland, Switzerland: IUCN, 1993).

completed. About 50 percent of the mammal species have been formally reviewed, and a much smaller percentage of the lower vertebrate species have been assessed (reptiles, 20 percent; amphibians, 12 percent; and fish, less than 10 percent). Merely a fraction of the invertebrates has been assessed to determine their status.

In the past few years IUCN has reexamined its categories of threat. The definitions that have been in use since the 1980s are very subjective in their interpretation, with discrimination among the categories depending to a great extent on whether the species does not meet the criteria of the next higher category of threat. For these reasons, a new set of categories is being considered that are more objective.[46] The definitions of the new categories are based on principles derived from population biology and rely on compilation of quantitative data. The proposed new definitions are provided in box 7-4, and an outline of the parameters to be considered in assigning a species to a particular category is provided in box 7-5.

I provide these definitions because future editions of the *IUCN Red List* will most likely follow them. Further, in November 1994, the parties to CITES adopted new criteria for assessing proposals to list species as threatened with extinction or the possibility of being threatened with extinction that are based on these definitions. It is likely that they will prompt more objective reassessment of the status of species that have been listed under national legislation.

Legislated Protection of Species

Conservation over the past thirty years has relied to a great extent on legislating controls over human behavior to protect species. In the United States twenty different federal laws and associated regulations control how wild species are used (or not used).[47]

The flagship of species protection laws is the United States Endangered Species Act of 1973. The central theme of this act is protection of species that are listed as "endangered" or "threatened." With very few exceptions, species listed as threatened are afforded the same level of protection as those designated as endangered. A total of 1,452 species (920 domestic species and 532 foreign species) have been listed on the U.S. Endangered Species Act through August 1994. The cumulative number of species listed under the act, from 1973 to the present, is shown in figure 7-9.

If a species is threatened, the listing of it should trigger remedial actions to correct or remove the causes of the threats. This is not what happens. Since 1973, only twenty-three species have been removed from the list: eight were listed in error, a court invalidated the listing of one, seven recovered, and seven

BOX 7-4   Proposed New Definitions of Categories of Threat

Extinct: There is no reasonable doubt that the last individual has died. A taxon is presumed extinct when exhaustive surveys in known and/or expected habitats, at appropriate times, throughout its historic range have failed to record an individual. Surveys should be over a time frame appropriate to the taxon's life cycle and life form.

Extinct in the wild: A taxon is known to survive only in cultivation, in captivity, or as a naturalized population (or populations) outside its historical range.

Critical: A taxon is facing an extremely high probability of extinction in the wild in the immediate future. A taxon is defined as critical by any of the quantitative criteria cited in the second column in box 7-5.

Endangered: A taxon is not critical but is facing a very high probability of extinction in the wild in the near future. A taxon is defined as endangered by any of the quantitative criteria cited in the third column in box 7-5.

Vulnerable: A taxon is not critical or endangered but is facing a high probability of extinction in the wild in the medium-term future. A taxon is defined as vulnerable by any of the quantitative criteria cited in the fourth column in box 7-5.

Susceptible: A taxon does not qualify for any of the categories above but is of concern because its range area is restricted (typically less than 100 km²) and/or it is found at few locations, which render it prone to the effects of human activities.

Safe/low risk: A taxon has been evaluated and found not to qualify for any of the threatened categories above.

Source: G. M. Mace, "The Status of Proposals to Redefine IUCN Threatened Species Categories," in Groombridge, *1994 IUCN Red List of Threatened Animals.*

BOX 7-5   Draft Criteria for Qualitative Calculations for Proposed New Categories of Threat

Standard	Critical	Endangered	Vulnerable
A: **Very small population size;** number of mature individuals equals:	≤ 50	≤ 250	$\leq 1,000$
B: **Small population size;** number of mature individuals equals:	≤ 250	$\leq 2,500$	$\leq 10,000$
and			
either all populations are small:	≤ 50	≤ 250	$\leq 1,000$
or			
there is a small number of subpopulations:	≤ 1	≤ 1	≤ 1
and			
the subpopulations are declining:	any rate	any rate	any rate
C: **Small distribution;** either in geographical extent:	$\leq 100 km^2$	$\leq 5,000 km^2$	$\leq 20,000 km^2$
or			
in range area:	$\leq 10 km^2$	$\leq 500 km^2$	$\leq 2,000 km^2$
and			
Two of the following:			
• The population is declining:	any rate	any rate	any rate
• The population is either fragmented:	+	+	+
or			
• There is a small number of populations:	≤ 1	≤ 2	≤ 5
• The populations are fluctuating:	>1 order of magnitude	>1 order of magnitude	>1 order of magnitude

(*continued*)

BOX 7-5 Continued

D: **Population declining**; at

rate of at least:	25%	50% in	50% in
	per year	5 years *or*	10 years
	over	2 gener-	*or* 3 gener-
	5 years	ations[1]	ations[1]

In any of:

- observed rates

- inferred rates

- projected rates

E: **Quantitative analysis**
    shows that the probability

of extinction in the wild is	50% in	20% in	10% in
expected to be at least:	5 years	20 years	100 years.
	or 2 gener-	*or* 5 gener-	
	ations[1]	ations[1]	

Note: To apply these criteria see definitions in box 7-4.
[1] Whichever is the longer.

went extinct. In other words, there is an equal probability that a species listed on the Endangered Species Act will recover or go extinct. Overall, the status of 270 species is improving or stable, while it is declining for 232 species, according to the U.S. Fish and Wildlife Service.

In twenty years only sixteen species have been "downlisted" from endangered to threatened.[48] Controls over the use of nine species were relaxed because their populations had recovered (three of them do not occur in the United States), two to allow controlled harvests, one because of a taxonomic change, and four due to a "data error."

Mandated recovery plans have been prepared for only 410 of the 711 native species under the jurisdiction of the Fish and Wildlife Service.[49] According to a report to Congress, the "primary objectives of the recovery plans include delisting, downlisting, or protection of existing populations for a specific time period or for the foreseeable future." The report notes that "the percentages of recovery objectives achieved are used as a measure of progress toward recovery." Figure 7-10 shows the percentage of recovery objectives that have been achieved in four categories defined by the service. Less than 25 percent of the objectives have been achieved for 588 species. More than 75 percent of the recovery objectives have been achieved for only thirty species (4 percent).

FIGURE 7-9   Cumulative Listings of Species on U.S. Endangered Species Act

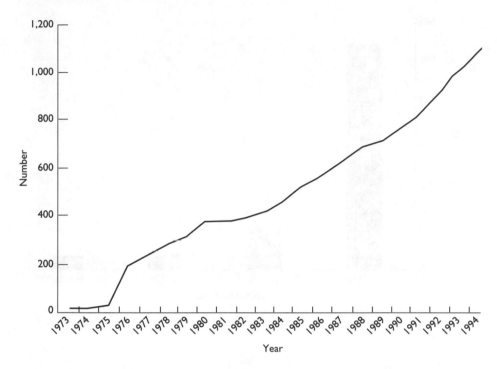

Source: George Drewry, U.S. Fish and Wildlife Service Division of Endangered Species, personal communication.

The Endangered Species Act is not achieving its goals. Based on these figures, listing a species "is a distinction without cause for celebration." In fact there is considerable support for the notion that "the government's listing of a species . . . is often the species' obituary notice."[50]

International Controls over Trade in Species

CITES is the principal international instrument to protect wild species. Since its inception in 1972, there has been a continuing debate over whether the treaty is an endangered species treaty, on the model of the U.S. Endangered Species Act, or a device to control trade in wild species so that they do not become endangered. According to the text of the convention, its purpose is the latter.

As of July 1994, 122 countries had acceded to CITES. Since CITES came into force in 1972, about 33,500 species have been listed on Appendixes I and

FIGURE 7-10   Recovery Objectives Achieved under the U.S. Endangered Species Act

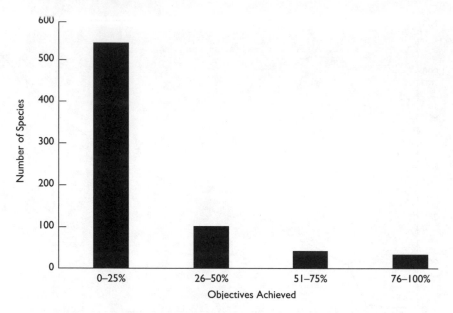

Source: U.S. Fish and Wildlife Service, *Endangered and Threatened Species Recovery Program*, report to Congress (Washington, D.C., 1992).

II as protected species. It is difficult to calculate the precise number of species on Appendix II because higher taxonomic categories, such as orders, families, or genera, are listed in addition to individual species. One family of plants alone, the orchids, listed on Appendix II may have as many as 27,000 species; however, the actual number is not known.

CITES lists species on three appendixes. Appendix I includes species threatened with extinction that are or may be affected by international trade. Appendix II includes species that may become threatened if trade in those species is not strictly controlled, or they look like other species that are listed on Appendix II. Appendix III includes species that are of concern within a country. Listing on Appendixes I and II is subject to approval of the parties to CITES. Listings on Appendix III may be made by any country without the approval of the other parties. International trade is allowed in Appendix II species according to various controls. No international trade is allowed in species or their products that are listed on Appendix I. Species listed on Appendix III are treated like species listed on Appendix I. The cumulative listings of species on Appendix I over the past twenty years are illustrated in figure 7-11.

FIGURE 7-11   Cumulative Listings on CITES Appendix I, 1975–1987

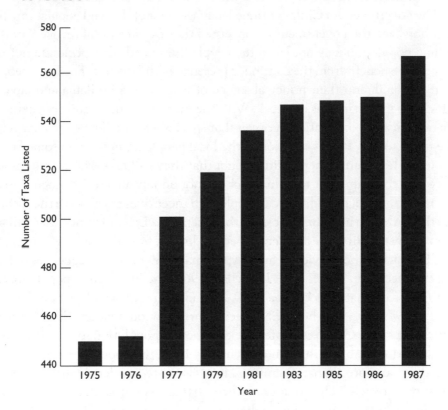

Data have not been maintained on transfers of taxa from one appendix to another since the convention came into force. However, very few species have been removed from the appendixes, even when it has been documented that the species has not been in trade for at least ten years.

When we look at the status of individual species, it is hard to determine if listing a species on Appendix I has been effective in its conservation. The vicuña may be one of the few examples in which populations of a species have recovered during the time that it was listed on Appendix I. In the vicinity of Lauca National Park in northeast Chile, the vicuña population was estimated to be about 2,000 animals in 1972. The species was listed on Appendix I in 1975. In 1992, twenty years later, the population was estimated to have grown to 28,000. In 1987, selected populations in Peru and Chile (including the Lauca population) were downlisted to Appendix II, allowing the harvest of their fiber and export of cloth woven in the country of origin. Only the government of Peru implemented a program to harvest vicuña fiber and export cloth, which subsequently ceased when civil conflicts broke out in the areas where the vicuña populations are located.

The government of Chile continued management of the wild population. Today, however, the government is concerned that the successful recovery of the vicuña population may not last if the people living with the species do not receive some benefit from the continued presence of the species. The local people own the land, and their principal source of income is from llama and alpaca, which are maintained as livestock. With the increase in the vicuña population, there is greater competition between that species and the llama and alpaca for the range forage. In this situation, if the local people do not obtain some benefits from the vicuña , there is little doubt that they will take whatever steps are necessary to ensure that their livestock are not unduly affected. Recognition of the importance of providing local people with incentives to conserve the species led the parties to adopt a proposal at the recent CITES conference that will allow rural people in Chile and Peru to benefit directly from the sale of vicuña fiber.

The black rhino provides a more typical pattern of the consequences of listing the species on Appendix I. In the late 1960s the black rhino population in Africa was estimated to be around 70,000 individuals. It was listed on Appendix I in 1975. By 1991 the IUCN/Species Survival Commission (SSC) African Rhino Specialist Group estimated the continental population to be 3,481. Zimbabwe had 1,400 black rhino in 1981, and by 1994, the population had been reduced to no more than 500 animals. It is also listed under Zimbabwe law as "strictly protected." The driving force behind this precipitous decline in the wild population is the demand for the horn in Asian medicines. The high demand coupled with the shrinking supply caused market prices for the horn to leap up. The actual wholesale prices paid for horn are not easily determined, but clearly they are high enough to motivate people to kill the animals, even at the risk of their own lives.

The Appendix I listing of the black rhino under CITES, or as "strictly protected" under Zimbabwe law, has not improved its conservation status. In fact, the listing may have precluded taking steps that could have had a positive influence on the black rhino's conservation. For example, because the species is listed on Appendix I, no international trade is allowed in any part of the species. Some people believe that if trade had continued, with greater controls on the market countries, there would have been strong economic incentives to conserve the species. But under CITES, the effectiveness of the treaty depends exclusively on the capacity of individual governments to enforce their own regulations. Opponents to continued trade in the horn argue that China, where most of the Asian medicines are prepared, is incapable of imposing ad-

International agreements that list both black and white rhinoceroses as endangered have not put a stop to poaching and trade in their horns. These agreements prevent local people from receiving some economic benefit from endangered species, so they have no incentive to protect them.

equate controls and therefore trade should not be allowed. They also argue that trade in back rhino products from Zimbabwe would encourage illegal harvests of other rhino species in other countries. However, given the precipitous decline in all rhino species, it is clear that the protectionist approach is not working.

At the recent meeting of CITES, the parties adopted a proposal to transfer South Africa's population of white rhino from Appendix I to Appendix II. This will allow international trade in live animals only, which will provide considerable economic incentives to conserve this important resource.

The African elephant is another species listed on Appendix I of CITES that requires more careful consideration of the factors contributing to the decline of the species. In the years prior to its listing in 1989, populations in East, West and Central Africa were being decimated by poaching for the ivory. Much of the population in southern African was increasing. Political pressure, not objective assessment of population data, determined the decision to list the African elephant. Factually, the poaching continues (albeit at a lower rate), and the habitat throughout many southern African countries is being destroyed by the growing populations of elephants. Halting legal trade in products from elephants dramatically reduced the incentive for Africans to manage this species responsibly. The African elephant is now the "poster child" for nongovernment organizations that have raised millions of dollars for its protection. Interestingly, at the recent CITES conference, South Africa withdrew its proposal to transfer its elephant populations to Appendix II because it did not believe it had sufficient support for the proposal to be adopted.

Over time, the decisions made by the parties to CITES have become more politicized. There is unquestionably a trend toward more and more species' being placed on the CITES appendixes. Nevertheless, the parties have a history of adopting creative solutions to specific conservation problems, such as ranching and farming of crocodilians. To my knowledge, because of the controls imposed under the convention, no listed species has gone extinct because of trade. Sadly, the black rhino may become the one exception to this claim. If the convention is to be responsive to the demands of the future, the parties must exhibit considerably more flexibility. Some of the more important challenges the parties must come to grips with are set out in box 7-6.

Lack of Legal Authority

Incentive systems to promote sustainable use of wild resources require that people be allowed to use them free from excessive government interference. This is not the case in most countries of the world where the land (and all of the resources on, or under, it) belongs to the state. Individual property rights are recognized in many countries; however, ownership of the land generally does not extend to the wild resources that reside on that property. Communal ownership of land is quite common in developing countries, but the wild resources on the land are most often owned by the state.

Exclusive government ownership of wild species prevents individuals from benefiting from uses of these resources. To accommodate sustainable use of wild species, laws must grant people "rights" to use the resources and to benefit from such uses. Without such benefits they have no incentive to conserve them. Rights to access and use wild species are essential if the users are to be accountable for their actions. Rights to use wild resources can be conveyed through property ownership or usufruct rights. To enhance the probability of sustaining the wild harvest or use, the people living with the resources must have sufficient incentives to conserve the resources.

BOX 7-6   Challenges Facing CITES

- Controlled trade in wildlife products is not a threat to their survival; however, such trade is still considered a negative or, at best, neutral influence on species survival.
- The convention must address basic principles of sustainable use, such as adaptive management, the rights of people who coexist with wildlife in developing countries to benefit from uses of wildlife, and the link between such benefits and the conservation of the resources.
- Individual parties must have the freedom, and encouragement, to try new, creative approaches to conservation of wild species.
- The "capacity gap" separating industrialized countries and developing countries must be addressed.
- Conservation of species is not assisted if markets for selected wildlife products are destroyed and demand reduced. On the contrary, such actions devalue the resources locally and often transfer the value of such resources to nongovernment organizations in industrialized countries.
- Greater emphasis is needed on increasing the compatibility of CITES with other treaties that aim to reconcile conservation and development.
- Harvests for international trade represent only a very small part of the actual offtake of wild species. Greater attention should be given to the impact of harvests for domestic consumption.

Protected Area Conservation: Big Is Not Necessarily Good

The aim of the protected area conservation movement is to set aside, by legal means, a significant percentage of representative ecosystems throughout the world as protected areas. Well over a billion hectares of the earth's surface have been set aside as protected areas, representing over 10 percent of the land area.

As of 1993, 8,619 protected areas had been created worldwide, representing about 800 million hectares and accounting for 5.9 percent of the total land area. A further 3,868 resource and anthropological reserves, of slightly less than 360 million hectares, and 977 marine and coastal zone areas, totaling another 200 million hectares, have been designated.

FIGURE 7-12   Cumulative Designation of Protected Areas in the World, 1900–1990

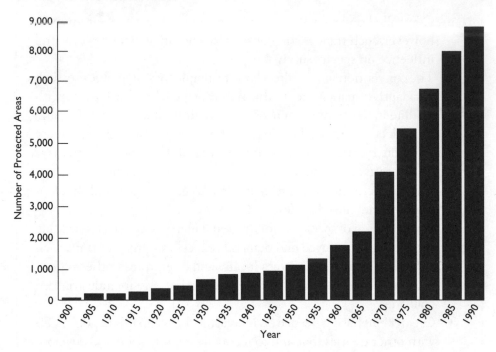

Source: IUCN, 1993 United Nations List of National Parks and Designated Areas (Gland, Switzerland: IUCN, 1994).

The 1993 United Nations List of National Parks and Protected Areas documents the growing trend toward creating designated protected areas between 1900 and 1990 (figure 7-12).[51] We are not, however, becoming more effective at conserving biodiversity. If governments lack the funds or the technical ability to manage the protected areas, they are no more than lines on paper.

Recently Rowan Martin examined the budgets for protected areas in a sample of thirty-two countries (table 7-4). He found that none of these countries spends more than a fraction of 1 percent of their national budget for protected area conservation. He notes that in "western countries, expenditure is primarily for management—not protection," which is the norm in most developing countries.[52] In North America and Europe, funding for protected areas was U.S.$1,283 per square kilometer.

Another problem that we must face is that protected areas are becoming "islands." When an area is first designated, it is likely that the habitat in the protected area was contiguous with the habitat outside the area. Where the parks

**TABLE 7-4   Protected Areas and Conservation Budgets in
Selected Countries**

| | | | Protected Area Budget | |
Region/Country	Number of Protected Areas	Area (sq. km.)	Total (U.S. $)	Per Square Kilometer (U.S. $)
Africa				
Cameroon	13	20,349	410,000	20.15
Central African Republic	12	58,374	480,000	8.22
Congo	10	13,338	40,000	3.00
Ethiopia	11	30,303	250,000	8.25
Gabon	6	10,439	420,000	40.23
Ghana	8	10,758	1,050,000	97.60
Kenya	36	34,726	18,200,000	524.10
Malawi	9	13,318	460,000	34.54
Nigeria	21	28,729	1,660,000	57.78
Senegal	10	21,758	620,000	28.50
Sudan	14	93,467	1,000,000	10.70
Tanzania	28	130,706	3,480,000	26.62
Zaire	8	85,842	1,000,000	11.65
Zimbabwe	25	50,736	6,700,000	132.06
Asia				
Brunei	5	778	36,060,000	46,349.61
Malaysia	51	14,740	4,980,000	337.86
Philippines	27	5,730	810,000	141.36
Taiwan	5	2,884	36,770,000	12,749.65
Thailand	90	55,152	15,100,000	273.79
Turkey	18	2,732	50,000	18.30
Europe				
Denmark	66	4,096	x	x
France	81	53,882	19,410,000	360.23
Germany	440	58,579	67,630,000	1,154.51
Netherlands	67	3,500	23,110,000	6,602.86
United Kingdom	140	46,271	162,680,000	3,515.81
Former USSR	213	244,184	32,700,000	133.92

Table 7-4   Continued

North America

Canada	426	496,812	282,990,000	569.61
United States	975	982,196	1,962,700,000	1,998.28

South America

Brazil	172	215,353	1,210,000	5.62
Ecuador	18	16,957	250,000	14.74
Peru	20	26,861	1,160,000	43.19

Oceania

Australia	748	468,129	12,220,000	26.10

Sources: Martin, "Should Wildlife Pay Its Way?"

catalyze development, commonly people move to the area around the parks. The continuity of habitat between the protected area and the surrounding area becomes disrupted, and eventually the protected area becomes isolated. When this happens, it becomes an ecological island, and it may not support all of the species that were once found in the area. Habitat disruption may "commit" species isolated in the parks to extinction before we become aware of the problem.

Divergence in Philosophies

The United States has unparalleled influence over international conservation policies. During the past thirty years, the conservation movement has sensitized the general public in both the United States and Europe to environmental issues. Organizations outside government are now the primary focus of citizens concerned about environmental issues. At one extreme, animal rightists, who have considerable influence on government policies, place the needs of individual animals above those of people, as epitomized by the recent statement by Favre: "Elephants are not turnips. . . . Even if a . . . [use] . . . is ecologically sustainable, it may be immoral. To kill elephants for the sole purpose of selling body parts like ivory is unacceptable."[53] At the other extreme are profiteers who harvest wildlife without any concern for the management needs of the resources they are exploiting. Neither extreme contributes to conservation in developing

The spotted owl became the focal point in a debate over how national forests in the northwestern United States would be used. Political management of resources often leads to zero-sum battles between activists and local people.

countries. Both confuse and obscure the fundamental issues that must be addressed if we are to conserve wild species. In particular, the growing influence of animal rightists is a major source of conflict between the North and the South.

Growth of the Conservation Movement

That governments own, or control access to, wild species in most countries "automatically . . . [provides the] . . . *raison d'etre* for the plethora of nongovernment organizations . . . to lobby the state."[54] The number of nongovernment conservation organizations has soared in the past ten years. Between 1983 and 1994 there was an increase of nearly 50 percent (388 to 576) in the number of organizations listed in the *Conservation Directory*.[55] The editor of the *Conservation Digest* notes that the environment is a growth field, with most organizations being formed within the past twenty-five years.[56] In addition to the increase

**TABLE 7-5   Membership in the Ten Largest
Conservation Organizations in the United States (thousands)**

	1985	1990	1994	Percentage Increase, 1985–1994
Conservation International	—	55.0	12.0	—
Environmental Defense Fund	35.0	150.0	250.0	714.3
Greenpeace USA	550.0	2,300.0	1,600.0	290.9
National Audubon Society	423.2	600.0	600.0	141.8
National Wildlife Federation	—	975.0	1,300.0	
Natural Resources Defense Council	62.0	168.0	170.0	274.2
Nature Conservancy	274.4	600.0	700.0	255.1
Sierra Club	326.6	560.0	550.0	151.7
The Wilderness Society	150.0	370.0	290.0	193.3
World Wildlife Fund	172.0	940.0	1,200.0	697.7
Total	2,029.2	6,718.0	6,672.0	328.8

Sources: Conservation Digest 2, no. 4 (1985); B. Gifford, "Inside the Environmental Groups," *Outside* (September 1990); Ned Martel, Blan Holman, et al., "Inside the Environmental Groups," *Outside* (March 1994).

in the number of conservation organizations, their revenues and memberships have also grown rapidly over the past ten years (tables 7-5 and 7-6).

Total membership in these organizations has increased by 328 percent, and reported annual income increased by over 300 percent to over a half billion U.S. dollars in ten years. At the same time, there has been a tendency for conservation organizations to become more specialized. Today, some work only on policy issues, some focus exclusively on legal matters, others on single species or particular habitats, such as marine or rain forest conservation.

For all of its growth and ability to attract funding, the conservation community has not done a very good job of monitoring its own activities. There are no programs to monitor how well different approaches are doing at conserving biodiversity. Results of conservation activity are generally not reviewed and evaluated to determine what works and what does not. Listing species to achieve legal protection is not working. Protected area conservation is succeeding in getting more land and marine areas designated as protected; however, it is difficult to assess how effective protected areas are at conserving biodiversity.

TABLE 7-6   Revenues Reported by the Ten Largest Conservation Organizations in the United States, 1985–1994 (millions of U.S. dollars)

	1985 Revenues	1990 Revenues	1994 Revenues	Percentage Increase, 1985–1994
Conservation International	—	$ 4.6	$ 11.2	—
Environmental Defense Fund	$ 3.4	12.9	21.9	644.1
Greenpeace USA	10.0	50.2	40.7	407.0
National Audubon Society	24.2	35.0	43.4	179.3
National Wildlife Federation	45.5	87.2	87.7	192.7
Natural Resources Defense Council	7.5	16.0	17.0	226.7
Nature Conservancy	43.3	156.1	216.2	499.3
Sierra Club	22.9	35.2	43.2	188.6
Wilderness Society	6.7	17.3	16.5	246.3
World Wildlife Fund	17.4	35.5	55.9	321.3
Total	180.9	450.0	553.7	306.1

Sources: Ibid.

What Is the Future?

The conservation of wild resources is on the agenda of every country in the world today. The World Bank, United Nations development agencies, the European Commission, and national development assistance agencies are all promoting better conservation of wild resources. Conservation organizations in industrialized countries are growing in terms of members and budgets, and they are becoming more diversified. They, along with the government development assistance agencies that they influence, are affecting policies in developing countries.

Incentive Systems

Sustainable conservation of biodiversity will be achieved only when the people who live with it recognize the importance of conserving. By and large, rural poor people are preoccupied with their daily survival; they live in the present. It is extremely difficult for people in this situation to plan their activities beyond tomorrow. They take what they have to to get through the day, and they do not discriminate between species that are considered "endangered" and those that are not. The term *endangered* has no meaning, except maybe in the context of their own situation.

In order to be effective, conservation activities in developing countries must emphasize incentive systems that are recognized by the rural poor. They must benefit from the investment they will have to make in order to conserve biodiversity. When species or habitats are used for economic purposes, it is essential that the people who live with the resource get an equitable share of the income. Further, there must be a link between the benefits and the management of the resource.

Change of Approach

Once we accept the premise that sustainable conservation depends on the rural people, it means that we have to understand their problems and concerns. We must ask them what they want. And we have to respond to those needs rather than to just the immediate conservation goals with which we may be preoccupied.

Letting rural people decide what conservation actions are taken is anathema to managers. Such an approach is open-ended. They could want to do anything. How can you plan? Calculate the costs? Schedule training and services? Order equipment? All of these questions are easily addressed if you know what you are going to do before you start. Yet success is dependent on meeting the needs of the people, which we can know only if they are part of the process.

Valuation of Wild Resources

In a free market system, the value of products is determined by the supply and the demand for a product. As the supply of a product decreases, its value typically should increase, assuming a constant or increasing demand. However, the value of wild-harvested products is determined by their availability in the marketplace, not in the wild. The sale price includes the cost of capture or harvest, manufacturing, shipping, and advertising, but, unlike raising chickens or cattle, the price does not normally include the cost of "producing" or managing the wild organism from which the product was derived. These costs have been covered by nature and are not factored into the valuation of products. If we continue to use wild resources, which I believe is a given, we must recognize that it costs something to keep them around. The "management costs" to sustain the resource will have to be factored into the retail price of wild-harvested products. Landowners, villagers, and others who manage wildlife must be compensated for their efforts to conserve wild resources and habitats. Given the vicissitudes of demand for wild products driven by markets outside the influence of the producer, it is not advisable (let alone sustainable) for a sustainable use program to be based on single species. Each additional use adds to the potential valuation of the resource and, hence, increases the incentive to maintain and manage it.

Governments and conservation organizations need to begin to promote using wild species rather than stopping such uses. Currently it is anathema for many of the mainstream conservation organizations to think of themselves as marketing agents for wild-harvest products, but that is exactly what is needed.

Some human development organizations are taking this step already. Jason Clay of Cultural Survival has identified several general principles for developing markets for wild-harvested products (box 7-7). His views are based on several years of experience of working with indigenous peoples in the Amazon rain forest to harvest and sell wild products to increase their

> **Governments and conservation organizations need to begin to promote using wild species rather than stopping such uses. Currently it is anathema for many of the mainstream conservation organizations to think of themselves as marketing agents for wild-harvest products, but that is exactly what is needed.**

incomes. The organization has also established Cultural Survival Enterprises "to create and test . . . alternative income generation . . . based on marketing nontimber rain forest products."[57]

Governments in developing countries must recognize the central role that sustainable conservation will play in achieving their economic development goals. To this end, they must encourage and support market development for wild-harvested products. They must also provide a framework within which responsible management and use of wild-harvested resources is possible.

Changes in Land Use Policies

In developing countries, governments commonly emphasize agricultural production to satisfy increasing domestic food requirements and to generate foreign exchange. Under such policies natural habitat is being converted to agricultural production. These governments do not generally place much emphasis on wildlife. Their wildlife departments are generally small agencies with little in-

---

BOX 7-7   Principles for Developing Markets for Wild-Harvested Products

1. Start with what is already on the market.
2. Diversify product and reduce dependence on a few products.
3. Diversify the markets for raw and processed . . . products.
4. Add value locally.
5. Capture value that is added further from the source.
6. Proposed solutions must equal the scope of the problems.
7. No single . . . group can provide enough commodities for even a small company.
8. Controlling a large market share of a commodity allows considerable influence over the entire market.
9. Make a decent profit in the market place, not a killing.
10. The markets . . . are for saving . . . [biodiversity] . . . rather than the . . . people [who harvest the products].
11. Certification of environmental sustainability is key.

Source: J. Clay, "Some General Principles and Strategies for Developing Markets in North America and Europe for Nontimber Forest Products," in M. Plotkin and L. Famolare, eds., *Sustainable Harvest and Marketing of Rain Forest Products* (Washington, D.C.: Island Press, 1992).

---

fluence and typically organized as police forces to stop people from taking wild species. It may be a reasonable way to protect species in economically advanced societies, but it is not a feasible option in most developing countries. Such efforts increase alienation between a government and its rural people. This approach will have to change.

National policies should recognize the important role wild resources play in sustaining the rural population, especially in developing countries. Once the politicians who run governments understand that it is in their best interest to support local decision making with regard to land uses, changes in the approach to biodiversity conservation will follow. This is already taking place throughout West Africa, where governments in several countries (Mali, Niger, Senegal) are pursuing a policy to decentralize authority over land use to regional committees and even the villages themselves.

Government agencies, conservation and development assistance organiza-

Parks have been created to preserve the gorilla habitat, but local people often ignore park boundaries and plow the forests for subsistence farming.

tions, and businesses must begin to support the management and use of wild resources by private landowners and rural poor people. This requires a major shift in the philosophy of government agencies and many nongovernment organizations. They need to recognize that rural people can be allies in conservation and not treat them as environmental criminals.

Legal Access to Wild Resources

One major sticking point in promoting local management and use of wild resources is the absence of legal authority, in most countries, for private citizens to do it. Access or usufruct rights for the people who live with wild resources are essential. At the same time the laws and government agencies must be structured to prevent overexploitation. In practical terms, local peer pressure will more than likely be the biggest factor in ameliorating the risk of overexploitation. This is true among private landholders in Zimbabwe as well as rural villages in other parts of the world that I have worked with.

In northern Pakistan, several villages have banned hunting in their valleys. One village went so far as to announce its decision on the government-run radio station. They let it be known that if they caught a poacher they would confiscate his weapons and fine him 10,000 Pakistan rupees, a sizable sum in local terms. I asked them if they had caught anybody. They had, and used the fine they collected to support their school and to buy trees to plant.

Human-based systems will never be perfect. Not all situations where individuals or rural villages have authority over wild resource uses will lead to optimal conservation of the resources. Sometimes shortsighted greed will win out over long-term sustainability, but faith in fundamental human respect for the environment leads me to believe that far more biodiversity will be conserved if people are given rights to use wild resources than if we continue our present course.

> **Conservation has been plagued with emotionality and subjectivity in its decision making. As a result, funds are wasted, it is difficult to evaluate success, and a lot of effort is being invested in activities that are not warranted.**

Capacity Needs

Decentralized authority over land uses and the right to use wild species is not an end point. Before rural people can effectively conserve the resources on which they depend, systems must be in place to augment traditional knowledge with technical skills. Governments and development assistance and conservation organizations must promote extension services to rural people.

Whenever I think about how daunting the task will be to improve the capacity of rural people to manage and use wild resources, I remember the villagers in the heart of the Korakorum in northern Pakistan who built a 20-kilometer road out of the side of a mountain by hand. Technical guidance was provided by the Aga Khan Rural Support Programme, but the villagers did the work. Surely if that village could build such a road, others have the capacity to manage wild resources, and we are obliged to help them.

Science in Perspective

Conservation has been plagued with emotionality and subjectivity in its decision making. As a result, funds are wasted, it is difficult to evaluate success, and a lot of effort is being invested in activities that are not warranted. Early efforts to conserve species and establish protected areas were driven to a great extent by subjective assessments and anecdotal information. Species were listed for legal protection because it was thought that they needed protection, not because surveys or other verifiable evidence dictated such an action. Protected areas were designated on a map by drawing a line around an area believed to be appropriate for protection, with little or no consideration given to the fact that people lived within the delineated area.

Greater attention is being given to methods and the results of controlled studies. More rigor is being placed on definitions and interpretation of the status of species. Protected area conservation recognizes that different types of areas serve different purposes, and their effectiveness should be assessed in the light of locally determined management objectives.

But science must be viewed in perspective. Science is about objective prediction.[58] Sustainable use of renewable natural resources involves a complex array of interdependent variables that span fields as diverse as the social sciences, economics, law, and biological sciences. Ecologists are of limited value in elucidating problems related to improving the sustainability of using wild resources if they limit their attention to discovering new information about ecological systems or new ecological theories.[59] What we need today are problem solvers who can work at various levels of resolution, assimilate information from the diverse factors affecting the sustainability of uses, and extract meaningful patterns or insights that can be applied to other use systems. This approach is particularly urgent according to Salwasser, who states that "the concept is expanding to embrace the complex relationships between human[s] . . . and social units and the ecosystem in which they occur."[60]

Webb notes that the most cost-effective way of assessing the impact of a harvest or use is to begin to harvest and use the resource, a point made by Walters, and Ludwig and associates.[61] Detailed scientific studies before a use is undertaken accomplish little because the conditions under which the studies are conducted will most likely not prevail in the real world in which wild harvests occur.

We must recognize that science will not give us the answers we are seeking. Science is not omniscient. It cannot be used to justify actions one way or another in conservation. Ultimately data obtained from scientific studies are subject to interpretations. Our problem today is that we have learned to believe that science has the answer, so it is politically correct to base our arguments for a particular action on scientific evidence.

This point was dramatically driven home to me when I recently chaired a one-day workshop at the first International Congress for Wildlife Management in San José, Costa Rica. The purpose of the workshop was to advise the Wildlife Society on how best to respond to international needs for wildlife management and conservation. To open the session, I asked the chairs of ten regionally oriented groups to tell me what they thought were the major wildlife problems in their regions. All the chairs of the groups representing developing countries told me that the number one problem was their lack of knowledge about the animals and plants they were working with. In contrast, the chair of the North American group said that the number one priority in wildlife management was immunocontraception of white-tailed deer. No wonder there is conflict between northern scientists and wildlife managers in developing countries.

There is no doubt that the trend is toward greater objectivity in the conservation community. Sustainable use provides the only approach that appears to be grounded in objective methodologies yet has the capacity to deal with the unpredictable and complex nature of human behavior. No one can guarantee that an activity will be sustainable forever; nevertheless, adaptive management techniques provide a means to monitor and assess progress and offer the necessary flexibility to alter management activities in the light of changing environmental and social conditions.

Sustainable conservation is sustainable development. They are one and the same. If we are to address the human needs on this planet, we must recognize that use of wild resources is an integral part of that process. Our job in conservation is to do what we can to ensure that those resources are there for the next generation, and the one after that. We cannot accomplish this goal in Washington, Brussels, or Tokyo. It can be done only by the people who live with the biodiversity.

NOTES

1. P. H. Raven, "Defining Biodiversity," *Nature Conservancy* (January-February 1994): 11–15.

2. A. Tiega, "Wildlife Use in Nigeria: Historical Trends and Future Perspectives," in *Proceedings of the First International Wildlife Management Congress* (San José, Costa Rica: Wildlife Society, 1993).

3. G. Webb, *Mining and Environmental Management: Some Current Trends*, Report to the Australian Mining Industry Council (Darwin: G. Webb Pty. Limited, 1990).

4. A. Kalland, "Seals, Whales and Elephants: Totem Animals and the Anti-use Campaign," in *Symposium on Responsible Management of Natural Resources* (Brussels: European Council for Conservation and Development, 1993).

5. Raven, "Defining Biodiversity."

6. P. Ehrlich and A. Ehrlich, *Extinction: The Causes and Consequences of the Disappearance of Species* (New York: Random House, 1975).

7. N. Meyers, *A Wealth of Species* (Boulder, Colo.: Westview, 1983).

8. B. Groombridge, ed., *1994 IUCN Red List of Threatened Animals* (Gland, Switzerland: IUCN, 1993); V. Haywood and S. Stuart, "Species Extinctions in Tropical Forests," in *Tropical Deforestation and Species Extinction*, ed. T. C. Whitmore and J. A. Sayer (London: Chapman & Hall, 1992).

9. A. Goudie, *The Human Impact on the Natural Environment*, 4th ed. (Cambridge, Mass.: MIT Press, 1994); W. V. Reid, "How Many Species Will There Be?" in *Tropical Deforestation*.

10. Groombridge, *IUCN Red List*.

11. Raven, "Defining Biodiversity."

12. J. D. Diamond, "Biogeographic Kinetics: Estimation of Relaxation Time for Avifaunas of Southwest Pacific Islands," in *Proceedings of the U.S. National Academy of Sciences*, vol. 69 (1972); E. F. Conner and E. D. McCoy, "The Statistics and Biology of the Species-Area Relationship," *American Naturalist* 113 (1979): 791–833; R. H. MacArthur and E. O. Wilson, *The Theory of Island Biogeography* (Princeton, N.J.: Princeton University Press, 1967).

13. Raven, "Defining Biodiversity."

14. P. R. Ehrlich and A. H. Ehrlich, *Extinctions* (New York: Random House, 1981); D. Simberloff, "Are We on the Verge of Mass Extinction in Tropical Rain Forests?" in *Dynamics of Extinction*, ed. D. K. Elliot (New York: Wiley, 1986); Raven, "Defining Biodiversity."

15. T. C. Whitmore and J. A. Sayer, "Deforestation and Species in Tropical Moist Forests," in *Tropical Deforestation*.

16. D. Simberloff, "Do Species-Area Curves Predict Extinction in Fragmented Forest?" in *Tropical Deforestation*.

17. Ibid.

18. Heywood and Stuart, "Species Extinctions."

19. World Resources Institute, *World Resources 1994–95* (New York: Oxford University Press, 1994).

20. Heywood and Stuart, "Species Extinctions."

21. D. Simberloff, "Are We on the Verge," and W. R. Reid and K. R. Miller, *Keeping Options Alive: The Scientific Basis for Conserving Biodiversity* (Washington, D.C.: World Resources Institute, 1989), in Heywood and Stuart, "Species Extinctions."

22. IUCN, UNEP, and WWF, *The World Conservation Strategy* (Gland, Switzerland: IUCN–World Conservation Union, 1980).

23. IUCN, UNEP, and WWF, *Caring for the Earth: A Strategy for Sustainable Living* (Gland, Switzerland: IUCN–World Conservation Union, 1991).

24. World Commission on the Environment and Development, *Our Common Future* (New York: Oxford University Press, 1987); United Nations Conference on Environment and Development, *Agenda 21* (New York: United Nations, 1992).

25. R. B. Martin, "Should Wildlife Pay Its Way?" (Keith Roby address, Department of National Parks and Wildlife, Perth, Australia, December 8, 1993).

26. R. Sandbrook, *Introduction to a Workshop to Review Sustainable Use of Natural Resources* (Epsom, Surrey: Africa Resources Trust, 1992).

27. R. B. Martin, "The Influence of Governance on Conservation and Wildlife Utilization" (plenary address to the Conference of Conservation of Queensland, Brisbane, Australia, February 8–11, 1994).

28. Ibid.

29. I. Bond, "The Economics of Wildlife and Land-use in Zimbabwe: An Examination of Current Knowledge and Issues," Project Paper, no. 35, WWF Multispecies Production Systems Project (Harare, Zimbabwe, 1993).

30. Martin, *The Influence of Governance.*

31. D. H. M. Cumming, "Developments in Game Ranching and Wildlife Utilization in East and Southern Africa," in *Wildlife Production: Conservation and Sustainable Development*, ed. L. A. Renecker and R. J. Hudson (Fairbanks: University of Alaska, 1991), pp. 96–108.

32. Ibid.

33. D. Hopcroft, "Nature's Technology: The Natural Land-use System of Wildlife Ranching," in *Third Biennial Woodlands Conference on Growth Policy* (Woodlands, Texas, October 29, 1979).

34. Personal communication.

35. G. W. Webb, "Criteria for the Sustainable Use of Wildlife," presented at the Workshop on Sustainable Use of Renewable Wild Resources, Nineteenth Session of the IUCN General Assembly, Buenos Aires, Argentina, January 1994.

36. R. H. V. Bell, "The Workshop Theme: Adaptive Management," in *The Proceedings of a Workshop Organized by the U.S. Peace Corps at Kasunga National Park*, ed. R. H. V. Bell and E. McShane-Caluzi (Washington, D.C.: U.S. Peace Corps, October 1984).

37. R. B. Martin, "Alternative Approaches to Sustainable Use: What Does and Doesn't Work," presented in Symposium 1: Conservation Benefits from Using Wildlife, Conservation Through Sustainable Use of Wildlife Conference, University of Queensland, Australia, February 8–11, 1994.

38. M. Albertson et al., "A Model and Pilot Project for Third-World Village Development" (paper presented at International Conference on Sustainable Village-based Development, Fort

Collins, Colo., September 25–October 1, 1993); A. O. Faulkner and M. Albertson, "Tandem Use of Hard and Soft Technology: An Evolving Model for Third-World Village Development," *International Journal on Applied Engineering Education* 2, no. 2 (1986).

39. S. Fitzgerald, *International Wildlife Trade: Whose Business Is It?* (Washington, D.C.: World Wildlife Fund, 1989).

40. R. Prescott-Allen and C. Prescott-Allen, *What's Wildlife Worth?* (London: Earthscan, 1982).

41. K. Makombe, ed., "Sharing the Land: Wildlife, People and Development in Africa," IUCN/ROSA Environmental Issues Series no. 1 (Harare, Zimbabwe: IUCN/ROSA, 1994).

42. G. W. Murdock, *Africa: Its Peoples and Their Cultural History* (New York: McGraw-Hill, 1958).

43. Tiega, "Wildlife Use in Nigeria."

44. S. Anstey, "Wildlife Utilization in Liberia," in *WWF/SDA Survey Report* (Gland, Switzerland: World Wide Fund for Nature, International, 1991).

45. Groombridge, *1994 IUCN Red List.*

46. G. M. Mace, "Guest Essay: The Status of Proposals to Redefine the IUCN Threatened Species Categories," in Groombridge, *1994 IUCN Red List,* pp. xlviii–lv.

47. R. Littell, *Endangered and Other Protected Species: Federal Law and Regulation* (Washington, D.C.: Bureau of National Affairs, 1992).

48. Data from Competitive Enterprise Institute.

49. U.S. Fish and Wildlife Service, "Endangered and Threatened Species Recovery Program," in *Report to Congress* (Washington, D.C.: U.S. Fish and Wildlife Service, 1993).

50. Littell, *Endangered and Other,* p. 30.

51. IUCN, *1993 United Nations List of National Parks and Protected Areas* (Gland, Switzerland: IUCN, 1994).

52. Martin, "Should Wildlife Pay Its Way?"

53. D. S. Favre, "Debate within the CITES Community: What Direction for the Future?" *Natural Resources Journal* 33 (Fall 1993): 875–918.

54. Martin, "The Influence of Governance."

55. National Wildlife Federation, *Conservation Directory,* 28th ed. (Washington, D.C.: National Wildlife Federation, 1983).

56. W. Henry, "Letter from the Publisher," *Conservation Digest* 2, no. 4 (1990): 2.

57. J. Clay, "Some General Principles and Strategies for Developing Markets in North America and Europe for Nontimber Forest Products," in M. Plotkin and L. Famolare, eds., *Sustainable Harvest and Marketing of Rain Forest Products* (Washington, D.C.: Island Press, 1992).

58. Webb, "Criteria for the Sustainable Use."

59. D. Ludwig et al., "Uncertainty, Resource Exploitation and Conservation: Lessons from History," *Ecological Applications* 3, no. 4 (1993): 547–549.

60. H. Salwasser, "Sustainability Needs More Than Better Science," *Ecological Applications* 3, no. 4 (1993): 587–589.

61. Webb, "Criteria for the Sustainable Use"; C. Walters, *Adaptive Management of Renewable Resources* (London: Macmillan, 1986); Ludwig et al., "Uncertainty."

Chapter 8

WATER OPTIONS FOR THE BLUE PLANET

Terry L. Anderson

HIGHLIGHTS

- *U.S. population has increased by 200 percent since 1900; per capita water usage grew 500 to 800 percent during the same period.*

- *Global water use has tripled since 1950.*

- *Water recycling increased more than 400 percent between 1954 and 1985 and is expected to double again between 1985 and 2000.*

- *Transferring just 5 percent of agricultural water to municipal uses would meet the needs of urban areas in the western United States for the next twenty-five years.*

- *Estimates show that a 10 percent increase in price would decrease agricultural water consumption in California by 6.5 percent and cut overall consumption by 3.7 percent in the seventeen U.S. western states.*

- *Current government policies do not adequately reflect the costs of operation, maintenance, and capital for water projects and fail to account for the environmental damage associated with water storage and delivery.*

- *Special interest groups constitute a formidable political force to block the efficient allocation of water.*

- *Inefficient water use prevails in urban areas because consumers do not have to pay the full cost of the water they use.*

- *Water overuse adversely affects the environment by reducing water in streams. Subsidized agricultural drainage water loaded with leached salts and trace elements was dumped into California's Kesterson Wildlife Refuge, causing the death and deformation of many animals, reptiles, and fish.*

- *Eliminating laws against water marketing and establishing private water rights would give consumers an incentive to use water more efficiently.*

- *Removing legal impediments to water markets would allow private firms to enter the water supply industry and take the burden off the public treasury. Market forces could pare demand, boost supply, reallocate water, and end the threat of water crises.*

For several years the U.S. Fish and Wildlife Service (USWFS), manager of the Kesterson National Wildlife Refuge in central California, was happy to take the inflow from the San Luis Drain, which constituted the main source of water for the refuge. Then in 1983 the USWFS noticed something seriously wrong: the once-productive refuge was turning into a grotesque sideshow of deformed fish, birds, and reptiles. Wings were missing, heads were not completely formed, beaks were misshapen, and animals were dying. Instead of being a wildlife refuge, Kesterson was more like a toxic waste site.

Studies showed that the toxic culprit was selenium, a trace element being leached out of soils by Bureau of Reclamation irrigation water in the state's Westlands Water District and carried to Kesterson via the San Luis Drain. Because farmers in the water district pay only about 15 percent of the cost of storing and delivering water and because the drainage system was also heavily subsidized by the federal government, farmers had an incentive to use large amounts of water to irrigate marginal lands, with little regard for the quality of wastewater. In the process, extensive irrigation leached the selenium and other trace elements from the soil and deposited them in the drain. In small dosages the selenium is necessary for life, but as it concentrated in the Kesterson Reservoir, it became a deadly pollutant.[1]

Although both federal bureaucracies involved in the Kesterson debacle were parts of the Department of Interior, solving the problem was not simple. Shutting off the water to the irrigation district and thereby halting drainage was an obvious solution, but the politically powerful farmers receiving the subsidized water wanted to continue business as usual, and the banks holding debt from the farmers could ill afford to see production in the area fall. Hence, costly pollution control alternatives costing millions of dollars have been implemented to solve a problem that would not have existed in the first place if the federal government had not engaged in a water policy that subsidized destruction of the environment.

The cause of the environmental disaster at Kesterson, distorted market signals to water users created by governmental bureaucracies, is typical of water problems around the world. Industrial, municipal, and agricultural water users seldom pay the true cost of water and therefore have insatiable demands. Without higher prices, users have no incentive to consider alternative technologies that could save tremendous quantities of water and reduce the pressure to build expensive, environmentally destructive dams and delivery systems. As economist Robert Repetto at the World Resources Institute emphasizes, the consequences of distorted water pricing are disastrous on many fronts: "The political economy of irrigation systems leads to poor use of water and invested capital. . . . The consequences are inefficient, inequitable, fiscally disastrous, wasteful of increasingly scarce water, and environmentally harmful."[2]

Low prices as well discourage the adoption of water-saving technology. The potential for water conservation is demonstrated by something as simple as the toilet, the largest user of water in a typical household. Compared to the typical American toilet that uses 3 to 5 gallons of water with each flush, the Japanese toilet is a technological marvel for saving water. The flushing mechanism allows the user to choose between two alternative surges of water: a small amount for smaller jobs and a larger amount for larger jobs. In addition, potable water that fills the tank of the Japanese toilet for the next flushing enters through a spigot and a wash basin on the top of the tank. Thus the user can wash his or her hands with clean water that will be reused on the next flushing. The Japanese use this technology because their high water prices provide an incentive to conserve.

Examples of how different technologies can change the amount of water used in production processes abound (table 8-1). The variance in the amount of water consumed in water-using activities around the world is a reflection of the potential for water saving. Gray water coming from sinks, showers, and tubs can be collected in a separate drainage system and used for irrigation. Such techniques have helped Israel recycle 35 percent of its wastewater for agricultural and industrial uses. Drip irrigation can save thousands of acre-feet of water by putting water near the roots of plants, where it is needed. Although technology exists to improve the water use efficiency in agriculture, irrigation systems remain only 40 percent efficient.[3] Dry cooling (working on a principle similar to that of a car's radiator) can cool thermal electric generating plants with a fraction of the water needed for wet cooling.

But does humanity really face a shortage of freshwater?

Water, Water Everywhere But in the Wrong Places

With the "blue planet" three-quarters covered by water, it seems impossible that water could constrain human progress. Indeed, as a global resource, water is not scarce. Over the entire globe, annual precipitation averages 525,100 cubic kilometers (ckm), or 264.2 billion U.S. gallons. Of this, only 20 percent occurs over land and freshwater, and of that only 9,000 ckm are useable for human consumption. The remainder runs off in floods, enters the ground and is irretrievable, or is locked in snow and ice.[4] In spite of the fact that human water consumption grew from 1,000 ckm in 1940 to 4,000 ckm in 1990, "the percentage of available resources actually used by man, including allowances for dilution of wastes, appears to be in the range of 38–64 percent of the readily available resource, though it is only 1 percent of the total precipitation."[5]

Water may not be scarce globally, but it is scarce at specific times and places around the world (table 8-2). Countries using less than 5 percent of total annual runoff can meet their demands with little problem. Those using 10 to 20 percent face quantity and quality problems. In Israel no new sources of freshwater remain untapped, thus requiring that all new demands be met by treating

TABLE 8-1   Water Use of Various Industries

Industry	Range of Flow (gal/ton product)
Cannery	
Green beans	12,000–17,000
Peaches and pears	3,600–4,800
Other fruits and vegetables	960–8,400
Chemical	
Ammonia	24,000–72,000
Carbon dioxide	14,400–21,600
Lactose	144,000–192,000
Sulfur	1,920–2,400
Food and beverage	
Beer	2,400–3,840
Bread	480–960
Meat packing	3,600–4,800[a]
Milk products	2,400–4,800
Whiskey	14,400–19,200
Pulp and paper	
Pulp	60,000–190,000
Paper	29,000–38,000
Textile	
Bleaching	48,000–72,000[b]
Dyeing	7,200–14,400[b]

Source: Peter Rogers, *America's Water: Federal Roles and Responsibilities* (Cambridge, Mass.: Twentieth Century Fund, 1993).
[a] Live weight.
[b] Cotton.

TABLE 8-2   Worldwide Water Scarcity

Country	Internal Water Availability (cm/person/year)[a]	Withdrawal as Percentage of Available Water[b]
Canada	109,510	1
Panama	60,760	1
Nicaragua	46,730	1
United States	10,060	19
China	2,520	16
India	2,270	21
Peru	1,840	—
Haiti	1,500	—
South Africa	1,400	18
Poland	1,290	30
Kenya	610	—
Tunisia	490	53
Israel	370	88
Barbados	200	51
Libya	170	374
Malta	60	92
Egypt	40	97

Source: World Resource Institute, *World Resources—An Assessment of the Resource Base That Supports the Global Economy* (New York: Basic Books, 1986), 1:67.

[a] Internal availability excludes river flow from other countries: renewable water only.

[b] Available water includes river inflow (renewable + net river inflow).

and reusing wastewater.[6] With about half of the world's land mass in arid or semi-arid regions, twenty-five countries are especially water stressed.[7] Jordan, Israel, Algeria, Egypt, Tunisia, and the countries of the Arabian peninsula use nearly all of their available water supply.[8] In Libya, total internal renewable water available each year is 0.7 ckm, but withdrawals are 2.62 ckm, or 374 percent of the available water supply.[9] Thus, significant mining of groundwater is occurring. At the other extreme, Canada withdraws only 1 percent of its internally available water. In between, India uses 50 percent of its total runoff, but it is estimated that India's total demand will rise to 92 percent of annual freshwater resources by 2025.[10]

Even in the United States where scarcity is not nearly as severe as in more arid parts of the world, there are predictions of shortages. The U.S. Water Resources Council estimated that

on the average, about 40,000 bgd (billion gallons per day) of water passes over the coterminous United States in the form of water vapor. Of this, approximately 10 per cent (about 4,200 bgd) is precipitated as rainfall, snow, sleet, or hail. The remainder continues in atmospheric suspension. Of the 4,200 bgd . . . about 2/3 (2,750 bgd) is evaporated immediately from the wet surfaces or transpired by vegetation. The remaining 1,450 bgd accumulates in ground or surface storage; flows to the oceans, the Gulf of Mexico or across the Nation's boundaries; is consumptively used; or is evaporated from reservoirs.[11]

Since 1900, the U.S. population has increased 200 percent; in the same period per capita water use shot up 500 to 800 percent. Between 1960 and 1970, water use in the United States increased 37 percent (270 bgd to 370 bgd), and between 1970 and 1980 22 percent (370 bgd to 420 bgd).[12]

Water consumption has increased steadily for most water uses. Table 8-3 shows how water use by continent rose between 1950 and 1990 and is expected to continue to increase until the next century. On average, about one-fourth of the water withdrawn in the United States is irretrievably consumed. The other three-fourths returns to surface-water or groundwater supplies. However, the growth in consumption appears to be slowing, increasing between 1970 and 1988 "by only 14%, compared with a 44% increase [during] the previous decade."[13] The slowdown in the growth rate of consumption is generally attrib-

TABLE 8-3 Water Use, by Continent (cubic kilometer/year)

	1950	1960	1970	1980	1990[a]	2000[a]
Africa	56	86	116	168	232	314
North America	286	411	556	663	724	796
South America	59	63	85	111	150	216
Europe	94	185	294	435	554	673
Oceania	10	17	23	29	38	47
Total	1,360	1,982	2,594	3,316	4,138	5,189

Source: M. Meybeck, D. Chapman, and R. Helmer, *Global Freshwater Quality* (Oxford: Blackwell, for UNEP and WHO, 1989).

[a] Estimate.

uted to a decline in industrial withdrawals resulting from improvements in production technology and tougher laws on water pollution.[14]

Although the difference between aggregate withdrawals and consumption
might indicate a plentiful water supply rather than a shortage, other factors contribute to shortages. Some of the water must be left to flow to Canada and Mexico. Furthermore, location-specific demands for water do not always coincide
with supplies. For example, the western United States receives less rainfall than
the East. Southern California receives fewer than 10 inches of rain each year,
compared with 33 inches in Chicago and 44 in New York.[15] Nevertheless, 80
percent of the nation's water is consumed in the West.[16] To make matters worse,
85 percent of the water in the West is used for irrigation. With California's population mushrooming from 20 million to 30 million in the past twenty years, the
demand for water west of the ninety-fifth meridian is not likely to decrease.

Normally arid conditions can be exacerbated by drought. Between 1987 and
1993, California's growing population was accompanied by less than normal precipitation, resulting in a severe water shortage, with the state's reservoirs falling
to as low as one-third of capacity. In 1991, the State Water Project cut off supplies to farmers, and the Bureau of Reclamation cut its supplies to farmers by 75
percent from its Central Valley Project, forcing the nation's largest supplier of
fresh vegetables to idle thousands of acres and causing farmers to pump what they
could from already-depleted groundwater supplies. Towns and cities rationed water under threat of similar cuts. The environmental toll was devastating.

Problems of water scarcity can be particularly important when considered in
the context of international relations. Globally, 47 percent of all land falls with
international river basins. This land is home to 40 percent of the world's people.[17] Within international water basins, the potential for conflict is reduced by
treaties, as between the United States and Mexico, but in others, disputes can
result in armed conflict, as they have in the Middle East.

A typical response to water scarcity is to tap underground sources. In Libya,
the Sarir Desert Project tapped groundwater for irrigating 15,000 hectares at a
cost of $30 million to $50 million for 1 ckm. The famous Ogallala aquifer in the
south-central United States, the world's largest groundwater basin covering
174,000 square miles, is also feeling the pressure of pumping. Withdrawals from
the Ogallala irrigate 15 million acres and account for 30 percent of total U.S.
groundwater used for irrigation. Overdraft (withdrawals in excess of recharge)
occurred in 95 percent of the Ogallala, sparking forecasts that the aquifer would
be 23 percent depleted by 2020.[18] The water tables beneath Dallas–Fort Worth
have dropped 492 feet since 1960. A similar situation exists in Arizona, where
the water tables around Phoenix have dropped 400 feet in the past fifty years.[19]

According to an estimate by the American Institute of Professional Geologists, Americans pumped 100 bgd of groundwater in 1985, a 12 percent increase over 1980 figures. In China, the water tables near two major manufacturing cities, Beijing and Tianjin, are decreasing by 3 to 12 feet annually, and in southern India, water tables have dropped 80 to 96 feet in ten years.[20] In Mexico City, groundwater pumping exceeds recharge by 50 to 80 percent.[21] While using half of its total runoff, India draws half as much again, 150 billion cubic meters per year, from groundwater sources.[22]

Groundwater mining per se is not a problem if the value of the water in its current uses is greater than any future value it might have if left stored in natural aquifers. It is a problem, however, to the extent that open access to the resource encourages overpumping, thus creating the "tragedy of the commons." Understanding why this is a problem is simple. Consider several children with drinking straws in a single soda. Each might like to drink leisurely, but an individual can garner more by drinking rapidly. Hence, there is no incentive to "conserve," and the soda is quickly depleted. Because most groundwater basins are pumped by autonomous individuals with an incentive to behave like children drinking from a common soda, groundwater mining can have important implications for resource allocation over time.

> **In short, although the blue planet has more than enough water to meet human demands, water is often found in the wrong place at the wrong time.**

When measured in terms of quality instead of quantity, water scarcity is much worse. Half of the world's population is estimated to suffer from infections caused by waterborne diseases such as yellow fever, malaria, and river blindness.[23] Diseases associated with water kill 5 million people per year,[24] and four of five child deaths in developing countries result from waterborne disease.[25] These diseases are highly correlated with income. The 25 percent of the world's population that enjoys centralized water and sewer systems avoids most waterborne diseases. Such diseases are more prevalent in developing countries, however, where 95 percent of the people discharge untreated sewage directly into surface water. In 1970, only 29 percent of the population in lesser developed countries had access to safe drinking water, and by 1980 this had risen to 43 percent.[26] In Poland the portion of river water suitable for drinking has dropped from 32 percent to less than 5 percent over the past twenty years. Three-quarters of Poland's river water is now too contaminated even for industrial use.

In short, although the blue planet has more than enough water to meet human demands, water is often found in the wrong place at the wrong time. Glob-

Old-fashioned flood irrigation systems like this one in India are responsible for making millions of acres of farmland unproductive because of salinization and alkalinization.

ally, water use has tripled since 1950 as the demand for water has outstripped population growth. Per capita water supplies are one-third lower than in 1970, due to the addition of 1.8 billion people to the planet since then. Almost a quarter of the world's nations lack sufficient freshwater to meet the needs of their burgeoning populations.[27] In the past century, water use for irrigation has increased tenfold and accounts for 70 percent of the world's freshwater use. By the turn of the next century, water shortages are likely to become even more widespread. Although the U.S. Water Resources Council predicts that withdrawals from streams will decrease by 9 percent between 1975 and 2000, during the same period, consumption of freshwater is expected to increase by almost 27 percent. The council estimates that seventeen subregions either have or will have a serious problem of inadequate surface-water supply by the year 2000.[28] Given these dire predictions, we must look carefully at how water supplies are allocated.

Trying to Fool the Invisible Hand

Every schoolchild learns the basic scientific principle that water runs downhill, but the economic and political principles that determine water flows are less well known. From economics we learn that more of any commodity will be supplied at higher prices and therefore *that water will run uphill to money*. California's recent experiment with an Emergency Drought Water Bank nicely demonstrates this principle. Facing a severe drought for nearly a decade, the state of California established a Water Purchase Committee that announced it would pay $125 per acre-foot for agricultural water. "By the end of June 1991, the Drought Water Bank had purchased about 750,000 acre-feet of water (400,000 from fallowed farmland, 210,000 from groundwater sources, and 140,000 from surface reservoirs). . . . It was a surprise to many people that such large quantities of water became available so quickly."[29] A lower price of $50 per acre-foot was set for water in 1992 after rains broke the drought, but still 154,000 acre-feet were delivered, much of it used for fish and wildlife purposes.

If water flows uphill to money, an even more important principle from political economy is that *water gushes uphill to politics*. This result occurs because politics allows costs to be diffused over a large population while benefits are concentrated on special interest groups that bear little or no cost. Not surprisingly, the average taxpayer footing the bill is not well informed about subsidies to water users because each subsidy costs him or her very little. The special interest group receiving the subsidized benefits, on the other hand, is keenly aware and politically active in support of the project. Politicians respond by spending millions of dollars to build infrastructure and deliver copious quantities of water to users at a fraction of the actual cost.

The insatiable demand for water has created a proliferation of massive storage projects. The foundation was laid for big projects with the Reclamation Act of 1902, but it was the New Deal projects that got the ball rolling. Under the Tennessee Valley Authority, the Public Works Administration, and the Columbia Basin Project, the federal government built hundreds of dams for flood control, irrigation, and hydropower production. Budgets for the agencies boomed, new employees were hired and entrenched in the bureaucracy, and land values soared in regions where the water was to be delivered.[30]

Subsidized financing of water projects is typical throughout the world, and since most of these schemes are touted as irrigation projects, the lion's share of the subsidies goes to farmers. The U.S. farmer, for example, typically pays only one-fifth of the true cost of irrigation from federal projects.[31] Table 8-4 shows similar estimates of subsidies for ten specific irrigation projects. From these data,

TABLE 8-4   Irrigation Subsidies

Project	Costs Allocated to Irrigation	Costs to be Repaid by Irrigators	Percentage of Irrigation Costs Subsidized
Central Valley, California	$687,152,000	$606,646,000	11.1
Chief Joseph Dam, Washington[a]	11,083,200	6,050,000	45.4
Collbran, Colorado	6,105,000	1,089,101	82.2
Columbia Basin, Washington	745,111,398	135,916,400	81.8
Fryingpan-Arkansas, Colorado	69,946,000	50,512,300	27.8
Rouge River, Oregon	18,064,000	9,066,500	49.8
San Angelo, Texas	8,853,904	4,000,000	54.8
The Dalles, Oregon	5,994,000	2,550,000	57.5
Venturia River, California	18,273,128	10,746,300	41.2
Washita Basin, Oklahoma[b]	10,403,011	8,221,000	21.0

Source: U.S. Department of Interior, Bureau of Reclamation, *Reclamation Payments and Payout Schedule* (Washington, D.C.: Government Printing Office, 1965).
Note: On some of these projects, a portion of the subsidy to irrigators came from industrial and municipal users.
[a] Includes costs and repayments from Foster Creek and Greater Wenatchee Divisions.
[b] Includes costs and repayments from Fort Cobb and Fass Divisions.

**TABLE 8-5   Economic Rents and Subsidies
in Bureau of Reclamation Irrigation Projects in the Western
United States**

Irrigation District	Actual Current Charges ($/acre-ft.)	Estimated On-Farm Water Value ($/acre-ft.)	Estimated Full Supply Cost ($/acre-ft.)	Actual Charges as Percentage of Supply Costs	On-Farm Value as Percentage of Full Supply
Black Canyon	1.41	24	15.7	0.09	1.52
Coachella	7.00	8	26.27	0.27	0.30
Columbia Basin East	4.19	20	41.16	0.10	0.49
Elephant Butte	6.45	67	24.43	0.26	2.74
Farwell	10.50	34	135.50	0.08	0.25
Glenn-Colusa	1.46	6	17.85	0.08	0.34
Goleta	59.24	35	263.12	0.23	0.13
Goshen	4.22	—	22.96	0.18	—
Grand Valley	1.18	4	31.10	0.04	0.13
Imperil	4.75	10	11.00	0.43	0.91
Lower Yellowstone	5.28	35	34.62	0.15	1.01
Milk River	7.79	80	119.13	0.07	0.67
Moon Lake	1.75	3	7.05	0.25	0.43
Oroville-Tonasket	11.47	90	21.33	0.54	4.22
Truckee-Carson	2.19	72	33.46	0.07	2.15
Wellton Mohawk	4.80	31	29.58	0.16	1.05
Westlands	15.80	27	67.56	0.23	0.40

Source: Richard W. Wahl, "Full Cost Pricing Option," in Department of the Interior, Bureau of Reclamation, *Westside Report on Acreage Limitation: Draft Environmental Impact Statement* (Washington, D.C.: Government Printing Office, January 1985).

economists Rucker and Fishback conclude that "these figures, which range from 11 percent to over 82 percent, demonstrate that this subsidy has been extremely valuable to irrigators on a number of federal reclamation projects."[32] Table 8-5 shows similar evidence for seventeen other federal projects throughout the West. Though all of the projects generate on-farm water values in excess of irrigation charges, only eleven have benefits in excess of the full supply cost. A U.S. General Accounting Office study of six irrigation projects concluded, "If water were priced high enough to recover the construction costs plus 7.5 percent interest charge, the potential customers for irrigation water could not generate enough extra agricultural yield to pay for the additional expense required by irrigated agriculture. The projects we studied failed the pragmatic test of economic viability for the irrigation facilities."[33]

President Carter's 1979 "hit list" of federal projects that would not be funded raised hopes that the economic irrationality of America's federal water system might be coming to an end, but good pork barrel does not die easily. In 1993 Congress authorized completion of the Central Utah Project (CUP). With storage facilities already in place, the project will deliver water to agricultural regions in southern Utah. Unfortunately, federal studies show that the economics is as bad as always. Roughly, the CUP will deliver water to farmers at an additional cost of $350 per acre-foot. The farmers, however, will pay only about $3 per acre-foot[34] for the water that will generate about $30 per acre-foot in additional farm value.[35] In short, farmers receiving water from the CUP will be paying one-hundredth of the cost of the water they receive. Despite the economic evidence that most water projects do not make economic sense, political pressure continues to allow these projects to proliferate because the interest groups that capture the benefits constitute a formidable political force.

The situation is much the same throughout the rest of the world, where revenues collected from farmers barely cover 10 to 20 percent of the construction and operating costs.[36] Water prices on government irrigation projects in South Africa average only 30 percent of operating and maintenance and make no provision for interest and redemption of capital.[37] Table 8-6 shows operating costs and total cap-

TABLE 8-6   Irrigation Project Cost Recovery

Country	Operation and Maintenance Costs as Percentage of Farmer Revenue	Capital and Recurrent Costs as a Percentage of Farmer Revenue (moderate estimate)	Farmer Benefits as a Percentage of Irrigation Charges (low estimate)
Indonesia	128	735	476
Korea	107	550	303
Nepal	181	1,388	200
Philippines	83	443	100
Thailand	362	1,818	111

Source: Robert Repetto, *Skimming the Water: Rent-Seeking and the Performance of Public Irrigation Systems,* Research Report 4 (Washington, D.C.: Water Resources Institute, 1986).

ital and recurrent costs as a percentage of farmer revenue and the economic benefits to farmers as a percentage of irrigation charges for selected countries.

In the light of these numbers it should not be surprising that farmers would provide the political support for politicians who are willing to pass the laws necessary to build the projects. The additional value of output generated by subsidized water projects (called "rents" by economists) is usually captured by the landowners to whom the water is delivered. The process whereby these recipients seek these rents (hence the term "rent seeking") in the political system is captured by economist Robert Repetto:

> Political manipulation, intimidation, and corruption replace economic efficiency as ways to get ahead. Inevitably, most of the available rents are captured by those with power, influence, and wealth. . . . Successful rent-seekers can well afford to spend a portion of their rents to safeguard, defend, and increase them. These defensive expenditures finance organizational efforts, political contributions and lobbying. . . . Those who control the allocation of rents, whether administratively or politically, are in a position of power . . . because they are dispensing rights to resources for which excess demand is chronic. They typically find ways to appropriate a share . . . for themselves—often through corruption and monetary gain, but also in other forms. Politicians gain votes and contributions, and public agencies gain expanded budgets, staffs, and authority.[38]

Given the strong incentive faced by recipients of subsidized water and by the politicians and bureaucrats who supply the water, it is difficult to eliminate the waste inherent in inefficient water projects.

The inefficiency created by keeping water prices artificially low is compounded by the fact that recipients have little reason to use water more effectively. Typically irrigation projects are less than 50 percent efficient, meaning that much of the water diverted for crops runs off, carrying with it pesticides, herbicides, and soil nutrients. Water logging also occurs as farmers apply generous amounts of cheap water to their crops. As a result, it is estimated that 150 million hectares—60 percent of the world's total irrigated area—need some form of upgrade to remain productive.[39] With cheap water, farmers tend to concentrate on high-valued, water-intensive crops like sugarcane, alfalfa, and fruits.

Inefficient water use is not confined to agriculture. Urban users also often do not face the full cost of water. Hence green lawns cover desert cities, faucets drip continuously, and antiquated distribution systems leak badly. "Losses as high as 40 percent have often been reported, with 20 percent commonplace and even well-maintained systems losing about 7 to 10 percent." Boston's delivery system lost nearly one-third of what was put into it until a leak detection program reduced losses to 12 percent between 1978 and 1983.[40] On average public water supplies in the United Kingdom leak 30 percent of their water. Overdraft of aquifers has been a serious problem in cities such as Tucson, Arizona; when the city council doubled water bills in 1976, the voters threw the entire council out of office in the next election. Like farmers, when urban people become accustomed to cheap water, their demands become insatiable, and they do not want to change their comfortable lifestyles.

Selling Our Lifeblood

It is true that "the energy crisis and the water shortage are inextricably linked," but the connection is more subtle than most environmentalists understand.[41] Rising energy prices have made supplying water more costly, but the general link between energy and water is more directly related to the extent to which prices are allowed to influence demand and supply. If water prices are kept too low, con-

> **Water prices have been kept at artificially low levels by governmental agencies, and shortages inevitably have followed.**

sumers will demand more and more water, but suppliers will have little incentive to meet the demands. This is exactly what happened during the "energy crisis" of the 1970s and precisely parallels what any water crisis is likely to be. So the question is, Why are water prices kept artificially low, and why don't prices adjust?

The 1970s answered this question in the context of the energy crisis: when the government used price controls to keep fuel prices artificially low, people wanted more cheap gasoline than oil producers were willing to supply. Shortages inevitably followed. Not until prices rose were companies willing to undertake exploration and production. Hence, the government was forced into the business of allocating scarce supplies by trying to restrict the demand for gasoline. Experience around the world has demonstrated over and over again that the only successful way to avoid shortages is to rely on free market pricing and allocation.[42] The same is true for water.

Water prices have been kept at artificially low levels by governmental agencies, and shortages inevitably have followed. Typically the government responds to these shortages by attempting to restrain demand, ration water, and increase the available supply. For example, in the face of the California drought in the late 1980s and early 1990s, municipalities implemented rationing, began constructing expensive desalinization plants, and paid thousands of dollars for cloud seeding; in 1991, the state cut off supplies to farmers.[43] Nevertheless, except in isolated cases where shortages have been caused by drought and where a cooperative community spirit has developed, efforts to ration water have not been very successful. Increased water supplies have been made possible through the construction of massive water projects that have dammed free-flowing rivers and built thousands of miles of aqueducts. Given that fiscal and environmental constraints will make it more difficult to meet growing demands with new sources of supply, water crises will become more common unless markets are allowed to play a greater role in water allocation.

Will consumers and suppliers respond to changes in water prices? Many people believe that because water is a necessity, there are few opportunities for responding to higher-priced water. Experience suggests otherwise.

Evidence that cities react to changes in the price of water is found all around the United States. For example, Tucson, Arizona, was able to reduce average daily demand by 27 percent in 1977 using a combination of water price increases and other forms of rationing. In their study of six subregions of the United States, economists Beattie and Foster found that a 10 percent increase in the price of water would reduce water consumption between 3.75 and 12.63 percent.[44] As might be expected, because high variability in precipitation influences water use, the northern California–Pacific Northwest region was the most responsive to higher water prices and the arid Southwest region the least responsive. Although these estimates may suggest that higher water prices could reduce consumption, it must be noted that in 57 percent of the twenty-three cities studied, real water prices declined between 1960 and 1976. Only three cities had real water rate increases of more than $1 per 1,000 cubic feet. Beattie and Foster concluded that "the water utility industry has done a good job for consumers. Unfortunately, because of this good job water users have adjusted their way of life so that needs for water are great."[45]

Similar data for the agricultural sector, which generally consumes 80 to 90 percent of all water, suggest that this demand is also price responsive. The dramatic voluntary response to the California Drought Water Bank that offered water owners a chance to sell conserved water for a profit is one example. Demand responsiveness varies by crop, of course, but some aggregate estimates by agricultural economist Del Gardner for California show that a 10 percent increase in price would bring about a 6.5 percent decrease in water consumption. The same price increase would cause an overall average consumption decrease of 3.7 percent for the seventeen western states. Gardner estimates that, starting from a price of $17 per acre-foot, a 10 percent increase in price would yield a 20 percent decrease in water use in California.[46]

Therefore, if water costs farmers more, they will use less on any particular crops, they will shift to different irrigation technology or water application practices, and they will change cropping patterns. Researchers at the University of California conclude that cutting water supplies would decrease most crop yields but that the reductions might not be as dramatic as expected because farmers could make other adjustments.[47] Flood irrigation techniques conserve on labor but use large amounts of cheap water. With higher water prices, it makes sense to substitute labor and capital for water and to use drip irrigation or similar techniques. In a simulation of a 640-acre farm in Yolo County, California, Hedges showed that the optimal cropping pattern at a zero water price would call for 150 acres each of tomatoes, sugar beets, and wheat; 47 acres of alfalfa; 65 acres of beans; and 38 acres of safflower. If the water price rose to $13.50 per acre-foot, water-intensive alfalfa acreage would drop out and safflower acreage, a crop that uses less water, would expand. The point is that many choices are available to water users, and they can adjust rationally to changes in water prices.

Modern drip irrigation conserves water by delivering it directly to plants and prevents salinization and alkalinization.

By motivating farmers to cut their consumption through improved irrigation techniques and modified cropping patterns, higher water prices would free up irrigation water for municipal and other uses. Transferring just 5 percent of agricultural water to municipal uses would meet the needs of urban areas in the western United States for the next twenty-five years.[48] Higher water prices would also reduce the need for supply projects and delivery systems that cause environmental harm through damming and diverting free-flowing streams. Higher prices would encourage private, profit-making firms to enter the water supply industry, taking the burden off the public treasury. If markets in water were permitted, demand would be reduced, supply would be increased, water would be reallocated, and the specter of water crises would vanish.

Industrial users also respond to higher water prices. The amounts of water used in production processes vary greatly (see table 8-1). In part, this variation is related to the amount of recycling that occurs. Water recycling has increased since 1954 and is expected to continue to do so (table 8-7). Much of the increase in recycling is occurring in response to environmental regulations that have raised the cost of treating water before it is returned to rivers or lakes, thus making recycling more economical. In Israel where water is especially scarce, 35 percent of all wastewater was recycled for agricultural and industrial use in 1986.[49]

The evidence shows that people will conserve water when they have an incentive to do so. Therefore, the key to avoiding water crises is to establish institutions that provide that incentive. There is historical precedent for such

TABLE 8-7   U.S. Water Recycling Rates in Major Manufacturing Industries

Year	Paper and Allied Products	Chemicals and Allied Products	Petroleum and Coal Products	Primary Metal Industries	All Manufacturing
1954	2.4	1.6	3.3	1.3	1.8
1964	2.7	2.0	4.4	1.5	2.1
1973	3.4	2.7	6.4	1.8	2.9
1978	5.3	2.9	7.0	1.9	3.4
1985 (est.)	6.6	13.2	18.3	6.0	8.6
2000 (est.)	11.8	28.0	32.7	12.3	17.1

Source: Sandra Postel, "Increasing Water Efficiency" in *State of the World 1986* (New York: W. W. Norton & Co., 1986).

institutions. Indian tribes living in the arid Southwest were among the early civilizations to recognize the link between incentives and efficient use. Although they collectively built dams to divert water and canals to deliver the water, the water delivered to the privately owned fields was also privately owned.[50] Similarly, indigenous Hawaiians defined private water rights on the basis of taro production. Until recent judicial interpretations and legislative enactments changed the traditional system, surface water rights could be transferred efficiently through markets with minimal bureaucratic intervention.[51]

Throughout the American West, the common law tradition of prior appropriation has established the foundation for water marketing.[52] This doctrine assigns water rights on the basis of "first in time, first in right," meaning that those who have the longest record of water use have first claim when water is scarce. In other words, the most senior user gets water first, the next most senior next, and so on until the total flow of a stream is completely claimed. Such rights can be bought and sold, transferred from one parcel of land to another, and trans-

This dam was built by the government of Saudi Arabia to supply irrigation water to its farmers. Nearly all governments subsidize farmers by providing below-cost irrigation water, which leads to wasted water rather than better farming.

ferred from one use to another, subject to judicial and bureaucratic regulations designed to prevent harm to third parties.

Unfortunately, these regulations often go beyond third-party problems and significantly restrict the potential for water markets. For example, many states require that water be put to beneficial use or the water rights are lost. This use-it-or-lose-it principle discourages water conservation because rights to unused water are lost. Moreover, instream uses of water are not recognized as beneficial uses; consequently, recent demands for instream flows to accommodate fish and wildlife cannot be met by buying water from willing suppliers.[53]

Today, pragmatic environmental groups recognize that markets promote water use efficiency and that, under the current system, protracted legal battles to preserve instream supplies for wildlife are all too often a case of too little, too late.[54] California, Colorado, and Montana have passed legislation allowing farmers to retain their rights to water they conserve. Therefore buyers can purchase water that would be diverted for irrigation water from farmers and dedicate it to instream uses. Montana law, for example, now allows the state's Department of Fish, Wildlife, and Parks (DFWP) to lease water from farmers for instream purposes. This means that when streams are dewatered by irrigation, the DFWP can negotiate with the owners of the irrigation water to find ways to increase stream flows at critical times. In one case, the DFWP agreed to help finance a more water-efficient sprinkler irrigation project in return for having the saved water left in the stream for spawning trout. Similarly, the Nature Conservancy has been particularly active in Colorado in obtaining conservation easements on irrigation water that allow it to be left in the stream without being jeopardized by the use-it-or-lose-it principle. And recently the Oregon Water Trust leased water rights from an eastern Oregon rancher to increase flows in a salmon spawning stream. As Janet Neuman, president of the Oregon Water Trust, put it, "These leases prove that the marketplace can be used to help address Oregon's environmental problems."[55]

Recent legal changes in New South Wales, Australia, suggest that interest in water marketing is not confined to the United States.[56] In the past, water users have had unrestricted access to water, but now rights are defined according to specified volumetric allocations for nondrought periods that specify how many megaliters each user is authorized. With rights so established, new users can obtain water only by purchasing existing rights. When there are costs associated with storage and delivery, users increasingly are being required to pay, where formerly water was free. In 1983 water sales were limited to twelve-month periods, but 1989 legal changes now allow permanent transfers.

As is often the case in the development of property rights systems, water transfers were not imposed from above; rather, they were the response of users to increasing resource scarcity. All the government did was to remove the legislative obstacles to transferability. With the arrival of transferable property rights, true water markets were born.[57]

Other obstacles to water marketing occur in international water basins, which encompass 47 percent of all land. Currently these basins generate envy, anger, and conflict between nations with differential water availability. The director of the United Nations Environment Programme, Mostafa Tolba, has concluded that "national and global security are at stake. Shortages of fresh water worsen economic and political differences among countries and contribute to increasingly unstable perception of national security."[58] This is especially the case in the Middle East, but water exports between the United States and Canada also generate animosity. International water basin controversies generally have had little to do with water marketing. Rather, the entire international water debate is couched in terms of whether one country, say Canada, should sell water to another, say the United States. In this context the question becomes one of whether the people of Canada should allow their water to be sold.

When water transfers are proposed, they are usually for massive projects to deliver water from remote northern regions to populated southern areas. For example, the infamous North American Water and Power Alliance (NAWAPA) would have diverted as much as 250 million acre-feet of water from northern Canada to southern Canada, the southern United States, and Mexico. In 1964 the construction cost alone of NAWAPA was estimated to be between $80 billion and $100 billion ($300 billion to $380 billion in 1990 dollars). Similarly, the Great Recycling and Northern Development (GRAND) Canal project would have pumped water from James Bay south to the Great Lakes at an estimated cost in 1984 of $100 billion. Such grandiose schemes capture the public's attention and create incorrect perceptions of what water transactions would be like under a free trade regime that included water. A fundamental problem with most proposals for international water transfers is that the people who would benefit would not have to pay the enormous costs of the projects. On the supply side, the citizens of the "selling" country gain little or nothing as individuals if exports are allowed. On the demand side, the "buying" country has an insatiable thirst because the real cost of water consumption is hidden in taxes or other fiscal illusions.

If these schemes were replaced with true water marketing, progress could be made in improving water allocation across international borders. Water compacts that would quantify water rights for each country could be followed by the establishment of private water rights defined in terms of quantity and quality and could provide a starting point for real continental water marketing. Innovative marketing federations that would transcend international borders could also reduce the political allocation that currently dominates international water transfers.[59] If water marketing based on willing buyers and willing sellers could be established with international water basins, subsidies could be reduced and disputes over water could be replaced with cooperation and efficiency. Law professor James Huffman summarizes the prospects for transboundary water marketing:

> [T]he challenges facing transboundary water marketing are similar to and different from those facing any transboundary commercial activity. To the extent that domestic law treats water as a marketable commodity or right, the usual trade barriers must be eliminated if water is to be efficiently allocated. To the extent that the domestic laws of Canada, the United States and Mexico treat water as a public good to be managed politically, the prospects for transboundary water markets are more distant.[60]

To solve the problem of groundwater overdraft caused by "tragedy of the commons," even more innovative solutions are needed. Unfortunately, defining groundwater rights is more difficult because recharge and discharge are hard to measure. Nonetheless, it is possible to define rights to stocks and flows and adjust those rights as more information becomes available regarding the constraints on the

A fundamental problem with most proposals for international water transfers is that the people who would benefit would not have to pay the enormous costs of the projects. On the supply side, the citizens of the "selling" country gain little or nothing as individuals if exports are allowed. On the demand side, the "buying" country has an insatiable thirst because the real cost of water consumption is hidden in taxes or other fiscal illusions.

aquifer.[61] Because groundwater and oil have similar characteristics, it would also be possible to develop a system of groundwater rights based on the idea of oil pool unitization. This system gives all owners in the pool a voice and a stake in the rate at which the resource is extracted and therefore reduces the chances of "the race to the pumphouse" found with the rule of capture.

Pollution problems could also be reduced with more reliance on property rights. In England, where fishing rights are privately owned, the owners have an

Federal projects that dam rivers to provide below-cost hydroelectric power and irrigation water have severely reduced salmon runs.

incentive to police water quality, identify polluters, and seek remedies when pollution adversely affects fisheries. Through the Anglers Cooperative Association, hundreds of legal actions have been taken against polluters and injunctions or damages obtained for municipal waste, industrial contaminants, and agricultural siltation. Economists Meiners and Yandle document how common law remedies in the United States prior to the Clean Water Act of 1972 effectively limited pollution damages. They conclude that

> common law rules have long been relied on for protecting property rights. This
> includes the right not to suffer from the damaging effects of pollution inflicted
> on those who have a right to clean water. . . . Like other areas of the common
> law, some aspects of water rights changed as technology and social conditions
> changed. Had state and federal statutes not been passed, we believe that com-

mon law would have taken these changes into account. By reviewing the fragments of common law that have survived despite legislative interference, we can build an image of how pollution could be managed effectively by common law rules.[62]

This image includes well-defined and enforceable private rights to a specific quantity and quality of water. Such rights would discipline users of water to economize on the water they withdraw and reduce pollutants returned to the water source.

Conclusion

The blue planet is unlikely to face global water shortages, but the chances that some areas could experience water crises are real. Droughts will occur as they have in the western United States in recent years, and new demands will arise with population growth and increased incomes. In particular, the demand for environmental amenities is likely to expand, as it has in the demand for increased flows in the Columbia River system to save endangered salmon stocks.

As with other natural resources, these changes need not create a crisis if individuals are allowed to respond through market processes. Indeed perhaps more than with other natural resources, water allocation has been distorted by politics under the notion that "water is different." Some would say that water cannot be entrusted to the market process because it is a necessity of life. To the contrary, because it is a necessity of life, it must be entrusted to the discipline of the market that can encourage conservation. Unless distortions created by governmental intervention are corrected, water shortages will become more acute and crises will be inevitable. When this happens, it will be difficult to suppress market forces. It would be better, however, if we would get legal impediments out of the way of markets before necessity becomes the mother of invention.

NOTES

1. For a complete discussion, see Richard W. Wahl, *Markets of Federal Water* (Washington, D.C.: Resources for the Future, 1989), chap. 7.
2. Robert Repetto, *Skimming the Water: Rent Seeking and the Performance of Public Irrigation Systems*, Report 4 (Washington, D.C.: World Resources Institute, December 1986), p. 37.
3. Sandra Postel, "Saving Water for Agriculture," in *State of the World 1990* (New York: W. W. Norton, 1990), p. 40.

4. M. I. L'vovich, *World Water Resources and Their Future*, trans. American Geophysical Society, ed. Raymond L. Nace (Washington, D.C.: American Geophysical Union, 1979).

5. Peter Rogers, *America's Water: Federal Roles and Responsibilities* (Cambridge, Mass.: MIT Press, 1993).

6. Sandra Postel, "Increasing Water Efficiency," in *State of the World 1986* (New York: W. W. Norton, 1986), p. 51.

7. GEMS Monitoring and Assessment Centre, *United Nations Environment Programme Environmental Data Report 1989/90* (London: U.K. Department of the Environment, 1989), p. 278.

8. World Resources Institute, "Freshwater," in *World Resources 1992–93: Toward Sustainable Development* (New York: Oxford University Press, 1992), p. 161.

9. GEMS, *United Nations 1989/90*, p. 236.

10. Robin Clarke, *Water: The International Crisis* (Cambridge, Mass.: MIT Press, 1993), p. 13.

11. U.S. Water Resources Council, *The Nation's Water Resources, 1975–2000* (Washington, D.C.: Government Printing Office, 1978), p. 12.

12. Henry C. Keski, *Saving the Hidden Treasure: The Evolution of Ground Water Policy* (Ames, Iowa: Iowa State University Press, 1990), p. 5.

13. Conservation Foundation, *State of the Environment: A View toward the Nineties* (Washington, D.C.: Conservation Foundation, 1987), pp. 226–227.

14. Ibid.

15. World Resources Institute, *Nation's Water Resources 1975–2000*, p. 35.

16. Conservation Foundation, *State of the Environment*, p. 229.

17. Clarke, *Water*, p. 91.

18. Jean Margat, "A Hidden Asset," *UNESCO Courier* 15, no. 4 (1993): 15.

19. Conservation Foundation, *State of the Environment*, p. 231.

20. "The Dry Season," *Northwest Orient* (August 1986): 13–19.

21. Sandra Postel, "Running Dry," *UNESCO Courier* (May 1993).

22. Margat, "A Hidden Asset," p. 15.

23. Clarke, *Water*, p. 126.

24. "The First Commodity," *Economist* 322 (1992): 11.

25. N. D. Jayal, "Destruction of Water Resources—The Most Critical Ecological Crises of East Asia," *Ambio* 14, no. 2 (1985): 96.

26. M. B. Fiering, "Models for Assessment of Water Resources," in *Resources and World Development*, ed. Digby J. McLaren and Brian J. Skinner (New York: John Wiley & Sons, 1987), p. 553.

27. "Facing a Future of Water Scarcity," *USA Today Magazine* 122 (September 1993): 68–71.

28. U.S. Water Resources Council, *The Nation's Water*, p. 2.

29. Rogers, *America's Water*, p. 9.

30. Randy T. Simmons, "The Progressive Ideal and the Columbia Basin Project," in *The Political Economy of the American West*, ed. Terry L. Anderson and Peter J. Hills (Lanham, Md.: Rowman and Littlefield Publishing, 1994).

31. Postel, "Increasing Water Efficiency," p. 57.

32. Randal P. Rucker and Price V. Fishback, "The Federal Reclamation Program: An Analysis of Rent-Seeking Behavior," in *Water Rights: Scarce Resources Allocation, Bureaucracy, and the Environment*, ed. Terry L. Anderson (San Francisco: Pacific Research Institute, 1983), p. 62.

33. U.S. General Accounting Office, *Federal Charges for Irrigation Projects Reviewed Do Not Cover Costs* (Washington, D.C.: Government Printing Office, March 1981), p. 40.

34. Utah Foundation, *The Utah Foundation Research Report*, report 572 (Salt Lake City: Utah Foundation, 1994), pp. 293–304.

35. Delworth B. Gardner and Ray G. Huffaker, "Cutting the Losses from Federal Water Subsidies," *Choices* (Fourth Quarter 1988): 24–26.

36. Repetto, *Skimming the Water*, p. 1.

37. Department of Water Affairs, *Management of the Water Resources of the Republic of South Africa* (Cape Town: CTP Book Printers, 1986), p. 1.33.

38. Repetto, *Skimming the Water*, p. 14.

39. Postel, "Saving Water," p. 43.

40. Rogers, *America's Water*, p. 110.

41. "The Browning of America," *Newsweek*, February 23, 1981, pp. 26–37.

42. Robert E. Hall and Robert S. Pindyck, "What to Do When Energy Prices Rise Again," *Public Interest* 65 (fall 1981): 68.

43. "Water Shortage Pits Man Against Nature," *Nature*, March 21, 1991, pp. 180–181.

44. Bruce R. Beattie and H. S. Foster, Jr., "Can Prices Tame the Inflationary Tiger?" *Journal of the American Water Works Association* 72 (August 1980): 444–445.

45. Ibid., p. 445.

46. Delworth B. Gardner, "Water Pricing and Rent Seeking in California Agriculture," in *Water Rights: Scarce Resource Allocation, Bureaucracy, and the Environment*, ed. Terry Anderson (Cambridge, Mass.: Ballinger Press, 1983), p. 88.

47. Trimble R. Hedges, *Water Supplies and Costs in Relation to Farm Resource Decisions and Profits on Sacramento Valley Farms*, Report 322 (Berkeley, Calif.: Gianinni Foundation, 1977).

48. Leslie Spencer, "Water: The West's Most Misallocated Resource," *Forbes*, April 27, 1992, pp. 68–74.

49. Hillel I. Shavel, "The Development of Water Reuse in Israel," *Ambio* 16, no. 4 (1987): 186.

50. Terry L. Anderson, *Sovereign Nations or Reservations? An Economic History of American Indians* (San Francisco: Pacific Research Institute, 1995).

51. Terry L. Anderson, "The Market Alternative for Hawaiian Water," *Natural Resources Journal* 25, no. 4 (1985).

52. Terry L. Anderson, *Water Crisis: Ending the Policy Drought* (Baltimore: Johns Hopkins University Press, 1983).

53. Terry L. Anderson and Ronald N. Johnson, "The Problem of Instream Flows," *Economic Inquiry* 24, no. 4 (1986).

54. Terry L. Anderson and Donald R. Leal, "Building Coalitions for Water Markets," *Journal of the America Water Works Association* 72 (August 1989).

55. Joan Laatz, "Rancher Leases Water Rights to Keep Stream Full for Salmon," *Oregon*, June 19, 1994.

56. Gary L. Sturgess and Michael Wright, *Water Rights in Rural New South Wales: The Evolution of a Property Rights System* (St. Leonards, Australia: Centre for Independent Studies, 1993).

57. Ibid., pp. 12–13.

58. Quoted in Clarke, *Water*, p. 92.

59. James L. Huffman, "A North American Water Marketing Federation," in *Continental Water Marketing*, ed. Terry L. Anderson (San Francisco: Pacific Research Institute, 1994).

60. Ibid., p. 25.

61. Terry L. Anderson et al., "Privitizing Ground Water Basins: A Model and Its Application," in *Water Rights: Scarce Resource Allocation, Bureaucracy, and the Environment*, ed. Terry L. Anderson (San Francisco: Pacific Research Institute, 1983).

62. Roger E. Meiners and Bruce Yandle, *Taking the Environment Seriously* (Lanham, Md.: Rowman and Littlefield Publishers, 1993), p. 88.

Chapter 9

RESCUING THE OCEANS

Kent Jeffreys

HIGHLIGHTS

- *The commercial harvest of ocean fisheries has risen from an estimated 18 million metric tons in the late 1940s to well over 80 million tons today. An additional 24 million tons annually is estimated to be taken by local fishermen.*

- *After thousands of years of extensive human use, many of the ocean's resources have been seriously damaged or depleted by shortsighted practices.*

- *Out of eighty-one marine fishery stocks examined, fourteen are still considered to be overexploited, and another thirty-six were being fished at the maximum capacity, according to the National Fish and Wildlife Foundation.*

- *The annual operating costs of the global marine fishing fleet in 1989 were U.S.$22 billion greater than the total revenues obtained from the sale of fish.*

- *The worldwide problem of overfishing has its source in the system of open access and free use, with its norm that only capture confers ownership of the fish.*

- *When oceans are part of the commons, users are more interested in extracting what they can than in stewarding what they cannot own.*

- *The government has contributed to the overfishing of marine fisheries by granting to all citizens the right to fish.*

- *The "open access" treatment of the oceans has allowed superefficient commercial fishing fleets to deplete the fisheries, in spite of state action to limit technologies, impose catch quotas, and restrict access to the fisheries.*

- *Drift nets, which indiscriminately net and kill any species they entangle, are an example of the inappropriate measures taken in the absence of property rights.*

- *As a first step, individual transferable quotas—a personal right or license, which may be sold or exchanged, to catch a given quantity of fish—can encourage better management of fisheries.*

- *When water-based resources are privately owned, the same legal rights that enable shopkeepers and homeowners to prevent trespass or property damage can be used to ward off water resource abuse.*

- *Attaching property rights to coastal zones would encourage better management of the fisheries and create legal rights against polluters.*

- *If coastal property rights are to be effective, they must be clearly defined, defended against challengers, and transferable.*

- *The coastal fisheries cooperative associations (FCA) system in Japan is exemplary of a property rights-based fishing industry. FCA property rights are sufficient to block potentially harmful or polluting coastal developments.*

- *In Washington State, where oyster beds may be privately owned, the threat of a damages suit against terrestrial sources of pollution has protected the state coastal zones.*

- *The lobster industry in Maine has developed a fairly effective, if informal, system of property rights management. Along the coast, small boat fishermen strategically place traps within their exclusively owned areas. Self-enforcement by these fishermen is the common response to interlopers and trap thieves.*

- *Aquaculture has great potential to relieve the impact of overharvesting fisheries while simultaneously increasing the food supply.*

Humanity derives enormous benefits from the global ocean: food, transportation, climate moderation, and mineral resources, to name just a few of them. Tens of millions of people depend on the resources of the sea for survival, and millions

> **If no one has responsibility for maintaining the resource, no one will do so.**

more respect its poetic beauty and intimidating vastness. Now, after thousands of years of extensive human use, during which most assumed the ocean's bounty was limitless, many of the ocean's resources have been seriously damaged or depleted by shortsighted practices. In particular, time is running out for certain fisheries that are being badly mismanaged.

Although there is a universal desire to protect the ocean from such threats as pollution, overfishing, and habitat loss, few agree on how to do it. Lack of accurate data is a problem, but the primary source of conflict is the competing interests of the various parties to the debate.

For several decades, concerned individuals have continued to dispute the relative merits of various marine resource management systems. A range of management approaches have been tried or suggested, including government ownership and management, comanagement by government and private entities, and outright private ownership. Part of the problem, as these proposed solutions suggest, stems from the "tragedy of the commons"; that is, if everyone has a right to exploit a resource, it will rapidly be degraded by overuse. If no one has responsibility for maintaining the resource, no one will do so. Oceans present the commons problem rather starkly.

The resolution of this debate is of more than academic importance. Marine fisheries are significant for both food production and economic vitality in many communities around the world. Although U.S. attention is often transfixed on the major fisheries of the Atlantic and Pacific oceans, countless small-scale fisheries are imperiled by the counterproductive practices of desperate people and encroaching development in the Third World.

The U.S. commercial fishing industry is huge. Consumers spent about $35 billion on seafood in 1992. That same year, commercial landings weighed in at more than 8.5 billion pounds, worth about $3.8 billion at the dock. U.S. imports added another 2.9 billion pounds, valued at almost $5.7 billion. Sport fishers contribute to the problem of overfishing in many areas, although they are not a major source of consumer products. Sport fishers are a potent political and economic force as well. Recreational (saltwater) anglers alone spent approximately $9.8 billion in 1990.[1]

The world's living marine resources are vast, but they are not limitless. According to the United Nations Food and Agriculture Organization (FAO) the commercial harvest of ocean fisheries has risen from an estimated 18 million metric tons in the late 1940s to well over 80 million tons today (figure 9-1).[2]

FIGURE 9-1   World Fish Catch

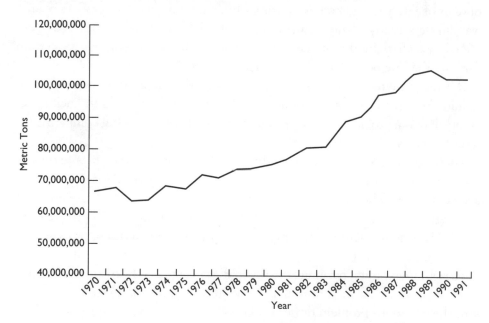

Source: World Resources Database, World Resources Institute, DSC Data Services, Washington, D.C. 1994.

An additional 24 million tons annually is estimated to be taken by local fishermen. Proper management of each resource is essential.

In recent years, the stated policy goal of the United States has been the maintenance of sustainable harvests.[3] Some observers have despaired of ever achieving such an elusive goal, given that naturally fluctuating levels of fish stocks compound human ignorance.[4] But calls for further research, while needed, are no substitute for corrective action.

The Ecology of the Ocean

Understanding marine ecology is vitally important if we are to develop effective management systems for the world's oceans. No one would ever suggest that humans could harvest—efficiently and effectively—the bounty of terrestrial ecosystems without an adequate understanding of the short- and long-term impacts of such activities. Unfortunately, key elements of such an understanding remain beyond current knowledge. Even for economically important species, we often know little about life cycles, feeding and migratory habits, or possible interactions with other species. Thus, much of the debate over fisheries management takes place in something of an information vacuum.

Marine ecology is complex and insufficiently understood, largely because of the difficulties posed in gathering data. Unique ecological niches abound, from the abyssal depths more than 30,000 feet below sea level to tidal marshes and estuaries. Researchers are only beginning to unravel the tangle of interconnected species, habitats, and life cycles. The majority of ocean life is found above or near the continental shelf regions that ring most land masses. Here, wind and wave action continuously mixes surface water with deeper layers, to perhaps 200 meters or so.[5] In addition, rivers discharge their nutrients directly into this zone. In some areas, upwelling of nutrient-rich bottom waters occurs, bringing vital materials back into the light zone where they can be utilized by photosynthetic plants. These primary producers serve as the food source for other microscopic organisms, which share the designation "plankton." These various organisms provide the oceanic food chain with its first link.

In fact, this link comprises a majority of the total biomass chain. Thus, primary and secondary production within the ocean contains most of the biomass. The estimated biomass of the world's oceans is approximately 190 billion tons.[6] Yet humans do not directly use a significant share of it. Instead, fishers target the species that consume the primary food producers. Because of the huge amounts of energy and human labor that are required to harvest the widespread primary or secondary marine biomass, it should be clear why such efforts will always have difficulty competing with more concentrated sources of protein, on land or in the sea.

It is the higher species that will remain the primary focus of human interest. While the estimates of total oceanic biomass range upward of 200 billion tons, almost all of this is in the form of noncommercial, even microscopic species. If this biomass is viewed as a pyramid, with the primary producers at the bottom, it becomes clear why the global catch of relatively conventional species is not unlimited. For at least twenty-five years, most experts have considered the maximum sustainable harvest per year for human use to be approximately 100 million tons.[7]

Human Impact on the Oceans

Human activities have had and continue to have significant impacts on marine resources. These impacts fall into three broad categories: pollution, habitat destruction or modification, and overfishing.[8] No management approach will be completely successful unless it can satisfactorily address these three categories of concern. While other categories exist—for instance, the introduction of exotic species to an ecosystem—these three, particularly severe in estuarine and coastal areas, will be the primary focus of this chapter.[9]

Pollution

There is significant, and well-placed, concern that human activities are reducing the productive capacity of many marine environments. River discharges and general terrestrial runoff contribute enormous quantities of material, from silt to toxic chemicals, to coastal zones. Marine transportation dumps major quantities of refined and crude oil products, including ballast discharges, and the crude oil spilled in tanker accidents and offshore well leaks. Air emission too can contribute significantly to marine pollution.[10]

According to at least one conference of experts, three contaminants are cause for the greatest concern: sewage, nutrients, and synthetic organic compounds.[11] Marine discharges of raw sewage are a growing problem in many developing countries and can lead to outbreaks of infectious diseases, particularly if shellfish beds are contaminated. Increasing reliance on organic pesticides is raising concerns about highly sensitive habitats, such as nearshore coral reefs. Widespread applications of agricultural fertilizers may also introduce excess nutrients into the marine environment. In the Chesapeake Bay watershed alone, farmers apply approximately 700 million pounds of commercial fertilizers each year. Manure from livestock also contributes enormously to the nutrients in runoff. Lancaster County, Pennsylvania, alone produces nearly 10 billion pounds of manure annually from its farm-raised cows, hogs, sheep, and poultry.[12]

In some cases, the nutrients are produced directly by human bodies. The average person excretes about 9 grams of nitrogen and 2 grams of phosphorus each day; in other words, 1 million humans produce about 9,000 kilograms of nitrogen and 2,000 kilograms of phosphorus every day. Such "overfertilization" of the oceans is the major threat to many coastal ecosystems, particularly bays and estuaries. Instead of directly poisoning species, the high nutrient levels trigger a "bloom" of algae and phytoplankton that eventually will die and decay. At this point, microbial decomposition will strip the oxygen from the local waters.[13] By depleting the available oxygen, fish and other species are suffocated or driven away.

In specific coastal areas, processing and discarding the fish catch can introduce enormous, if localized, amounts of nutrient pollution. For example, in Louisiana, over 100 million pounds of crustacean shells are discarded each year.[14]

Natural events, such as floods or shifts in normal ocean currents, can also wreak havoc with marine habitats by either denying or delivering additional nutrients and suddenly altering the balance of life in the region. One

> **Lancaster County, Pennsylvania, alone produces nearly 10 billion pounds of manure annually from its farm-raised cows, hogs, sheep, and poultry.**

result of such shifts can be the occasional and sudden increase in the populations of certain species of algae. Most of these events are not especially harmful to species of most interest to fishers but still are capable of producing remarkable changes in the local environment. The water's very color can be altered by the presence of vast numbers of these organisms, from yellow to brown to red. Indeed, the Red Sea earned its name from such blooms of algae.[15]

Some species that may undergo a population explosion produce a toxin so potent that it can kill all of the fish in the immediate vicinity. Furthermore, predators, including marine mammals such as dolphins, can succumb to these same toxins by consuming the stricken fishes.[16] Humans can receive dangerous dosages of these toxins if the bloom occurs above shellfish beds. Debate continues over the relative contribution human activities may make to periodic algal blooms, or red tides. Although human sources of increased nutrients are probably not accountable for many of the deadly blooms, in some cases there appears to be a correlation with human effluence.[17] Yet despite the fact that scientific suspicions were raised decades ago, major uncertainties remain about the origins of these events.[18]

Although the Clean Water Act and other legislation has greatly reduced the pollution of U.S. waters, significant amounts continue to be generated annually. A particularly difficult question has been the huge number of extremely small pollution sources, from backyards to barnyards, that cumulatively contribute perhaps 50 percent of the total pollution in some areas. In response, the Coastal Zone Act Reauthorization Amendments of 1990 required coastal states to develop plans to reduce and eliminate nonpoint sources of marine pollution.[19] Additional actions can be expected as more information is generated by these regulations.

No other marine pollution events capture the public's attention quite like oil spills. And indeed, even fairly small spills of crude oil can have a devastating impact on local wildlife and habitat. Unfortunately, the major media concentrate on the immediate effects of a spill and do little to convey the remarkably rapid—and completely natural—recovery process.

The public outcry in response to any major spill often stimulates a desire to "do something" about the tarred beaches or the struggling sea life. However, in recent years, spill experts have begun to realize that aggressive spill cleanup measures—that is, those beyond containment and skimming operations—may do more harm than good.[20] For example, most efforts to remove oil from rocky or sandy shores also disturb or destroy surviving plants and microorganisms that, if left undisturbed, would eventually repopulate the oiled zones. Experts now admit that even the largest oil spills are "relatively modest and, as far as can be determined, of relatively short duration."[21] Certainly it is preferable to avoid all spills, but oil spills are actually temporary disasters.[22]

In other contexts, dealing with the problem of marine pollution is a complex process.[23] The difficulty of persuading sovereign nations to restrict pollution with impacts on common fishery resources is readily apparent. Some commentators have argued that governmentally directed investments in pollution control can create sufficient benefits to satisfy the investing society even in the face of free riders who take advantage of the resulting improvements without contributing to the effort. However, without careful coordination among nations, admittedly a difficult process, environmentally destructive behavior may be beneficial to uncooperative countries in the short term.[24]

It can also be demonstrated that the lack of strongly defined and defended property rights has permitted pollution of the coastal ocean regions to continue. In practice, where water-based resources can be privately owned, private defenders of the resource can establish legal precedents to ward off future resource abuse. In this way, vital ecological habitat can be protected by the same legal mechanisms that enable shopkeepers and homeowners to prevent trespass or property damage.

For example, in the state of Washington, oyster beds may be privately owned.[25] Because of these private rights, individuals are able to protect the environment and their financial investment simultaneously. In at least one instance, managers of terrestrial sources of pollution have agreed to eliminate sewage discharges because of the threat of a damages suit from private oyster bed owners.[26] Treating coastal zones as private property has important implications for fisheries management, as well as recreational and development activities.[27]

For another example, consider water pollution in the context of a river drainage basin.[28] One of the most celebrated of such instances was the Pride of Derby case in England, which was resolved in 1951. The English common law had long permitted private legal actions against polluters, the logical extension of common law protection against trespassing and the creation of nuisances (which interfere with the "quiet enjoyment" of property ownership): "The Common Law entitles every riparian [riverbank] owner to have the water flowing

A demonstration channel catfish pond is harvested at Virginia State University's aquaculture facility. Aquaculture is growing rapidly worldwide.

past his land in its natural state of purity, and every fishery owner is entitled to the free movement of fish up and down a river from the sea to the source. Any denial of rights can be restrained by injunction and be the subject of a claim to damages."[29]

The costs of these civil legal actions, however, were often prohibitive, and nearly 100 years passed during which no cases were taken to the British High Court to stop water pollution. Criminal statutes against water pollution had been passed by Parliament as early as 1876, but these required the local government authorities to prosecute the polluters. Since the local governments were, in fact, almost always the largest polluters (generally because they owned and operated ineffective sewage plants) there was more than a little reluctance to seek criminal sanctions.

The Anglers' Co-operative Association (ACA), a group of fishing club members, was formed in 1948, and among its first acts was a request for an injunction against a water polluter under the common law. The resulting case included a damages award to the ACA and began a trend toward private enforcement of the common law against polluters. The ACA represented the Pride of Derby Angling Club and the Derby Angling Association against three major corporations that had been polluting the Rivers Derwent and Trent. The suit was successful, and the ACA has continued to represent local member organizations in over 1,500 cases. Today most cases are settled out of court. The ACA has noted that it was fighting pollution twenty years before the public and politicians became aware of the threat from water pollution.

It should be noted that the property rights of riparian owners are not absolute. Although the effect is similar to private ownership, technically the water in a stream and the fish in the water are public property. The riparian owner may have title to the land that forms the banks and bed of a river or stream but is allowed to possess the water and the fish only under important restrictions. For example, any use of the water must not appreciably diminish the quality or the quantity of the water flow downstream. This applies to all uses: industrial, agricultural, and municipal. As with marine species, the fish in the river are considered wildlife until they are caught, at which time they become the property of the fisherman.[30]

Still, the owner of the land along a river may legally limit access to the water. Fishermen privately contract with landowners to secure the exclusive right to exploit this limited access in order to fish. Under this approach, for generations large landowners in Britain have hired "river keepers" to manage their fishery resources. In principle, this practice could be extended to the oceans.[31] Indeed, in Denmark, farmers adjacent to the coastline have an ancient right to lay eel traps.[32] This right enables them to control access and secure rent. Thus, they are encouraged to practice good stewardship of the resource (and to require it of lessees) to ensure a lifetime income.

Since property rights have been historically beneficial along rivers, a property rights approach merits serious consideration as a solution to the problem of marine resource depletion. (This is not to imply that there is no role for government. Government involvement may be necessary to enforce private contracts and property allocations, even when it played no significant role in their negotiation.)[33]

Habitat Loss

In many coastal zones around the world, the most significant threat to fisheries arises from habitat losses.[34] Without sufficient breeding grounds and areas for young creatures to find refuge and food, species cannot sustain themselves. Habitat losses can take many forms; for example, sea grasses may be destroyed by silt and shore runoff or stripped away by periodic bottom trawling, and highly sensitive coral and other reefs may be damaged by anchors, shipwrecks, pollution, divers, and even dynamite fishing. Most of these events are unintentional, but some habitat alteration is by design. Harbors and shipping channels are constantly being dredged, for example, and the direct scouring of the bottom is exacerbated by the dumping of the spoil elsewhere.

Shoreline development may have a major impact on fisheries. In prosperous countries and resort areas, the result is often a series of high-rise buildings that attract a seasonal influx of tourists and lead directly to increases in recreational fishing, diving, pleasure boating, and nutrient loading of the surrounding sea. In addition, shoreline crowds or construction may displace critical habitat for juvenile marine species as well as shore birds and plants.[35]

In Third World countries, many economic development schemes have eliminated miles of mangroves and other coastal wetlands with little concern for the ecological consequences. Even logging operations on land can significantly affect marine ecosystems, as silt runs into the coastal waters.[36] As a result, fish populations that depend on these habitats for all or part of their life cycles have dwindled. The result is a decline in economic well-being and a scramble for the remaining fish resources.

Overfishing

While debate continues to swirl around possible resolutions to the previous two problems, at least one aspect of the issue of overfishing has long been agreed upon: "The worldwide problem of overfishing has its source in the regime of open access and free use, with its norm that only capture confers ownership of fish."[37] A general consensus has evolved with respect to the problems raised by open access.[38] In brief, the problem with fishing is that anyone can do it. Fish are considered fair game for everyone; if you catch one, you are thereby richer.

In numerous island and coastal communities, overfishing has been driven by rising human population and, especially, improvements in fishing technologies. To make matters worse, the larger, older fishes are generally the first to be taken, with a negative impact on the sustainability of the fishery because the more mature individuals of the species are likely to be far more prolific when they are reproducing.

Even the capital-intensive commercial fisheries of the developed nations have not been immune from major depletion and even collapse. Government regulators in Canada and the United States have responded with moratoriums and bans on fishing certain stocks.[39] After years of official denial, the trend of closures accelerated in 1994.[40]

There is no longer any real dispute over the fact that the major fishing fleets are simply too effective and too numerous. It is estimated that many fleets could be reduced by up to two-thirds and still reap the same harvest with more catch per boat.[41] In one tentative study of the subsidies provided by governments to national fishing fleets, it was estimated that the annual operating costs of the global marine fishing fleet in 1989 were as much as $22 billion greater than the total revenues. The total figure may have declined in recent years due to the collapse of the Soviet Union and the resulting inability of the Russian government to maintain previous levels of subsidies to the fleet, but the difference is still staggering.[42] It is bad enough that all those fish are free for the taking. Even worse, the world seems to be subsidizing an army to take them.

Obviously the fish populations are not the only ones to suffer. Many traditional fishing communities are threatened with extinction even if the fish do survive.[43]

A Case Study: Sea Turtles

I could list many other examples that illustrate the three primary problems of overfishing, habitat losses, and pollution, but a single case study of sea turtles highlights the issues.

There are eight species of sea turtle, and all are considered to be threatened or endangered.[44] Although each of these air-breathing reptiles exploits a par-

Not only are some of the world's major fisheries endangered by overfishing, but so are many smaller local fisheries.

ticular niche in the ocean, all share one important reproductive trait: the females return to sandy beaches to dig shallow nests and lay their eggs. Thus, sea turtles are exposed to every hazard humans can pose: the nesting beaches may be developed; the eggs may be taken; many coastal cultures prize adult sea turtles for their meat, shell, and leather; and large numbers of sea turtles are killed accidentally by fishing nets and other gear or die in collisions with powerboats.

In response to a major recent dispute concerning the high turtle mortality rate from shrimp trawlers, a study of turtle excluder devices for shrimp trawling nets was performed from 1990 to 1993. The U.S. shrimp industry responded by reducing turtle mortality rates some 97 percent. Witnessing this success, several environmental organizations filed suit against the U.S. government in the U.S. customs court seeking to force a ban on imported shrimp caught.[45] Still, the human damage is virtually impossible to contain by such piecemeal manuevering.

Even without the intervention of humans, the odds are steeply stacked against each individual sea turtle. Most female sea turtles lay around 100 eggs in a nest. If left undisturbed, perhaps 85 percent will hatch.[46] But of those that reach the sea, only a fraction will survive to adulthood. The additional pressures introduced by humans make the survival of the sea turtles more precarious.

One of the more interesting experiments in preserving sea turtles was the Cayman Turtle Farm. The green sea turtle, like all other sea turtles, is considered to be threatened by a combination of fishing pressures and loss of egg-laying habitat due to coastal development. The green sea turtle had long been a target of fishermen throughout the Caribbean but had never been farmed until a group of entrepreneurs established Mariculture, Ltd. in the Cayman Islands.[47] Their goal was to satisfy part of the world demand for turtle meat, shells, and leather through a sustainable farm-raising program. Unfortunately, the U.S. Fish and Wildlife Service (FWS) denied the farm an exemption from the ban on trade in sea turtle products.[48]

Shrimping in the Gulf of Mexico endangers sea turtles, which get caught in the nets and drown.

This bureaucratic edict effectively eliminated the possibility of profit on a sufficient scale to permit a major breeding program. Sadly, at the time of the decision, the turtle ranch seemed well on its way to sustainability. By 1982, the Cayman Turtle Farm had built up a population of almost 80,000 captive turtles and for years had been returning yearlings to areas of depleted wild populations. Survival rates for hatchlings at the farm were exceeding 50 percent prior to the FWS decision. In the wild, survival rates are estimated to be only a few percent.

Because of the denial of permission for shipment through the United States, the turtle hatchery's operations were limited to local markets. Ironically, legislation was introduced in the U.S. House of Representatives in 1990 to create an American turtle farm under FWS control.[49] (The bill died in Congress.)

Other nations have tried similar state-sponsored approaches. Mexico has a long-established sea turtle research and protection program.[50] The first turtle refuge, or "camp," was created at a natural nesting site at Rancho Nuevo in the state of Tamaulipas in 1966.[51] Yet in the early years of this project, most beaches were unprotected, and vast numbers of turtle eggs were taken for human consumption each year. The more recent response to the continual decline of turtle populations has been that conservation programs have been taken more seriously. In one unique effort to engage the private sector in the battle to save sea turtles, *hueveros*, or egg collectors, are granted a monopoly along one beach in Costa Rica. Revenues from the strictly limited egg-gathering season are divided among the collectors and a government research program. Granting similar rights to villagers elsewhere could save the sea turtle. In fact, two conservationists witnessed a dispute in a Mexican cantina over the origin of some turtle egg snacks. They later observed, "If the plight of sea turtles was being discussed in bars, then the conservation ethic really was getting out."[52]

Interim Steps toward Improved Management

Simply calling for conservation measures will not suffice in the major commercial fisheries. Before there can be an industry-wide agreement to conserve the resource, a method for allocating rights among the many claimants must be developed.

Certainly the most discussed innovation in this regard is the individual transferable quota (ITQ), a personal right or license, which may be sold or exchanged, to catch a given quantity of fish. The most advanced example of this approach (for a national fishing industry) is found in New Zealand. By October 1986, New Zealand's inshore fisheries were all placed under a management system consisting of ITQs.[53] In most instances, the quota rights are similar to a perpetual share of the permissible harvest of a given species.[54] The New Zealand government sought to establish quotas on overfished species that were sufficiently low to permit gradual recovery. But to satisfy the political realities of the fishing industry, the government instituted a "buy-back" program that assigned quotas based on historical catch levels and then paid fishermen to relinquish the amount above the design quota.[55]

The New Zealand ITQ system has undergone periodic modifications as more is learned about how the fishers and the fish respond to the new situation. For example, bycatch—any unwanted species (illegal, inedible, or immature) caught by the fishing gear—is a serious concern in many commercial fisheries. When New Zealand tried to confront the issue of bycatches under its quota management system, it first declared that any nonquota catch (bycatch) must be forfeited to the government. In response, fishers dumped any part of the catch that would be forfeited. Thereafter, the New Zealand government established a catch-quota trade-off scheme, which allowed fishers to trade selected quota species for bycatch species. If the quota for a particular species is exceeded or bycatch is taken, no dumping need occur because it has become a tradable commodity.

The New Zealand system is complex, but most economic analysts feel that it was a step in the right direction.[56] Although few would argue that the lessons have been easily learned, if at all,[57] overall, the ITQ approach seems to provide the incentives that encourage better management of the fisheries by the fishers themselves.[58]

Several ITQ systems now exist in the United States. In September 1991, the North Pacific Fishery Management Council adopted a draft plan for individual, transferable quota shares for halibut and sablefish in the Alaskan fishery. The northern Pacific fisheries have remained highly productive under these various fishing rights allocation schemes. Although direct comparisons are not possible, it may be important to note that the recent closure of the once highly productive Georges Bank fishing region off the Atlantic coast of New England and Canada occurred in a region that does not have a similar fishing rights allocation system. Instead, the managers of this region have relied primarily on seasonal and technological restrictions. Yet over the years, the total catch in this management region has plummeted for haddock, cod, and flounder. For example, in the 1960s, the annual haddock catch was approximately 50,000 metric tones. By 1993, the haddock harvest had collapsed to less than 1,000 metric tons.[59] While much of the difference in trends between the Pacific and Atlantic regions may be the result of superior biological productivity in the northern Pacific, the existence of distinct management systems merits further investigation by researchers and policymakers.

Other fisheries with ITQs in place include the South Atlantic Wreckfish fishery, the Atlantic Surf Clam and Ocean Quahog fishery, and Canada's Pacific Halibut fishery.[60] Because each fishery is unique, it can be difficult to compare results without careful qualification. And an ITQ in one fishery may lead to an increase in fishing pressure for nearby fisheries that remain open access.[61]

The ITQ concept has met with perhaps somewhat greater enthusiasm among economists than among fishers or administrators. In testimony before the U.S. House of Representatives, officials of the Environmental Defense Fund (EDF) endorsed a careful conception of ITQs. One of the most important features of ITQs, according to the EDF, was that ITQs would reduce the management burden on federal fisheries agents "by eliminating the need to engage in contentious debates about the allocation of fishing privileges after the initial allocation is made. Market forces will take over this function."[62]

Yet the initial allocation of the quota is certain to be a strongly contested process. Many fishers suspect that it is subject to political influence, and not without reason.[63] The result may produce a windfall profit for the fortunate few. Some observers have recommended various methods for avoiding such a result, among them use of an auction to allocate initial shares, imposition of user fees, or various taxes applied to catch amounts or transfer of quota rights.[64]

ITQs are not perfect. If the initial quota catch is set too high, the fishery will continue to decline. This was thought to be the case with orange roughy in New Zealand, and the catch quotas were severely cut back in response. In fact, 17 out of 169 quota-managed species were considered to be overfished by more than 10 percent during the 1987–1988 season.[65] However, successfully designed and implemented ITQ approaches can offer important benefits to both fishers and fisheries, primarily by reducing the total fishing effort through buyouts or other consolidations within the fishery. The economic rewards to the fishers can be significant. In Canada's halibut fishery, the market value of an individual license increased rapidly, soaring from $80,000 in 1990 to about $150,000 in 1992. Even with the increased prices, these fishers were able to reduce total catch to a level below the program's total allowable catch.[66]

In any event, it can be argued that ITQs should be only a first step toward more complete property rights, as one of the leading theorists on the economics of fisheries, Anthony Scott, notes:

An ITQ harvesting regime, requiring continued regulation, is best seen as only a brief stage in the development of management. Its evolution can be expected to continue until each owner has a share in the management decisions regarding the catch, and, further still, until he has an owner's share in management of the biomass and its environment.[67]

Whether ITQ systems will be fully successful in the effort to prevent overfishing is an unanswered question. Other methods, short of pure property rights, are often employed in conjunction with or as alternatives to ITQs. Restrictions on net type, boat sizes, and other technological aspects of the fishing industry are widespread. Selection of appropriate technologies is an important control mechanism for any resource manager. Protection of the future productivity of an area can be facilitated by careful determination of permissible equipment. However, politics has often played a role in the selection of technologies.[68] Once in place, politically endorsed technologies can attract overinvestment by fishermen, especially when access to the fishery remains largely uncontrolled or subsidies are made available to replace banned technologies.

Drift nets are an example of the kind of technological response encouraged by an "open commons" treatment of the oceans. Drift nets, as the name implies, are released by the fishing boat and allowed to float until retrieved. They indiscriminately net and kill many species, including marine mammals and sea birds, that become entangled. Japan, Taiwan, and South Korea have primarily relied on large-scale drift nets, especially in the North Pacific. United Nations Resolution 44/225 instituted a ban on South Pacific drift-net fishing and, in the absence of conservation measures, a total ban on drift-net fishing by July 1, 1992. In response, the U.S. Congress adopted legislation, the Driftnet Moratorium Enforcement Act of 1991, to apply trade sanctions against nations that fail to comply with the UN resolution.[69]

One problem with technological limitations is that they can significantly reduce the income received by fishermen per hour of labor. Technology restrictions seek to make the harvest less efficient in order to prevent overfishing, but mandating inefficiency can lead to other social and economic problems.

A common alternative or supplement to limits on technology is a limit on the total allowable catch. One means of accomplishing this is through quotas on individual species (although there are often difficulties with enforcement). Another approach to prevent overfishing is to limit legal access to the fishery.[70] Licensing of boats is a common method for enforcing this concept, as is restricting licenses to local or national citizens. Still another approach is to restrict the duration of activity through the establishment of seasons for target species. The Alaskan halibut fishing season has been reduced to only a few twenty-four-hour frenzies each year.[71] It has even been reported that the Alaskan season for herring roe was a brief forty minutes.[72]

Of course, commercial fishers confront more than just attempts to make their harvest less efficient. Recreational fishermen and organizations have successfully closed many commercial fisheries through political action.[73]

Even with all of these tools available, U.S. fishery management has had significant difficulties. For years, many of the largest factory vessels targeted the highest-value product and dumped the rest back into the water. This was common in the Alaskan pollock fishery, where for years the valuable roe was kept for sale, and millions of pounds of fish were discarded. The U.S. Commerce Department (which has oversight of U.S. fisheries) prohibited this practice in an effort to encourage conservation of a valuable food resource. Although the roe is no longer a specific target, many regulations still work to promote economic waste by encouraging overinvestment in boats and equipment, and some regulations still promote waste of the fishery resources.[74] For example, to protect small-scale halibut fishermen, all halibut caught by trawlers must be returned to the sea. Unfortunately, most of the netted fish are already dead.[75] This, and other types of bycatch, result in some 2 billion pounds of potential food being thrown back each year.[76] The worst example of such waste is found in the Gulf of Mexico. It is estimated that 10 pounds of bycatch are discarded for every pound of shrimp caught in the southeastern U.S. Gulf shrimp fishery.

According to one observer, "Our use of marine resources represents a prime example of the 'tragedy of the commons.' Users are more interested in extracting what they can than in stewarding what they cannot own."[77] Each competing interest, whether governmental or private, will seek to take as much of the resource as possible before it can be taken by others. Simply demarcating international boundaries within the ocean is insufficient if a property rights treatment is lacking within those boundaries.

The Property Rights Concept

Territoriality is common in the animal kingdom, but property rights are a uniquely human characteristic. In essence, property rights comprise a spectrum of rules that govern human resource use. The literature on property rights has focused on two basic approaches to ownership: ownership in common and ownership by individuals. However, it is important to note that a single line cannot divide the various forms of property ownership into neat units. Although "ownership" implies complete control over a resource, that is not always the case.

When discussing marine resources, perhaps the crucial property right is the ability to control access to the resource. Failure to concentrate on the question of access has led to an unfortunate degree of confusion over property rights. Most of the existing management schemes for marine resources fail to deal adequately with the question of access to the resource. However, observers are approaching near unanimity on the issue of limiting access in some form.[78]

And indeed, many researchers have long recognized the importance of control over access. If access is carefully regulated by a responsible party, the technical ownership of the resource becomes relatively less important. For example, Francis T. Christy, Jr., considers the term "common property" in the light of the nature of control over access rather than the identity of the owners. He points out that much of the "public" land in America is treated as private property through contractual arrangements with the governmental owner.[79] S. V. Ciriancy-Wantrup and Richard C. Bishop have importantly distinguished between "common property" and property held by "tenancy in common." As to the former, multiple owners will have equal rights to use the resource, but this does not imply that each owner will hold an identical percentage interest in the resource. These authors have also criticized the practice of equating an unowned resource (such as the high seas) with the "common property" concept.[80]

More recently, Elinor Ostrom examined the traditional methodology, which assumes that resource management must be entirely governmental or entirely private, and found that this does not correspond to real-world situations. She found that permitting individuals to contract privately over resource use can lead to equitable and efficient arrangements utilizing an array of enforcement mechanisms.[81]

The reason that some resources suffer from overuse or abuse is that control over access to the property resource is somehow denied or rendered ineffective. In the ocean context, this has been the result of both political policy and the physical impracticality of enforcement of rules governing restricted access. Nevertheless, some researchers have begun to document the results of private property approaches to the management of living marine resources. One comprehensive overview of the concept concluded that "private forms of sole ownership—common property or individual private property—appear more likely than limited entry, ITQs, or other forms of state ownership to conserve living marine resources and to secure significant net economic

> **The reason that some resources suffer from overuse or abuse is that control over access to the property resource is somehow denied or rendered ineffective.**

benefits." This report observes that merely limiting access to the fisheries, although an improvement, is a weak first step. Essentially this method divides people into two groups: fishers and nonfishers. Establishing individual quotas, however, distinctly improves limited entry schemes by dividing the resource itself into shares owned by individual fishers. Sole ownership further extends the rights and obligations of the fishers to include management and conservation actions so long as they do not interfere with the resource rights of others. The key to success is the concept of ownership rather than the identity of the owner. Although sole ownership is a relatively new concept for living marine resource managers, it offers a way to provide individuals with the incentive to respect the rights of others.[82]

Nevertheless, perfect economic efficiency will not be found in any human endeavor, regardless of the property arrangements. The goal should be to select (or permit) the economic arrangements that most fully and efficiently provide for society's needs. Although traditional environmentalists might shudder at such words (wondering at the same time what happened to "respect for mother nature" or "awe and wonderment"), the key link between traditional environmentalists and the new thinking is this: society's needs include a healthy marine environment. We can satisfy both groups if we can allow society a natural way to meet its needs. In this regard current economic debate has shifted from a focus on guidance through government action toward a recognition of the efficacy of markets. Markets are a necessary yet insufficient precondition. As Milton Friedman has said, "Markets exist even in Albania."[83] Crucial to their effectiveness is the type of governmental infrastructure within which markets are allowed to operate. In addition to a sound political infrastructure, efficient markets rely on privately owned goods (property) that can be transferred to others through voluntary exchange.[84]

Three critical aspects of property rights must be present for there to be efficient and equitable market interactions: the rights must be clearly defined, defended against challengers, and transferable, in whole or in part, to others through voluntary exchange.[85] The detailed terms of these transfers to others are the basis of contracts—legally enforceable agreements.[86]

Although resources vary across space and time, human interactions have been remarkably consistent for centuries. Anthropologists have been able to document the attitudes and practices of coastal cultures around the globe.[87] Interestingly, most have used a property rights approach to managing coastal resources. The property rights are not necessarily limited to an identified individual but may be owned by kinship or tribal groups. Societal structures operate to identify the decision maker for the particular resource in question. In this manner they have been able to deal with many of the same problems—albeit on a smaller scale—that face marine resource managers today. It is the modern world that has created an ocean problem, but not by its inexorable growth—rather, by departing from age-old traditions governing the ownership of coastal waters.

The Japanese Example

Only recently have most American scholars begun to recognize that the theoretical case for private property rights in fisheries has a real-world example in traditional coastal societies. (Most such societies have been overwhelmed by central government authorities.) Perhaps the best example of a surviving property rights–based coastal fishing industry is found in Japan, where "tracts of sea and the resources they contain adjacent to Japanese communities are fishermen's bona fide property."[88] This is not a new development; evidence indicates that some form of marine property rights existed in Japan at least 2,000 years ago. The rights of coastal villages evolved through Japan's feudal period; by the period after the American occupation following World War II, fisheries rights had a "legal status equal to that of land ownership."[89]

Although the overall system in Japan is complex and based on centuries of cultural development, it is no more complex than alternative systems elsewhere that have failed to produce a similarly sustainable fishery resource. In Japan, coastal fisheries cooperative associations (FCA) generally hold the rights to the small-scale coastal fisheries and distribute these rights among the members.[90] The local government, or prefecture, interacts with the community-based FCAs, and most large-scale or open sea fisheries are licensed by the government to private organizations or individuals.[91]

The property rights of the FCAs are sufficient to block potentially harmful or polluting coastal development in many cases. While the central government can take fisheries rights from FCAs for development purposes (and compensate for the loss), private developers are required to negotiate a settlement with the FCA in advance. This may require purchase of all fishing rights that would be destroyed or diminished as a result of the proposed development. If the FCA refuses to accept a private offer, even the central government is powerless to impose an exchange.[92]

Nonetheless, Japan's rapid postwar development led to numerous pollution events along the coast. Under a series of court cases, it was established that victims of this pollution were entitled to compensation. As with the English Pride of Derby case, pollution victims are now often able to settle out of court.[93]

The History of Law and Custom

Although many societies historically implemented property rights regimes for coastal resources, through most of human history the open ocean was unowned, unexplored, and unimportant to most individuals. With the advent of ocean exploration and commerce, the legal treatment of the sea became an important issue for many nations.

Today's international treatment of the oceans as a resource owned in common by all nations is a result of the 500-year-old struggle among European interests striving to control the ocean's resources. Prior to that time, it was not uncommon for naval powers to assert ownership of

> **Through most of human history the open ocean was unowned, unexplored, and unimportant to most individuals.**

coastal waters and even entire seas. Indeed, in 1493 the Vatican went so far as to endorse officially the division of most of the Atlantic and Pacific oceans into Spanish and Portuguese possessions. However, as superior sailing (and warring) technologies diffused through the maritime nations, the Iberian partners could no longer enforce the full extent of the papal edict. Piracy and other forms of challenge were supported by England, France, and the Netherlands in an effort to wrest control of valuable trading routes from the Spaniards and the Portuguese.[94] It was in this climate that Hugo Grotius argued in the seventeenth century that the ocean was incapable of being reduced to ownership.[95] This was essentially a political argument by the Dutch against the Portuguese, in their contest for supremacy of Far Eastern trade. Nevertheless, this common ownership approach has dominated international law until very recently.

The current attitude toward possible ownership of ocean areas is also a result of conflict over limited resources. As improved technologies made fisheries more susceptible to overharvesting, coastal nations began a push to extend exclusive economic rights farther to sea. Today the 200-mile exclusive economic zone (EEZ) is accepted as the international norm.[96]

America extended its EEZ to 200 miles by presidential proclamation in 1983. With over 90,000 miles of coastline, the U.S. EEZ covers approximately 2 million square miles, making it the largest EEZ of any other coastal nation. More than eighty other nations now claim 200-mile EEZs. The result is that over 90 percent of the living marine resources are now found within EEZs, and almost all fishing takes place within the internationally recognized waters of sovereign nations as opposed to the high seas. One interesting outcome has been the reduction of distant-water fleets and thus a reduction in the energy expended per unit of catch within the EEZs.[97] However, the total catch continued to climb after imposition of the EEZ. For example, in the U.S. EEZ, the foreign fleet catch "declined from about 3.8 billion pounds in 1977, to zero in 1992." Simultaneously, the U.S. catch grew from less than 1.6 billion pounds in 1977 to well over 6 billion pounds in 1992. The total catch from the U.S. EEZ grew approximately 24 percent in the first ten years after the 200-mile limit was established, but overfishing has so depleted the stocks that it has since declined slightly from its peak.[98]

> **The trend in almost all nations is away from open access toward more restrictions on entry to exploit the fishery.**

Within this 200-mile zone marine resources (both the ocean and the fish) are considered the property of the coastal state. Regardless of the historical origins, the justification for state control over marine resources still can be based on two theories. First is the belief that government should own all things that have not been otherwise reduced to private possession. However, as technology advances, this justification will become less important, as the relative costs of private ownership are likely to decline substantially. The second justification is that government must play a role as trustee over marine resources for the benefit of all its citizens. Thus, it may be said that "the people" own the ocean, with the state as a managing trustee. For example, the state of Rhode Island's supreme court declared in 1958 that the state constitution "gives the benefits of fishery to all the people in equal measure."[99] Nonetheless, this legal theory can imply a utilitarian duty for the state to maximize societal benefits from the resource.[100] Thus, distribution of the resources to private individuals may become a duty of government where it can be shown that this would result in resource management superior to the existing arrangements.

The trend in almost all nations is away from open access toward more restrictions on entry to exploit the fishery. This can be seen as a necessary first step toward more extensive private ownership opportunities. Unfortunately, restrictions within national 200-mile EEZs can be circumvented by overfishing unrestricted or "open" fisheries. For example, a little over 200 miles off the coast of Alaska in the central Bering Sea, over 150 foreign vessels aggressively trawl for pollock and other bottom-dwelling species. (This region is known as the "doughnut hole.") Most of the major fishing nations are represented: Japan, South Korea, Russia, and others. Occasionally some fishermen cross the 200-mile line into U.S. waters, whether by accident or design. The fines imposed for such violations can be severe. In 1990, one Japanese boat owner paid a $900,000 fine for illegally fishing in U.S. waters.[101]

This and similar situations stimulate a push for uniform restrictions through international agreements. However, the complexity of satisfying every potential marine resource interest group around the globe is staggering. Hence, the progress toward more rational management of fishery resources can be exceedingly slow.

Why Private Ownership Can Yield Public Benefits

In a market-oriented economy, prices tend to rise when a good or service becomes relatively more scarce, and marine resources are no different in that regard. An extreme example is the case of the bluefin tuna, which can sell for $30,000 per fish, or more, in Japan.[102] An argument is often presented that higher prices lead to more pressure to deplete the resource. But it is equally true that with a property rights system in place, higher prices would provide the incentive to produce more fish through aquaculture or other stock enhancement techniques. Currently no one has any incentive to bolster the supply side; everyone has an incentive to deplete.

In any event, there is little evidence to imply the superiority of political over private management of renewable resources. Government officials seeking to manage fisheries are confronted by the Sisyphean task of reconciling the disparate claims of competing interest groups. Too often, they must come down on one side or the other: either they overprotect or they underprotect the resource. This is not necessarily due to incompetence. Bureaucrats, and government officials in general, are often exceptionally well-trained, highly skilled individuals. But because the bureaucrat will not receive rewards commensurate with effort, there is less of an incentive structure—reward for success, punishment for failure—than in the private sector. The incentives that do exist generally lie in the direction of bigger bureaucracy, not better environmental quality. In addition, the difficulty of gathering accurate and timely data in a central location and then distributing them effectively further complicates the workings of government.[103] Over time, the superiority of property rights–based markets begins to show results. Partially, this is due to the fact that with a widely held private property infrastructure, mistakes are more likely to be balanced by accurate forecasts from competing property owners.

However, two important points should be noted here. First, the owner need not be the same person who actually manages the resource.[104] Second, many economists have pointed out that an inept private resource manager is likely to be bought out by more efficient managers, so long as there is a private right in the resource that may be exchanged voluntarily.[105] Some environmentalists fear that this process must lead to the rapid depletion of resources as owners attempt to make as much money as fast as possible. Yet it may be argued that this fear is becoming reality in marine fisheries today precisely because of the lack of transferable property rights. Without the right to exchange contractually, better resource managers are excluded from all or part of the fisheries.[106]

In an important respect, marine fishery management resembles the treatment of oil and gas resources in America. To be owned, the fish must be captured. This is analogous to the rule of capture applied to fluid underground resources. However, in the American oil and gas industry, the mineral resource itself may be privately owned. Almost all onshore oil reservoirs in the United States are held by several, often hundreds of, individual owners. Because of the threat of inefficient use or overexploitation of the oil resource, conservation-oriented contractual arrangements were developed years ago among these otherwise competing owners. These arrangements were then codified in many states.

As we have seen, coastal regions (now often referred to as "coastal oceans," in contrast to the high seas) generally have been regarded as private property of one sort or another.[107] Primarily, this has been the result of proximity to the claimant and the relatively easier task of enforcing the claim and a reflection of the fact that the cost of watching over the resource (as well as the cost of exploiting it) is much lower close to shore. Yet even out of sight of land, reefs and other identifiable features, such as upwellings of nutrient-rich water, have permanent locations, attracting numerous species of fish. These fishing spots are often treated as the private property of a single individual or select group. Even in the absence of governmental recognition of a property right, the privately held knowledge of an experienced fisherman can act to give exclusive control of the resource to an individual. As one observer commented, "As much as we may yearn for the traditional mode of fishing, the development of technology, in the presence of finite stocks, is driving us in the direction of privatization, away from a common property resource."[108]

Indeed, the lobster industry in America (at least along the coast of Maine) has developed a fairly effective, if informal, system of private property rights management.[109] The particular circumstances of lobstering have created the incentives for property rights–style management, and total catch has been stable for decades. Crucial to success in lobster fisheries is an intimate knowledge of the unique conditions of the local area and the response of the lobsters to these conditions. Almost all lobstering is conducted from small boats, often by a single fisherman. Hundreds of wooden traps, or "pots," must be strategically placed within a fisherman's area to ensure an adequate catch. Custom dictates that these areas are the exclusive property of individuals, with self-enforcement by the owner the common response to interlopers or trap thiefs.[110] In a study of the lobster and herring fisheries off Matinicus Island, Maine, Francis and Margaret Bowles suggest several factors that must be present before a property rights approach will be adopted in a given fishery. The abundance of the target species should be sufficient to merit specific effort by the fisherman, and its distribution and habits should be predictable, thus rewarding the local fishermen who have "invested" in superior knowledge. In addition, the fishing sites themselves should be close enough to enable particular individuals (or a discrete group) to gain a competitive edge in time and cost. The authors conclude that "as the distribution of the species becomes less predictable, territories will become less beneficial and more costly to maintain."[111]

Federal Policies

As the domestic fishing industry attracted more fishermen and more capable technologies, total catch continued to rise through the 1950s and into the 1960s. Unfortunately, such levels could not be maintained indefinitely. The total catch statistics disguised a profound shift within most fisheries. As the initially abundant resource was depleted, fishers shifted their effort toward other, often less commercially desirable species in a process called "serial depletion."[112]

One of the most important federal responses to the conflicts over marine resource management was the Fishery Conservation and Management Act of 1976, now usually referred to as the Magnuson Act.[113] The passage of this legislation was encouraged by domestic American fishing interests' seeking to restrict the access of foreign fleets to U.S. coastal waters and to increase the authority of the federal government to regulate the U.S. industry. Under the Magnuson Act, the federal government is moving to gather more reliable data on fishery stock conditions, the effects of fishing on target and other species, and restricting harvest levels to the calculated optimum sustainable yield. However, according to the National Fish and Wildlife Foundation, out of eighty-one marine fishery stocks examined, fourteen are still considered to be "overexploited," and another thirty-six were being fished at the maximum capacity.[114]

Whales

The problems of resource management can be greatly magnified when the target species is slow to mature, produces few offspring, or migrates long distances (thereby crossing many international boundaries). The best-known example that fits this description is not a fish at all: the whale.

> Whales provide a classic example of the wanton waste generated by a common property resource system. Part of the "common heritage of mankind" and owned by no individual, whales have been rapidly exploited, to the near extinction of some species. Under the system of non ownership, even the most conservation-minded whaler is unlikely to stop overharvesting; if he didn't take the whales, someone else would.[115]

For years, it has been widely assumed that the only way to prevent the extinction of cetaceans was to ban their hunting or harvesting. The data would seem to indicate that the largest whales have been hunted to excess, while many species of smaller whales and porpoises are sufficiently numerous to permit some harvesting. However, despite recent improvements, significant bycatch of whales continues. The UN Food and Agriculture Organization has estimated that the global bycatch for all cetaceans is between 300,000 and 1 million.[116]

Beginning in 1985, the International Whaling Commission (IWC), representing thirty-seven countries, established a five-year moratorium on commercial whaling for large species, but its police powers were fairly weak, and for many years there was a thriving pirate whaling industry.[117] The continued improve-

> **According to the National Fish and Wildlife Foundation, out of eighty-one marine fishery stocks examined, fourteen are still considered to be "overexploited," and another thirty-six were being fished at the maximum capacity.**

ment of alternatives to whale oil and the decline of demand for whale products, coupled with active opposition to the pirates by private and public groups, has reduced piracy as a threat to larger species of whales. Despite the problems with the killing ban, the IWC met in Reykjavik, Iceland, for its annual meeting in May 1991, and the moratorium on whaling was extended.[118] The moratorium was once again extended following the 1993 meeting. In response, Norway unilaterally renewed its small-scale harvest of minke whales, taking a total of 157 the first season.[119] Because the estimated population of minke whales worldwide approaches 1 million, Norway has always objected to, and therefore been exempted from, enforcement of the IWC's designation of North Atlantic minke whale populations as "protected stocks."[120]

To protect against the threat of extinction, ownership arrangements could be extended to whales. Currently, whale watching by boat is a lucrative tourist trade, similar to the elephant-watching safaris in Kenya.[121] Specific watching areas could be leased to individual tour companies (or individuals), and efforts to protect the whales could be extended through commercial ventures.

Despite the obvious difficulties, whales could be owned privately in at least a limited fashion. Tracking, by satellite, sonar, or radio transmitters, is already feasible. Whales communicate with complex and distinctive sounds, and these sounds could be used to track whale pods or even individual whales. Unfortunately, the study of private property rights in whales primarily has been concerned with hunting and how to establish ownership over a wounded or dead whale.[122] The concept now needs to be extended to living whales with an eye toward preserving particular threatened or endangered species.

Other Methods of Protection Through Ownership

If establishing property rights in mobile creatures is theoretically possible, then immobile resources should be much easier to convert to private ownership. Reefs typically are home to a significant diversity of marine life. Thus, "artificial" reefs may offer the potential to expand stocks of reef fish.[123] Artificial underwater structures rapidly attract fish and other ocean species, much like their natural counterparts. This is particularly true in warmer waters. In the United States, artificial reefs have been used to improve fishing prospects for over a century.[124]

The artificial reef aspect of oil and gas drilling rigs along the outer continental shelf of the United States is an example of easily defined private areas in the middle of the coastal ocean. Private firms already harvest some of the mussels and other delicacies found on rigs, and sport fishermen find the best fishing is often adjacent to these artificial reefs. Exploring methods of capturing a portion of the wealth created by these reefs could create incentives to expand the use of artificial reefs. Japan is already spending huge sums on ocean habitat construction and research; the United States is only beginning to recognize the opportunities to any significant degree.[125]

In essence, human junk can be turned into reef structures along much of the world's coastline. For example, old automobile tires have been used to create underwater habitat along otherwise sandy, barren coastlines. Broward County, Florida, even passed a law requiring that used tires be turned over to the county for this purpose.[126] In the state of Texas, coastal areas appropriate for artificial reefs have been mapped,[127] and federal laws have been passed to provide revenues to the states to assist reef-building efforts.[128]

Private incentives to create underwater habitat are virtually nonexistent. Unless an individual stands to benefit personally from the creation of an artificial reef, the costs and regulatory hurdles may be overly discouraging. In the absence of defensible property rights to the fishing bounty created by the reef's structure, there are few benefits beyond altruistic satisfaction.[129] More action would be encouraged if private benefits could be collected by those who expend money and effort to improve the local fishery resources. In this manner, private property rights for marine resources could guarantee an increase in both biodiversity and total fish populations.

Unfortunately, natural coral reefs are under severe pressure from human activities in many coastal regions. For example, in 1989 the world harvest of corals exceeded 3.4 million pounds, most of which consisted of the stony, reef-building corals. In addition, natural stresses abound for coral-based communities, particularly storms and predation. Perhaps the most famous incidence of predation on corals occurred in the late 1970s when a population explosion of the coral-eating crown-of-thorns starfish resulted in significant losses of corals throughout the Indian and Pacific oceans. The eastern coast of Malaysia suffered 70 to 90 percent losses during this time. Because corals grow so slowly, decades must pass before the ecosystem fully recovers, if ever. Even when violent storms do not damage coral (or other reef) systems, extreme fluctuations in water temperature can devastate a reef. The most important reef-building corals can survive in only a narrow range of water temperatures of at least 68°F, largely confined to the equatorial region between 30 degrees south and 30 degrees north. During the 1980s, many researchers noted that high water temperatures (often associated with El Niño events) in the Pacific caused many corals to weaken and die. These shallow water corals, which normally rely on a symbiotic relationship with certain algaes, expelled the algae due to high water temperatures, eventually dying as a result.[130] This sequence of events is known as coral bleaching because the corals lose much of their color without the algae.

The largest single coral reef system in the world is the Great Barrier Reef system extending for 1,200 miles along the northeast coast of Australia. Most of this enormous system, approximately 135,000 square miles, is included within the Great Barrier Reef Marine Park. This is an innovative, multiple-use arrangement that establishes separate zones for numerous activities such as fishing, scientific research, tourism, and oceanic shipping.[131] This approach could well serve as the model for many other marine ecosystems around the globe.

Aquaculture and Mariculture

There are numerous examples of farming marine resources, from oysters in the Chesapeake Bay to Japanese pearl farms. Abalone, salmon, and shrimp are all successfully farmed.[132] Along with the creation of new reefs, which increase the size of the fishery resource, it is also possible to treat the oceans as a fertile farm. Cultivating the sea has a great potential to relieve the impact of overharvesting fisheries while simultaneously increasing the food supply. Whereas

But if demand continues to grow for fish products and if existing fisheries continue to decline, the fishing industry may have little choice but to become more of a fish farming industry.

reefs develop into complex, interdependent colonies of organisms, aquaculture normally seeks to concentrate a single (or a few) species in one spot, much as a terrestrial farmer does with chickens or cattle. Until recently, aquaculture was not favored by the traditional fishing industry, which saw it as a potential competitor. But if demand continues to grow for fish products and if existing fisheries continue to decline, the fishing industry may have little choice but to become more of a fish farming industry. Already much of the mariculture industry is being developed by the traditional fishing and processing industries. Production is significant: approximately 12 million tons (of all species) per year and growing (figures 9-2 and 9-3).[133]

The state of Texas has passed a "rigs-to-reefs" bill through the state legislature. Generally an international treaty requires that an abandoned production rig be moved within a matter of months. Additionally, a major oil company will soon provide research and development money to establish a pilot open-sea "net

FIGURE 9-2   World Aquaculture Production, 1984–1991

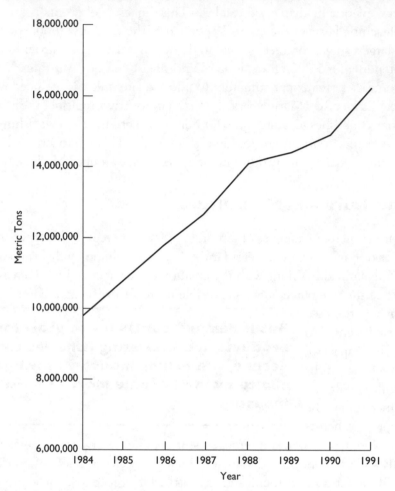

Source: Ibid.

pen" aquaculture project.[134] This project will initially focus on redfish (or red drum), which cannot be caught commercially in Texas because of overfishing, and eventually on other species as well. Texas domesticated the redfish over ten years ago and now has a small fish release program for several bays.

Currently aquaculture in America is dominated by freshwater catfish farming. Catfish alone accounts for 45 percent of the total output by aquaculture in the United States. Dozens of other species are being studied in order to develop sufficient knowledge to be able to raise them in ponds and net pens. In other parts of the world, tilapia and shrimp are common farm products.

FIGURE 9-3   Regional Aquaculture Production, 1984–1991

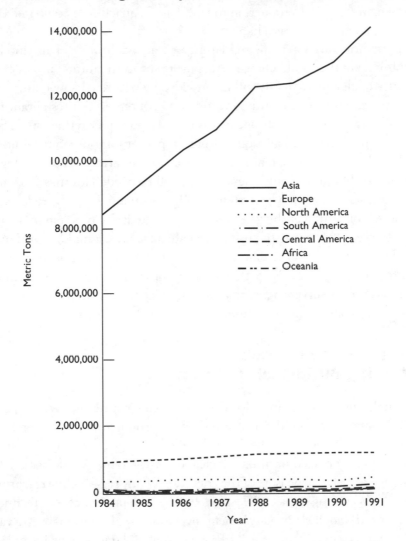

Source: Ibid.

Texas has even begun investigating the possibility of aquaculture in West Texas, far from the Gulf of Mexico. Although West Texas has only limited fresh-water supplies, it has abundant sources of saline groundwater. It is estimated that over 300,000 acres of arid lands in West Texas are underlain by saline water with the potential for aquaculture applications. Such inland farms would avoid the risks associated with hurricanes and accidental releases of nonnative marine species. Research into raising redfish and shrimp in these waters holds out the potential for utilizing saline groundwater for food production around the world. Shifting marine species production to land-based aquaculture could reduce the stress on ocean resources and habitat.

Laws traditionally have favored land-based aquaculture over marine-based mariculture, but this could change. At some point in the future, it may be that a majority of all aquatic foods will be provided by advanced aquaculture, leaving the commons to the sport and recreation economies. All restaurant trout, almost all catfish and crawfish, and nearly half of the oysters in the United States already are produced through aquaculture. In poorer nations, aquaculture may hold the key to preventing malnutrition or even starvation. As technology develops and techniques improve, less acreage will be needed for these aquafarms, and the impact on existing ecosystems (such as wetlands) may be reduced.

Another concern over the growth of aquaculture is the problem of the waste produced by concentrating the fish in a small area. For example, some have estimated that a twenty-acre salmon farm can produce as much organic waste as a town of 10,000 people. Nonetheless, these problems must be overcome because almost all observers agree that any significant increase in future marine harvests must come from fish farming.[135]

Pulling It All Together: Protecting Biological Diversity

Traditionally, most studies have focused on a single marine species. Yet perhaps the most important consideration from an environmental (and therefore an economic) point of view is the health of the entire marine ecosystem. Today, more emphasis is being placed on the biological diversity found within large marine ecosystems (LME) as a measure of the condition of any region's living marine resources.[136] According to a 1987 study by the federal Office of Technology Assessment, "Biological diversity benefits everyone, is valued by many in a variety of ways, and is owned by no one."[137] Protection of biological diversity, or simply biodiversity, is therefore an important concern in marine resource management.[138]

Little is known about how one species decline affects other species. The Consultative Group on Biological Diversity organized a workshop in Monterey, California, to address this topic.[139] According to one report from the workshop, "access to marine resources has not been restricted, and this 'open access' has aggravated the 'tragedy of the commons' that has so often depleted natural resources and reduced the diversity of life on earth." The conference went on to call for an examination of human behavior toward the marine environment; "indeed, fisheries management has more to do with managing people than it does with managing fish." One of the recurrent recommendations of the Monterey participants was to improve the economic valuation of marine resources. It was noted that a "review of current economic literature/analysis bearing on topics discussed in this Workshop would show a slim bibliography and a thin roster of economists who are focusing their research on marine resource choices."[140]

At an October 1990 international conference on large marine ecosystems held in Monaco, approximately 140 participants from diverse disciplines discussed research and management approaches to the LME concept. Although the advantages of considering entire ecosystems are apparent, monitoring and data gathering applicable to this purpose have been inadequate. It was

> **Traditionally, most studies have focused on a single marine species. Yet perhaps the most important consideration from an environmental (and therefore an economic) point of view is the health of the entire marine ecosystem.**

generally agreed that further international efforts were necessary to develop the LME into a useful tool for regional management of living marine resources.[141]

In this regard, an option that must be considered is the creation of marine refugia.[142] Because our knowledge of marine resources is yet insufficient to inform our actions fully, a cautious approach may be in order. Marine reserves have been established along the U.S. coastline, primarily where particularly significant concentrations of species occur, such as Monterey Bay, California, and the Florida Keys. However, many commercial fisheries would benefit from the establishment of reserves where all fishing was prohibited. Ideally, these should be established around known breeding grounds and other habitats vital to the perpetuation and proliferation of target species. As one expert observed, marine refugia "that are carefully planned and grounded in good scientific understanding of ecosystem dynamics can be an effective tool to complement other forms of fisheries regulation."[143]

Conclusion

The development of property rights is an evolutionary process.[144] No single individual can have sufficient knowledge to anticipate all contingencies in a field as complex as marine fisheries.[145]

There is no theoretical limit to the ways in which title to a given volume of ocean may be divided. For example, transportation on (or under) the sea may be left open to all. The sea floor may be separated into plots much like terres-

trial farms. Any particular range of depths may be selected for private ownership. The only true limits are the extent of technology and human ingenuity.

The oceans, along with the atmosphere, still present the most difficult (con-
ceptually and physically) problems for a property rights–based analysis. Clear
and defensible private ownership has been either technologically impossible or
prohibitively expensive, an impasse that has led many observers to conclude
that property rights approaches are inappropriate for global environmental issues—especially the so-called pure property rights approach of private, individual ownership of portions of any global resource.[146]

No one expects universal private property rights to marine resources to be
embraced overnight. But there are several avenues along the road to private ownership that hold promise for improving our treatment of marine fisheries. The
examples of individual transferable quotas, the Japanese coastal fishing culture,
the Pride of Derby type of private protection mechanisms, and others point to
the value of incorporating property rights approaches in marine fisheries policies.

At least as far back as 1969, it was said that "fisheries in the United States
are beset by senseless restrictions and marked by obsolescence, waste, and
poverty. . . . And there is little hope for change—unless dramatically new institutions and new forms of management can be developed and adopted."[147] If the
current trends are any indication, new institutions and forms of management
are necessary to maintain the vitality of living marine resources.

The development of oceanic property rights must be an evolutionary process.
There will not be an overnight shift to a new management regime. Nonetheless, an attempt should be made to improve the regulatory program through
greater reliance on individual decision makers. A property rights approach can
empower individuals to protect and to utilize fishery resources simultaneously
and thereby allow sustainable resource utilization into the future. It is time to
determine who should own the ocean.

NOTES

1. Eugene H. Buck, "Marine Fisheries Issues," Congressional Research Service Issue Brief (June 21, 1994), p. CRS-2.

2. Francis T. Christy, Jr., "Marine Fisheries and the Law of the Sea: A Decade of Change," FAO Fisheries Circular 853 (Rome: FAO, 1993), p. 4.

3. World Commission on Environment and Development, *Our Common Future* (New York: Oxford University Press, 1987).

4. Gary E. Davis and Jenifer E. Dugan, "Applications of Marine Refugia or Replenishment Zones to Fisheries Management," in Karyn L. Gimbel, ed., *Limiting Access to Marine Fish-*

eries: Keeping the Focus on Conservation (Washington, D.C.: Center for Marine Conservation and World Wildlife Fund, 1994).

5. R. S. K. Barnes, *An Introduction to Marine Ecology*, 2d ed. (Boston: Blackwell Scientific, 1988), p. 8.

6. Peter Weber, *Net Loss: Fish, Jobs, and the Marine Environment*, Worldwatch Paper 120 (July 1994), p. 10.

7. John P. Wise, "The Future of Food from the Sea," in Julian Simon and Herman Kahn, eds., *The Resourceful Earth* (New York: Basil Blackwell 1984), p. 119.

8. Boyce Thorne-Miller and John Catena, *The Living Ocean: Understanding and Protecting Marine Biodiversity* (Washington, D.C.: Island Press, 1991), pp. 14–20.

9. William G. Gordon, "Fisheries," in R. Neil Sampson and Dwight Hair, eds., *Natural Resources for the 21st Century* (Washington, D.C.: Island Press 1990), pp. 222–230.

10. Ronald Mitchell, "Intentional Oil Pollution of the Oceans," in Peter M. Haas et al., eds., *Institutions for the Earth: Sources of Effective International Environmental Protection* (Cambridge, Mass.: MIT Press, 1993), pp. 183–247.

11. "Intergovernmental Meeting of Experts on Land-Based Sources of Marine Pollution," Halifax, Canada, May 6–10, 1991, pp. 1–7.

12. Tom Horton, "Chesapeake Bay: Hanging in the Balance," *National Geographic* (June 1993): 2–35.

13. Barnes, *Introduction to Marine Ecology*, pp. 307–316.

14. Eugene H. Buck, "Waste from Fish Harvesting and Processing: Growing Environmental Concerns," Congressional Research Service Report for Congress (December 9, 1990), p. CRS-7.

15. Rachel L. Carson, *The Sea Around Us* (New York: Oxford University Press, 1951), p. 20.

16. Donald M. Anderson, "Red Tides," *Scientific American* (August 1994): 62–68.

17. J. M. Burkholder, E. J. Noga, C. H. Hobbs, and H. B. Glasgow, *Nature*, July 30, 1992, pp. 407–410.

18. Wesley Marx, *The Frail Ocean* (New York: Ballantine Books, 1967), pp. 9–23.

19. U.S. Department of Commerce and U.S. Environmental Protection Agency, *Coastal Nonpoint Pollution Control Program* (Washington, D.C.: Government Printing Office, October 1991), p. 1.

20. John Lancaster, "Weighing the Gain in Oil-Spill Cures," *Washington Post*, April 22, 1991, p. A-3.

21. James E. Mielke, "Oil in the Ocean: The Short- and Long-term Impacts of a Spill," CRS Report 90-356 SPR (Washington, D.C.: Library of Congress, July 24, 1990).

22. Donald R. Leal and Michael D. Copeland, "A Second Look at Oil Spills," *PERC Viewpoints*, no. 10 (April 1991).

23. U.S. Interagency Subgroup on Marine Pollution, "Marine Pollution: Policy Options for the 1992 Conference on Environment and Development" (March 18, 1991).

24. Brian R. Copeland, "Strategic Enhancement and Destruction of Fisheries and the Environment in the Presence of International Externalities," *Journal of Environmental Economics and Management* 19 (1990): 212–226.

25. Washington State recognizes private property rights down to the low-tide mark, whereas most other states traditionally limit private ownership to the high-tide line. In certain areas, several hundred feet of tidal mud flats can end up in private hands.

26. Personal conversation with Charles Roe, attorney.

27. See, for example, Bell et al. v. Inhabitants of the Town of Wells, Maine Superior Court, 1987, Civil Action Docket No. CV-84-125.

28. Jerome W. Milliman, "Can Water Pollution Policy Be Efficient?" (and following comments), *Cato Journal* (Spring 1982): 165–204.

29. Quote from Bryan Williams, "The A.C.A. and the Common Law" (speech to the Anglers' Co-operative Association Water Protection Officers' seminar, National Water Sports Center, Holme Pierrepont, Nottingham, n.d.).

30. In the case of isolated lakes or ponds, the fish and the water are the property of the landowner, but the right to be free of the pollution of others still applies.

31. Karen Wigan, "Shifting Control of Japan's Coastal Waters," in John Cordell, ed., *A Sea of Small Boats* (Cambridge, Mass.: Cultural Survival, 1989), pp. 388–410.

32. Hannes H. Gissurarson, "Evolution of Private Property Rights: The Story of the Icelandic Fisheries" (unpublished paper, Faculty of Social Science, University of Iceland, n.d.).

33. Elinor Ostrom, *Governing the Commons: The Evolution of Institutions for Collective Action* (New York: Cambridge University Press, 1990).

34. Elliott A. Norse, ed., *Global Marine Biological Diversity: A Strategy for Building Conservation into Decision Making* (Washington, D.C.: Island Press, 1993), pp. 106–117.

35. Jerome A. Jackson, "Terns on Tar Beach," *Natural History* (July 1994): 47–52.

36. Norse, *Global Marine Biological Diversity*, pp. 114–115.

37. Seyom Brown et al., *Regimes for the Ocean, Outer Space, and Weather* (Washington, D.C.: Brookings Institution, 1977), p. 104.

38. Gimbel, ed., *Limiting Access*.

39. Christopher B. Daly, "Fishermen Beached As Harvest Dries Up: Rules Imposed to Let Atlantic Restock Itself," *Washington Post*, March 31, 1994, p. A-3.

40. "Salmon Fishing Banned Along Washington Coast," *Washington Post*, April 10, 1994, p. A-4.

41. "The Tragedy of the Oceans," *Economist*, March 19, 1994, pp. 21–24.

42. Christy, "Marine Fisheries," pp. 12–13, 20.

43. Paul Schneider, "Breaking Georges Bank," *Audubon* (July–August 1993): 84–91.

44. Anne Rudloe and Jack Rudloe, "Sea Turtles: In a Race for Survival," *National Geographic* (February 1994): 96–120.

45. Eugene H. Buck, "Marine Fisheries Issues," CRS Issue Brief (June 21, 1994), pp. CRS-1, CRS-4.

46. National Research Council, *Decline of the Sea Turtles: Causes and Prevention* (Washington, D.C.: National Academy Press, 1990), p. 4.

47. Mark Fisher, "Conservation through Commercialisation," *Economic Affairs* 14, no. 3 (April 1994): 15–17.

48. Sam Kazman, "Petition to the Fish and Wildlife Service, et al., for a Trade Exemption for Maricultured Green Sea Turtle Products" (on behalf of the Association for Rational Environmental Alternatives and the Pacific Legal Foundation), January 22, 1982.

49. The original proposal, Sea Turtle Conservation Assistance Act of 1989 (House Resolution 3842, introduced November 21, 1989), was essentially rolled into the Trawlers Relief and Working Livelihood Act of 1991 (House Resolution 1478, introduced March 19, 1991).

50. Instituto Nacional de la Pesca, "XXV Years of Research, Conservation and Protection of Marine Turtles" (May 1990).

51. Secretaria de Pesca, "National Program for the Research, Conservation and Protection of Marine Turtles" (January 1990), p. 7.

52. Rudloe and Rudloe, "Sea Turtles," pp. 101–106.

53. Gary Bevin et al., "Economic Review of the New Zealand Fishing Industry 1988–89" (New Zealand Fishing Industry Board, April 1990). See also Ian N. Clark, Philip J. Major, and Nina Mollett, "Development and Implementation of New Zealand's ITQ Management System," *Marine Resources Economics* 5 (1988): 325–349, and Christopher M. Dewees, "Assessment of the Implementation of Individual Transferable Quotas in New Zealand's Inshore Fishery," *North American Journal of Fisheries Management* 9 (1989): 131–139.

54. For most regulated fisheries, the appropriate government agency establishes a total allowable catch, or annual quota. This concept is generally based on a calculation of the maximum sustainable yield of the fishery.

55. Ian Clark et al., "Development and Implementation of New Zealand's ITQ Management System," *Marine Resources Economics* 5 (1988): 329.

56. B. D. Shallard, "New Zealand Management Profile" (paper presented at the Australian and New Zealand Southern Trawl Fisheries Conference, Wellington, New Zealand, n.d.).

57. Grant Monk and Grant Hewison, "A Brief Criticism of the New Zealand Quota Management System," in Gimbel, ed., *Limiting Access*, pp. 107–119.

58. "A Case Study—Establishing Property Rights to Chatham Islands' Abalone (Paua)," in Alan Moran et al., eds., *Markets, Resources and the Environment* (Melbourne: Tasman Institute, 1991), pp. 188–206.

59. Christopher B. Daly, "New England Fishermen Crying Foul," *Washington Post*, December 14, 1994, p. A3; Timothy Egan, "Hook, Line and Sunk," *New York Times Magazine*, December 11, 1994, p. 77.

60. See in Gimbel, *Limiting Access*: John R. Gauvin, "South Atlantic Wreckfish Fishery: A Preliminary Evaluation of the Conservation Effects of a Working ITQ System," pp. 169–183; Bonnie J. McCay, "ITQ Case Study: Atlantic Surf Clam and Ocean Quahog Fishery," pp. 75–93; and Bruce R. Turris, "Canada's Pacific Halibut Fishery: A Case Study of an Individual Quota Fishery," pp. 132–151.

61. Gauvin, "South Atlantic Wreckfish Fishery," p. 181.

62. Rodney M. Fujita and D. Douglas Hopkins, "Individual Transferable Quotas for Fish Harvest Privileges," Committee on Merchant Marine and Fisheries, U.S. House of Representatives, February 9, 1994, p. 6.

63. Al Ota and Don Hamilton, "Fish Ruling Aids Clinton Backer," *Seattle Times*, April 17, 1993.

64. Francis T. Christy, Jr., "The Value of Fishing Privileges" (paper delivered at the Conference on Problems and Strategies in the Scientific Management of Fisheries and Marine Mammals, April 12, 1994).

65. Monk and Hewison, "A Brief Criticism," pp. 111, 114.

66. Turris, "Canada's Pacific Halibut Fishery," pp. 148, 149.

67. Anthony Scott, "Development of Property in the Fishery," *Marine Resource Economics* 5 (1988): 289–311.

68. Robert Higgs, "Legally Induced Technical Regress in the Washington Salmon Fishery," *Research in Economic History* 7 (1982): 55–86.

69. National Coalition for Marine Conservation, *Currents* (Savannah, Ga., September 1991), p. 2.

70. Eugene H. Buck, "Limited Access for Commercial Fishing: Individual State Programs," Congressional Research Service Issue Brief (May 15, 1986).

71. Bill Shapiro, "The Most Dangerous Job in America," *Fortune*, May 31, 1993.

72. "The Tragedy of the Oceans," *Economist*, March 19, 1994, p. 22.

73. Eugene H. Buck, "Fishery Resource Allocation Conflicts," Congressional Research Service Issue Brief 87-182 ENR (March 3, 1987).

74. For example, by restricting the duration of the fishing season while leaving access open to all, regulations may induce fishermen to invest in faster boats and more expensive equipment that will give them a competitive edge.

75. Alaska's halibut fishery is scheduled to convert to an ITQ regime in 1995.

76. Hal Bernton, "Halibut Fleet Caught Up in Short Fishing Frenzy: Alaska May Expand Day-long 'Seasons,'" *Washington Post*, May 18, 1991, p. A-2.

77. Carl Safina, *A Primer on Conserving Marine Resources* (New York: National Audubon Society, 1991), p. 4.

78. Gimbel, ed., *Limiting Access*.

79. Francis T. Christy, Jr., "Property Rights in the World Ocean," *Natural Resources Journal* 15 (October 1975): 695–712.

80. S. V. Ciriancy-Wantrup and Richard C. Bishop, "Common Property as a Concept in Natural Resources Policy," *Natural Resources Journal* 15 (October 1975): 713–727.

81. Ostrom, *Governing the Commons*, p. 23.

82. Steven F. Edwards, Allen J. Bejda, and R. Anne Richards, "Sole Ownership of Living Marine Resources," U.S. Department of Commerce, NOAA Technical Memorandum NMFS-F/NEC-99 (May 1993), pp. vii, 8, 18.

83. Milton Friedman, "Private Property: Is the Market Moral?" *National Review* (November 1990): 56.

84. Arguments have been made in favor of the private provision of this infrastructure, including courts and enforcement entities, but that is beyond the scope of this chapter.

85. Terry L. Anderson and Peter J. Hill, "Constitutional Constraints, Entrepreneurship, and the Evolution of Property Rights," in James Gwartney and Richard Wagner, eds., *Public Choice and Constitutional Economics* (Washington, D.C.: Cato Institute, 1988), pp. 207–227.

86. Steven N. S. Cheung, "The Structure of a Contract and the Theory of a Non-exclusive Resource," *Journal of Law and Economics* 13, no. 1 (April 1970): 49–70.

87. Bonnie J. McCay and James M. Acheson, eds., *The Question of the Commons* (Tucson: University of Arizona Press, 1987).

88. Cordell, *A Sea of Small Boats*, pp. 333–410.

89. Kenneth Ruddle and Tomoya Akimichi, "Sea Tenure in Japan and the Southwestern Ryukyus," in ibid., pp. 337–370.

90. As with the fishing rights of England, the Japanese property right is to the fishery, not the ocean itself.

91. Ruddle and Akimichi, "Sea Tenure," pp. 345–349.

92. Ibid., p. 364.

93. Karen Wigen, "Shifting Control of Japan's Coastal Waters," in Cordell, *A Sea of Small Boats*, p. 392.

94. Marx, *Frail Ocean*, pp. 181–191.

95. Hugo Grotius, *The Freedom of the Seas*, trans. Ralph van Deman Magoffin from the Latin text published in 1633 (New York: Oxford University Press, 1916).

96. The value of these 200-mile claims is hard to overstate. Consider that Japan has been building up a tiny speck of sinking land in the western Pacific in order to maintain its claim over thousands of square kilometers of open sea.

97. Ivar E. Strand, Jr., et al., "Trade, Institutions, and Preference for Living Marine Resources," *American Journal of Agricultural Economics* (December 1992): 1150–1154.

98. Buck, "Marine Fisheries Issues," p. CRS-5.

99. Dennis W. Nixon, "Managing Marine Fisheries Through Individual Transferable Quotas," in Gimbel, ed., *Limiting Access*, p. 69.

100. To an extent, this duty to maximize benefits is reflected in the major piece of U.S. fisheries legislation, the Magnuson Fishery Conservation and Management Act.

101. Hal Bernton, "Trawling Free-for-All Threatens Alaska Catch," *Washington Post*, May 26, 1990, p. A-11.

102. John Seabrook, "Death of a Giant: Stalking the Disappearing Bluefin Tuna," *Harper's* (June 1994): 48–56.

103. Ragnar Arnason, "Minimum Information Management in Fisheries," *Canadian Journal of Economics* (August 1990): 630–653.

104. Very few private mineral owners have adequate personal knowledge to be able to exploit their resource successfully. They contract with others who do have such knowledge. The same is true of almost every item of property in modern commerce. Markets make possible the efficient exchange of information, materials, and capital, thus permitting individuals to improve their living standards without knowing, for example, how to build an automobile or catch a tuna.

105. Milton Friedman and Rose Friedman, *Free to Choose* (New York: Harcourt Brace Jovanovich, 1980), pp. 9–37.

106. Terry L. Anderson and Donald R. Leal, *Free Market Environmentalism* (San Francisco: Pacific Research Institute for Public Policy, 1991), pp. 121–134.

107. John Cordell, "Sea Tenure," in his *A Sea of Small Boats*, pp. 1–21.

108. S. Fred Singer, "Fisheries Management: Another Option," in Jon G. Sutinen and Lynne Carter Hanson, eds., *Rethinking Fisheries Management*, Proceedings from the Tenth Annual Conference, Center for Ocean Management Studies, University of Rhode Island, June 1–4, 1986, p. 203.

109. See, for example, Francis P. Bowles and Margaret C. Bowles, "Holding the Line: Property Rights in the Lobster and Herring Fisheries of Matinicus Island, Maine" in Cordell, *A Sea of Small Boats*, pp. 228–257.

110. See, for example, James M. Acheson, "The Lobster Fiefs Revisited" in McCay and Acheson, *The Question of the Commons*, pp. 37–65.

111. Bowles and Bowles, "Holding the Line," p. 254.

112. Davis and Dugan, "Applications," p. 227.

113. Fishery Conservation and Management Act of 1976. Public Law 94-265, approved April 13, 1976, last reauthorized in 1990. It was later renamed in honor of the late Senator Warren G. Magnuson.

114. "Living Marine Resources: Background Paper for the U.S. Response to the 1991 Geneva PrepCom" (Preparatory Committee Working Group II—Oceans and Coastal Areas, United Nations Conference on Environment and Development, 1991), p. 6.

115. Robert J. Smith, "Private Solutions to Conservation Problems," in Tyler Cowen, ed., *The Theory of Market Failure: A Critical Examination* (Fairfax, Va.: George Mason University Press, 1988), p. 341.

116. Almost all of these are smaller species not subject to International Whaling Commission jurisdiction.

117. The IWC still permits whaling for research purposes. Not surprisingly, some nations conduct a significant amount of research on dead whales.

118. "Whaling Commission Continues Hunting Ban," *Washington Post*, June 1, 1991, p. A-24.

119. Peter Weber, "Safeguarding Oceans," in Lester Brown, ed., *State of the World* (New York: W. W. Norton, 1994), pp. 41–60, 55.

120. *Analyses of Proposals to Amend the CITES Appendices*, prepared by the IUCN (World Conservation Union) Species Survival Commission and the TRAFFIC Network for the Ninth Meeting of the Conference of the Parties to CITES (Convention on Trade in Endangered Species) (Gland, Switzerland: IUCN, August 1994), p. 64.

121. For example, the Whale Adoption Project of Cape Cod, Massachusetts, provides a newsletter and photographs identifying individual whales for its members. Allied Whale of the College of the Atlantic in Bar Harbor, Maine, has cataloged more than 6000 individual whales.

122. Robert C. Ellickson, "A Hypothesis of Wealth-Maximizing Norms: Evidence from the Whaling Industry," *Journal of Law, Economics, and Organization* 5, no. 1 (Spring 1989): 83–97.

123. R. B. Stone, H. L. Pratt, R. O. Parker, Jr., and G. E. Davis, "A Comparison of Fish Populations on an Artificial and Natural Reef in the Florida Keys," *Marine Fisheries Review* (September 1979): 1–11.

124. Frank M. D'Itri, ed., *Artificial Reefs: Marine and Freshwater Applications* (Chelsea, Mich.: Lewis Publishers, 1985), p. 3.

125. See, for example, Richard B. Stone, comp., *National Artificial Reef Plan*, NOAA Technical Memorandum NMFS OF-6 (Washington, D.C.: National Marine Fisheries Service, November 1985).

126. Cowen, ed., *The Theory of Market Failure*, p. 350.

127. See, for example, C. D. Stephan et al., *Texas Artificial Reef Fishery Management Plan*, Fishery Management Plan Series 3 (Austin: Texas Parks and Wildlife Department, n.d.).

128. See, for example, Dingell-Johnson Federal Aid in Fish Restoration Act and the National Fishing Enhancement Act of 1984 (P.L. 98-623, Title II), which established the National Artificial Reef Plan.

129. "Artificial" reef is perhaps a misnomer. While the hard substrate is provided by humans rather than nature, the newly created biological niches are accepted by reef building organisms as readily as any natural outcrop. While research continues on this point, it is likely that most of the organisms that come to populate these reefs are not "stolen" from other reefs. The net result is an increase in marine biomass, since the new reef saves countless organisms from dying due to a lack of sufficient reef expanses.

130. Eugene H. Buck, "Corals and Coral Reef Protection," Congressional Research Service Report for Congress (September 9, 1991), pp. CRS-3, CRS-7, CRS-12, CRS-13.

131. Norse, *Global Marine Biological Diversity*, p. 271.

132. Smith, "Private Solutions," pp. 341–360.

133. "Tragedy of the Oceans," p. 21.

134. A net pen is constructed from large fishing nets and is used both to keep the aquaculture species in and potential predators out.

135. Hal Kane, "Growing Fish in Fields," *WorldWatch* (September–October 1993): 20–27.

136. Boyce Thorne-Miller and John Catena, *The Living Ocean: Understanding and Protecting Marine Biodiversity* (Covelo, Calif.: Island Press, 1991).

137. Office of Technology Assessment, "Technologies to Maintain Biological Diversity," Report PB87-207494 (1987).

138. Norse, *Global Marine Biological Diversity*, pp. 204–208.

139. The workshop was held November 14–15, 1990, at the Monterey Bay Aquarium. The purpose was to "identify and illuminate current and emerging marine environmental issues which must be engaged in the 1990s in order to achieve acceptable levels of 'sustainability' for living marine resources." Marine Conservation Workshop, "Report on Marine Environmental Issues" (New York: Consultative Group on Biological Diversity, March 28, 1991), p. 5.

140. Ibid., pp. 5, 9, 13.

141. *Conference Summary and Recommendations*, Large Marine Ecosystem (LME) Concept and Its Application to Regional Marine Resource Management, October 1–6, 1990 (Monaco: January 2, 1991).

142. William J. Ballantine, *Marine Reserves for New Zealand*, Leigh Laboratory Bulletin 25 (Auckland: University of Auckland, October 1991).

143. M. Tundi Agardy, "Closed Areas: A Tool to Complement Other Forms of Fisheries Management," in Gimbel, *Limiting Access*, pp. 197–204.

144. For an excellent overview, see Terry L. Anderson and P. J. Hill, "The Evolution of Property Rights: A Study of the American West," *Journal of Law and Economics* 18, no. 1 (April 1975): 163–179. See also Ross D. Eckert, *The Enclosure of Ocean Resources* (Stanford, Calif.: Hoover Institution Press, 1979).

145. See, for example, F. A. Hayek, *The Fatal Conceit: The Errors of Socialism*, in *The Collected Works of F. A. Hayek*, ed. W. W. Bartley III (Chicago: University of Chicago Press, 1988), 1:88.

146. Consider the scale of the global environment. It is not merely unlikely but impossible for an individual or even a quite large collection of corporations to control a significant proportion of the earth's environmental resources without the aid of government (international, in this example) actions.

147. James A. Crutchfield and Giulio Pontecorvo, *The Pacific Salmon Fisheries: A Study of Irrational Conservation* (Baltimore: Johns Hopkins Press, 1969), p. v.

Chapter 10

RICHER IS CLEANER

Long-Term Trends in Global Air Quality

Indur M. Goklany

HIGHLIGHTS

- *Richer is cleaner. Anything that retards economic growth also retards ultimate environmental cleanup.*

- *Traditional pollutants like smoke and particulate matter, sulfur dioxide, and fecal coliform bacteria begin to decline when per capita incomes reach $3,280, $3,670, and $1,375, respectively.*

- *Empirical evidence indicates that projections using the famous I = PAT equation, widely publicized by biologist Paul Ehrlich, are dead wrong. I represents the cumulative impact on the environment; P, the population; A, affluence; and T, technology. Greater economic and technological progress are necessary preconditions for the solution, rather than the causes, of environmental degradation. They make cleaner environments more affordable, reduce the impacts per unit of production, and help moderate population growth.*

- *In the United States, sulfur dioxide emissions per capita are down 60 percent from the peak in 1920. Both particulates and carbon monoxide emissions per capita have declined more or less continuously since World War II—inhalable particulates down 79 percent since 1940 and carbon monoxide down 53 percent since 1945. Volatile organic compounds emissions per capita (contributors to smog) stayed roughly constant between 1925 and 1970, before they declined; in 1992 they were down 36 percent from the peak. Nitrogen oxide per capita peaked in 1980. It is now down about 10 percent from its peak.*

- *Ambient sulfur dioxide and particulate matter have been declining for at least the last two decades in most developed countries.*

- *Ambient sulfur dioxide levels in many world megacities have already declined significantly from the 1970s. Cities in the developed world have relatively low levels. Cities in the developing world have the highest levels.*

- *Ambient particulate matter is still a problem in most megacities. Among a sample of twenty world megacities, the three having low levels were all located in the developed world (New York, London, and Tokyo).*

- *The gravest air quality problem in the world is indoor air pollution due to heating and cooking with coal, wood, and dung or other biofuels in developing countries. Indoor particulate levels in poor countries rival those of the worst acute industrial episodes of the past.*

Environmental Quality and Affluence

The Environmental Transition and Affluence

One of the more celebrated relationships in the environmental sciences is the equation publicized by biologist Paul Ehrlich: $I = PAT$, where I is the cumulative impact on the environment, P is the population, A is affluence, and T, denoting technology, is a measure of the environmental impact per person per level of affluence.[1] Under some formulations, affluence is replaced by consumption (C). It has been reasoned that with time, the resource base will be depleted and people will turn to lower-grade resources, which will mean greater impact per unit of production. Based on this equation, many environmentalists have concluded that population and/or affluence are the roots of environmental degradation and that sooner or later the human race will face disaster. According to their analyses, one or more of these factors—P, A, or T—will increase inexorably until the environmental impact becomes overwhelming. They conclude that human population growth and/or economic growth must stop. But in fact the actual history of the planet's various societies over the past 100 years reveals a completely different outcome: anything that retards economic growth generally also retards environmental cleanup and consigns millions to squalid and untimely deaths.[2]

Countries undergo an environmental transition as they become wealthier and reach a point at which they start getting cleaner. For example, ambient sulfur dioxide (SO_2) and particulate matter (PM) concentrations in the air and fecal coliform in river water drop significantly as a country becomes wealthier (figure 10-1). (SO_2 and PM can affect respiratory and cardio-

> **Anything that retards economic growth generally also retards environmental cleanup and consigns millions to squalid and untimely deaths.**

vascular systems; at high enough levels they could, at least acting together, cause premature deaths. PM included ash and soot from combustion in motor vehicles, boilers, and other heating equipment; dust from industrial operations such as mining and smelting; and chemically transformed products of SO_2 dioxide and nitrogen oxide (NO_x). Fecal coliform is a measure of fecal matter in water, generally due to untreated or poorly treated human or animal wastes. It is associated with waterborne diseases.) Consider figure 10-1 and the general shape of the SO_2 and PM curves, which go up and then come down (like a camel's back). The reason for the turning point is complex, but essentially the wealthier a nation is, the more it values and the more it can afford to pay for a healthier environment and environmental amenities.

FIGURE 10-1   Affluence and the Environment

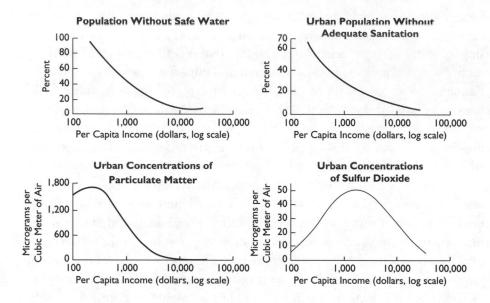

Source: World Bank, *World Development Report, 1992: Development and the Environment* (New York: Oxford University Press, 1992).

The level of affluence at which a pollutant level peaks (or environmental transition occurs) varies. A World Bank analysis concluded that urban PM and SO_2 concentrations peaked at per capita incomes of $3,280 and $3,670, respectively.[3] Fecal coliform in river water increased with affluence until income reached $1,375 per capita.[4]

Other environmental quality indicators (e.g., access to safe water and the availability of sanitation services) improve almost immediately as the level of affluence increases above subsistence. For these indicators the environmental transition is at, or close to, zero. In effect, the environmental transition has already occurred in most countries with respect to these environmental amenities because most people and governments are convinced of the public health benefits stemming from investments for safe water and sanitation. In fact, the vast majority of the 3 million to 5 million deaths each year due to poor sanitation and unsafe drinking water occur in the developing world.[5]

Other indicators apparently continue to increase, regardless of gross domestic product (GDP) per capita. Carbon dioxide and NO_x emissions and perhaps dissolved oxygen levels in rivers are in this third category. (Dissolved oxygen is a measure of the oxygen available in the water. It enables aquatic species to breathe. NO_x contributes to acid precipitation. At high enough levels, it can irritate the lungs. It is still controversial whether carbon dioxide [CO_2] should be considered a pollutant. See chapter 3.) On the surface, these indicators seem not to improve at higher levels of affluence, but their behavior is quite consistent with the notion of an environmental transition. The transition is delayed in these cases because decision makers have only recently realized the importance of these indicators, or the social and economic consequences of controlling them are inordinately high relative to the known benefits, or both.[6] Moreover, dissolved oxygen levels appear to have been improving since the 1970s for several affluent countries.[7] The delay in the environmental transition for this indicator is to be expected since it is related more to improving environmental, rather than public, health. Also, NO_x levels have begun to drop in the United States and in Japan.[8]

All the evidence indicates that, ultimately, richer is cleaner, and affluence and knowledge are the best antidotes to pollution.

Environmental Transition and Technology

The environmental transition for any pollutant depends on the state of technology. Technology refers to both "hardware" (e.g., scrubber, catalytic converter or sewage treatment plant) and "software" (e.g., knowledge of the effects or costs of pollution or its control, incentives or disincentives to control pollution, management practices, or operating rules).[9] Using technologies created by the United States and other developed nations, the newly industrializing countries (NICs) and newly developing countries (NDCs) should go through their environmental transitions at earlier stages in their development, that is, at lower levels of affluence relative to the currently developed nations. In fact, this explains why, with respect to safe water and sanitation services, there has been greater progress in NICs and NDCs compared to the developed nations at equivalent stages. Developed countries began improving their sanitation and drinking water in the late nineteenth and early twentieth centuries once the relationship between them and public health became clear. One would expect that over time, with wider knowledge and new technology, we should obtain better environmental quality at lower levels of affluence. And this indeed is the case for sulfur dioxide and sanitation (figure 10-2).[10]

FIGURE 10-2 Urban Population Without Adequate Sanitation

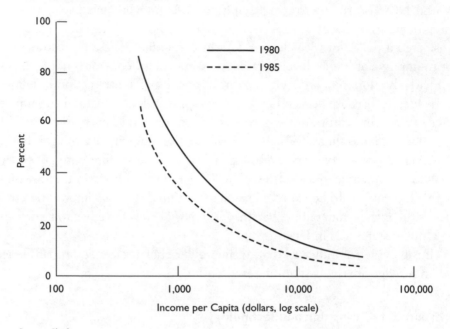

Source: Ibid.

Population, Affluence, and the Environment

The reasons that projections based on the $I = PAT$ equation break down in the real world are, first, that the population growth rate itself is a function of affluence.[11] As demographic studies have indicated, the more affluent a nation is, the greater is the likelihood that its population will grow at a lower rate. Second, technology too is a function of wealth. Wealthier populations can afford newer technology even if it costs more initially—as many indeed do. They can also invest more in research and development, which further stimulates new technology. More important, the legal and economic framework that leads to affluence, that is, free markets, and the respect and the protection of property rights (including intellectual property) stimulate innovation, including environmental innovation.[12] Technology and affluence make more efficient processes possible, whether it is for the production of energy, materials, food, or services. This means that the environmental impact for each unit of production goes down rather than up. Therefore, it is no surprise that processing facilities, such as steel mills and coal mines, are generally more efficient in developed countries and emit less effluent per unit of production. Agriculture offers a striking example of how technology has lowered environmental impacts. In the absence of any technological improvements such as pesticides, fertilizers, and other agricultural chemicals, the United States would need at least four times as much cropland as it used in 1910 to produce the same amount of food per capita as it does now. This would mean that all productive land in the United States would be under the plough, with disastrous consequences for biological diversity and forests.[13]

The fact that richer countries are also cleaner ones is often coincidental with political interventions for environmental purposes, but politics is not the chief reason that richer equals cleaner. It is no accident that rich countries have adopted more stringent clean air legislation: they are the ones that can afford such regulations. But as we shall see, the legislation is only part of the reason that the rich countries became clean.

The $I = PAT$ equation is misleading at best, and at worst, it suggests erroneous solutions. In particular, $I = PAT$ wrongly presumes that affluence is the problem rather than a necessary condition for solutions to environmental problems. The problem is not that there is too much affluence but that there is not enough affluence. *Our Common Future*, the report issued by the World Commission on Environment and Development, headed by Norway's prime minister Gro Brundtland, agrees that poverty hinders sustainable development.[14] It would be unconscionable to insist that communities and countries that have yet to go through the environmental transition for certain pollutants live with lower levels of affluence. Increasing the affluence of poorer countries will accelerate their environmental cleanup.

Air Pollution: A Historical Perspective

Air pollution is as old as the first fire. The first recorded complaint against air pollution in England came from Queen Eleanor while visiting Nottingham in 1257.[15] But like the weather, little was done about air pollution until this century, when air pollution in the growing cities of the industrialized world rose to dangerously high levels due to increasing amounts of coal and wood combustion by homes, businesses, and industries, and lower mortality and morbidity risks from infectious diseases meant that the public health risks of high levels of air pollution became relatively more important and more readily apparent.

TABLE 10-1   Air Pollution Levels During Historic Episodes

| Location | Date | Pollutant Concentrations ($\mu g/m^3$) (24-hr.) | |
		Sulfur Dioxide	Smoke-shade[a]
Meuse Valley, Belgium	December 1930	25,000[b]	12,500[b]
Donora, Pennsylvania	October 1948	1,800[c]	5,320[c]
London	December 1952	3,830	4,460
	January 1955	1,200	1,750
	January 1956	1,500	3,250
	December 1956	1,100	1,200
	December 1957	1,600	2,300
	January 1959	800	1,200
	December 1962	3,300	2,000
New York	November 1953	2,200	1,000
	November 1962	1,800	800(6.3)
	January 1963	1,300	800(6.3)
	February 1963	1,260	900(7.0)
	March 1964	1,730	520(4.8)

Sources: F. W. Lipfert, "Sulfur Oxides, Particulate, and Human Mortality: Synopsis of Statistical Correlations," *Journal of the Air Pollution Control Association* 30 (1980): 366–371; H. H. Schrenk et al., *Air Pollution in Donora, Pa.: Epidemiology of the Unusual Smog Episode of October 1948, Preliminary Report,* PHS Bulletin 306 (Washington, D.C.: PHS, 1949); R. E. Waller and B. T. Cummins, "Episodes of High Pollution in London 1952–1966," in *Proceedings: Part I. International Clean Air Congress, London, 4–7 October, 1966,* International Union of Air Pollution Prevention Association (London: National Society for Clean Air, 1966), pp. 228–231.
[a] The numbers in parentheses represent coefficient-of-haze (COH) values. COH is a unit of measurement for smoke-shade. The London data are British smoke measurements, which give generally lower values than corresponding U.S. figures.
[b] Estimated from calculations and assumed emission factors.
[c] Based on retrospective measurements.

Between 1920 and 1960, several air pollution episodes struck areas on both sides of the Atlantic (table 10-1). The 1930 Meuse Valley episode (in Belgium) increased mortality by 9.5 times the "background" level.[16] Twenty-four-hour SO_2 and PM concentrations were estimated at 25,000 micrograms per cubic meter ($\mu g/m^3$) and 12,500 $\mu g/m^3$ respectively.[17] By comparison, the twenty-four-hour

U.S. public health–related ambient air quality standards adopted in 1971 for SO_2 and PM were 365 $\mu g/m^3$, and 260 $\mu g/m^3$, respectively.[18] Based on these, the levels of SO_2 and PM both exceeded levels safe to breathe by about fifty or more times.

In October 1948 eighteen people died due to a four-day weather inversion in Donora, Pennsylvania, a town of about 14,000 people, 30 kilometers from Pittsburgh and the site of a steel plant, a wire plant, and a zinc plant. That death toll is 15 percent of all deaths we would expect today from all causes in a full year in a population of equivalent size. Retrospective measurements indicated that daily SO_2 and PM concentrations in Donora may have been at least 1,800 $\mu g/m^3$ and 5,320 $\mu g/m^3$, respectively.[19] In December 1952 one-day mean SO_2 and smoke concentrations in London reached 3,830 $\mu g/m^3$ and 4,460 $\mu g/m^3$, respectively.[20] Visibility was down to between 1 and 5 meters.[21] Four thousand "excess" deaths were attributed to that five-day episode out of a population of 8.5 million. For a population equal to that of the United States today, this would translate into about 100,000 excess deaths.

These episodes galvanized public support for air pollution control. The U.K. Clean Air Act was passed in 1956. In the United States, the initiative was taken by numerous local and state agencies. By 1956, there were eighty-two local air pollution control programs.[22] The first state program began in 1951, and by 1954, there were fourteen.[23]

State and local programs generally concentrated on controlling the density of smoke visible to the naked eye, open burning, incinerators in buildings and apartments, and rudimentary controls on industries. Some areas had sulfur-in-fuel regulations and lower limits on stack heights to ensure that concentrations at or near ground level, where most of the people spend much of their time, were not inordinately high.

But much of the cleanup of air pollution was voluntary. Cleaner energy sources, such as natural gas, oil, and electricity, were becoming increasingly available as substitutes for coal and wood in homes, businesses, and industries. New technologies that entered the marketplace substantially increased the efficiency of all types of combustion equipment, reducing the amount of soot produced and fuel burned for a given amount of usable energy. These included more efficient and cleaner furnaces and boilers for homes, businesses, industries, and power plants and diesel locomotives to replace coal-fired ones. Most important, increasing affluence made it possible for households and businesses to purchase these new technologies voluntarily. Urbanization accelerated the adoption of cleaner energy sources because higher population densities reduce access to

Pittsburgh has experienced some of the highest levels of outdoor air pollution ever recorded.

wood and other biofuels and ensure that distribution systems for natural gas and electricity are more cost-effective and economical. Thus, the number of residential natural gas consumers increased from 11 million in 1944 to 39 million in 1970. Bituminous coal consumption in railways dropped from 135.6 million tons in 1944 to about 1,000 tons in 1970, and in homes, from 117 million tons in 1940 to 12 million tons in 1972.[24]

Switching to oil and diesel also facilitated sulfur-in-fuel controls. It is easier to desulfurize these fuels and to enforce controls since authorities need only focus on a few fuel distributors, and even fewer refiners, rather than on millions of consumers. Moreover, desulfurization of natural gas is self-enforcing since any significant sulfur in the gas corrodes the distribution system.

As a result, SO_2 emissions from the residential sector declined from about 2.52 million tons in 1940, to 0.49 million tons in 1970, and 0.17 million tons in 1992.[25] Residential particulate emissions over this period declined from 2.24

million tons in 1940, to 0.39 million tons in 1970, and then increased slightly to 0.46 million tons in 1992. Residential carbon monoxide (CO) emissions declined from 14.8 million tons in 1940, to 3.6 million tons in 1970, and then rose slightly to 5.0 million tons in 1992.

The resulting gains in air quality might have been lost due to increased generation and use of electricity. Once again, technological progress came to the rescue. More efficient plants burned less coal per unit of electricity generated, and new technology enabled plants to be located farther from consumers and closer to coal mines. In addition, power plants installed tall stacks, a practice now frowned upon officially in the United States, to reduce ground-level concentrations in their immediate vicinity. Finally, pollution control technology improved dramatically for all types of processes. For example, older technologies reduced particulates in electric generating plants by 60 to 75 percent; now modern technologies (electrostatic precipitators and baghouses) cut particulates by as much as 99 percent.

In 1965, the federal government extended California's vehicle emission standards for hydrocarbons (HC) and carbon monoxide nationally, starting with the 1968 model year. The Clean Air Act of 1970 asserted federal responsibility for stationary sources whether or not they had an interstate impact. The Clean Air Act of 1970 set national ambient air quality standards (NAAQS) and deadlines to meet primary (health-related) NAAQS. The act also established (1) a series of technology-forcing stationary source pollution control requirements (new source performance standards and national emission standards for hazardous air pollutants), which are applied regardless of any need based on ambient air quality and, therefore, structurally divorced—to a large extent—from benefits; (2) the concept of prevention of significant deterioration (PSD); and (3) the requirement that states obtain approval from the federal government for their programs and policies to ensure compliance with all the previous requirements. It also established new motor vehicle emission standards (for HC, NO_x, and CO) and, of course, federal enforcement authority. The 1977 amendments codified and tightened the PSD requirements, which were not based on any balancing of costs and benefits; replaced the old set of rigid deadlines with a new set, because many areas had missed the previous ones; and established sanctions against states and areas that failed to meet those deadlines or did not have approved pollution control plans.[26] These regulations stepped up a trend, accelerated by the oil stocks, that was occurring on its own as part of a process in which engineers constantly seek cost-effective ways to improve efficiency. More efficient cars and factories are cleaner cars and factories.

Air pollution in Los Angeles in 1972. Since then air pollution has been dramatically reduced there, even though the population of the city has skyrocketed.

Cumulatively these technological, economic, social, and regulatory developments substantially improved U.S. air quality. The first improvements came from total suspended particulates (TSP), CO, and SO_2 emission reductions from ground-level sources in urban areas. The public health benefits of those reductions were high because urban areas have higher population densities and ground-level emissions magnify the local effects of air pollution. As a result, severe air pollution episodes and their very high mortality rates are now a part of history in the developed countries. For example, dustfall in Pittsburgh, measured at 166 tons per square mile each month in 1923–1924, had fallen by an order of magnitude by the mid-1970s.[27] TSP levels (annual arithmetic averages) declined from about 275 $\mu g/m^3$ in 1959 to 110 $\mu g/m^3$ in 1971. By 1993, the highest annual reading in Allegheny County, where Pittsburgh is located, had dropped to 81 $\mu g/m^3$.[28] Similarly, daily TSP levels in New York City, which were as high as 1,000 $\mu g/m^3$ during the November 1953 acute pollution episode (see table 10-1), declined to 280 $\mu g/m^3$ in 1972 and are now below 207 $\mu g/m^3$—a reduction of about 80 percent.[29]

With respect to SO_2, although national emissions kept on climbing until about 1970, the situation in urban areas improved significantly prior to that. SO_2 concentrations in New York City fell dramatically between 1963 and 1972 after peaking in 1964 (figure 10-3).[30] Annual levels dropped from about 570 $\mu g/m^3$ in 1964 to about 81 $\mu g/m^3$ in 1972, mainly due to fuel switching and sulfur-in-fuel requirements, and in 1993 they were below about 50 $\mu g/m^3$.[31] Daily SO_2 levels in New York City, once as high as 2,200 $\mu g/m^3$ during the November 1953 episode (see table 10-1), declined to 392 $\mu g/m^3$ in 1972 and are now below 168 $\mu g/m^3$—an overall reduction of more than 90 percent.[32]

Nor were the improvements in air quality restricted to the United States. Particulate (smoke) concentrations and emissions in London declined by about two-thirds between 1952 and 1965, largely due to the reduction in domestic coal consumption.[33] At that time, highest concentrations for PM and SO_2 were in high-density residential areas, exceeding those in industrial areas. Between 1952 and 1965, SO_2 concentrations declined about 10 percent due to a combination of decreased domestic and railway coal consumption and taller stacks at the new thermal power plants despite increased emissions of about 30 percent, mainly

FIGURE 10-3   New York, Ambient Sulfur Dioxide, 1963–1972

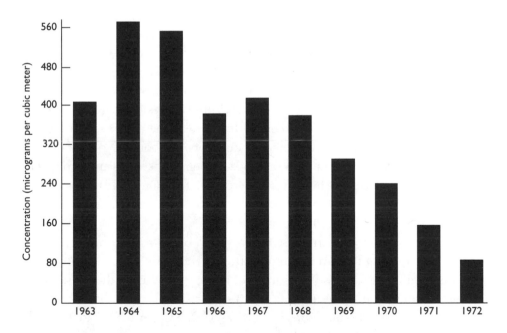

Source: H. Schimmel and T. J. Murawski, "SO_2—Harmful Pollutant or Air Quality Indicator," *Journal of the Air Pollution Control Association* 25 (1975): 739–40.

due to increased generation of electricity in thermal power plants. SO_2 emissions in the Greater London area were reduced by over 90 percent between 1962 and 1988.[34] Annual SO_2 concentrations in the City of London declined from about 300 $\mu g/m^3$ to about 40 $\mu g/m^3$ between the mid-1960s and mid-1980s.

While most of the developed world focused on reducing particulates (smoke) and SO_2 from stationary sources, beginning in the late 1940s, Los Angeles discovered its own brand of smog, a mixture of ozone and other oxidants, caused by vehicle emissions. One-hour ozone concentrations in excess of 1,200 $\mu g/m^3$ were reported and often exceeded 1,000 $\mu g/m^3$ in the 1960s.[35] In response, California promulgated vehicle emission standards for the 1967 automobile model year that were later adopted nationally by the federal government for the 1968 model year.

Human Exposure to Air Pollutants: The Role of Indoor Air Quality

Box 10-1 indicates the potential health impacts of traditional air pollutants: particulate matter, nitrogen dioxide, carbon monoxide, sulfur dioxide, and volatile organic compounds. To address these impacts, governments and international bodies such as the World Health Organization (WHO) define healthy air in terms of the air quality at a fixed point outdoors. However, indoor air quality, particularly in the home, is a far better indicator of the impact of air pollution on public health. First, virtually no one spends an entire day, let alone an entire year, rooted at the same spot outdoors. In fact, most people spend the vast majority of their time indoors, and generally at home. Studies of human activity patterns in the United States indicate that the average person spends about 93 percent of his or her time indoors, 5 percent in transit, and the remainder (2 percent) outdoors.[36] About 70 percent of the time is spent indoors at home. The average homemaker spends an even greater amount of time indoors at home (88.7 percent). The numbers are less stark in developing countries, but there too more time is spent indoors than out, and much of that time is in the home.

The quality of air is often worse indoors than outdoors. We have spent too much time studying the latter and too little on the former. Heating and cooking equipment that use fossil fuels and biofuels (e.g., wood and dung), smoking, solvents, and various cleaning solutions used or stored in the home all cause more problems than do outdoor smokestacks. CO, NO_x, and TSP are generally higher in homes employing natural gas than outdoors, while SO_2 and ozone are higher outdoors, by almost a factor of two to five.[37]

Not surprisingly, empirical studies of human exposure show that the concentration of pollutants in outdoor air contributes only a small amount to the total dosage received by human beings. One U.S. study showed virtually no correlation between CO levels in the blood, the physiological route by which CO affects people, and outdoor monitored levels; the latter explained less than 3 percent of variation in blood CO levels.[38] Similarly, outdoor concentrations of nitrogen dioxide (NO_2) are relatively poor predictors of total population exposure, while average indoor concentration explains 50 to 60 percent of total exposure.[39] Finally, calculations for the U.S. indicate that 1 gram of indoor PM emissions can have a greater effect on total exposure to the population than 1 kilogram (1,000 grams) released by a power plant from a relatively high stack.[40]

Indoors versus Outdoor Levels around the World

The most important air pollution problem worldwide is the indoor particulate levels in developing countries where people rely on coal and biofuels for home heating and cooking.[41] In some developing countries, in-home air quality rivals levels seen in the developed world during the acute pollution episodes in Donora or London. These data indicate:

- In developed countries, the average urban and rural inhabitant spends about 90 percent and 85 percent of his or her time indoors, respectively. The corresponding numbers for inhabitants of NICs and NDCs are 75 percent (urban) and 65 percent (rural) (table 10-2).

- Consistent with the notion that richer is cleaner, indoor air quality improves with affluence. Affluence and the Human Development Index (HDI), which takes into account quality-of-life factors such as longevity and literacy, in addition to per capita income, are crudely related. (See table 10-2.) This is because the less affluent populations cannot yet afford cleaner commercial energy sources such as natural gas and oil and instead have to rely on coal, wood, or other biofuels.

- For most groups of nations, indoor particulate concentrations are higher than those outdoors (see table 10-2). The apparent exception is for urban areas in developing nations with HDIs in the midrange. These countries are well on the way toward industrialization but have not yet instituted broad-scale controls or switched to cleaner fuels for industrial, commercial, or residential sectors.

BOX 10-1 Effects of Traditional Air Pollutants at Sufficiently High Levels

PARTICULATE MATTER (PM) AND SULFUR DIOXIDE (SO_2) EFFECTS

Health concerns include effects on breathing and respiratory system, aggravation of existing respiratory and cardiovascular disease, alterations in the body's defense systems against foreign materials, damage to lung tissue, carcinogenesis, and premature mortality. Past episodes (e.g., in Donora and London in the late 1940s and 1950s) of simultaneously high PM and SO_2 levels were responsible for a number of "excess" deaths. A legacy of the episodes is that even today, epidemiologists are unsure as to whether the excess mortality was due to PM, SO_2, or both. One reflection of this controversy is that while the U.S. primary (public health related) standards are specified separately for SO_2 and particulate matter, the World Health Organization's guidelines are specified in terms of the joint occurrence of SO_2 and PM levels in excess of specified amounts. For developing nations, whose resources are extremely limited and have not yet invested billions in air pollution control, this is a critically important question. The correct answer will help them reduce the public-health-related impacts of air pollution more cheaply and therefore more quickly. Sensitive subgroups include individuals with chronic obstructive pulmonary or cardiovascular disease, or influenza; asthmatics; the elderly; and children.

PM also causes soiling and unsightly smoke. SO_2 is also a precursor to acidic deposition (acid rain), which is associated with a number of effects, including acidification of lakes and streams, accelerated corrosion of buildings and monuments, and visibility impairment.

CARBON MONOXIDE (CO) EFFECTS

Carbon monoxide enters the bloodstream and reduces the delivery of oxygen to the body's organs and tissues. Sensitive individuals include those with cardiovascular disease, particularly angina or peripheral vascular disease. Healthy individuals also are affected but at higher levels. Exposure to elevated carbon monoxide levels is associated with impairment of visual perception, work capacity, manual dexterity, learning ability, and performance of complex tasks.

(continued)

OZONE (O_3) EFFECTS

Health concerns include damage to lung tissue, reduced lung function, and sensitivation of lungs to other irritants. Ambient levels of ozone not only affect people with impaired respiratory systems, such as asthmatics, but healthy adults and children as well. Exposure to ozone for six to seven hours at relatively low concentrations can reduce lung function in normal, healthy people during periods of moderate exercise. Repeated exposure in animal studies for months to years can damage the lungs. It can reduce agricultural crop yield and causes foliar damage in many crops and trees at current ambient levels.

NITROGEN DIOXIDE (NO_2) EFFECTS

Nitrogen dioxide can irritate the lungs and lower resistance to respiratory infection (such as influenza). The effects of short-term exposure are still unclear, but continued or frequent exposure to concentrations higher than those normally found in the ambient air may cause increased incidence of acute respiratory disease in children. Nitrogen oxides are precursors to both ozone and acidic deposition. In some western areas, it is an important precursor to particulate matter concentrations.

Source: U.S. Environmental Protection Agency (EPA), *National Air Quality and Emission Trends Report, 1992*, EPA 454/R-93-031, Office of Air Quality Planning and Standards (Research Triangle Park, N.C.: EPA, 1993).

Clearly total human exposure is dominated by indoor air concentrations (table 10-3). In developing countries, 75 to 92 percent of total human exposure is due to indoor concentrations. In developed countries, the corresponding figures are 90 to 93 percent.[42]

Trends in Human Exposure to Air Pollutants in the United States: Exposure in the Home

The good news is that the average nonsmoking household's in-home concentrations of traditional pollutants over the last half-century have declined (that is, improved) 85 percent for CO, 97 percent for SO_2, and 90 percent for TSP. Residential NO_x emissions per occupied household declined 40 percent between 1970 and 1992. Since no long-term indoor air quality measurements are avail-

TABLE 10-2   Concentrations and Time Spent Indoors Around the World

Group of Nations	Annual PM Concentrations (µg/m³) Indoors	Outdoors	Percentage of Time: Indoors	Outdoors
Developed				
Urban	100	70	90	10
Rural	60	40	85	15
Developing				
High HDI				
Urban	175	110		
Rural	150	70		
Mid-HDI				
Urban	190	300	75	25
Rural	400	80	65	35
Low HDI				
Urban	400	280		
Rural	700	100		

Source: K. R. Smith, "Fuel Combustion, Air Pollution Exposure, and Health: The Situation in Developing Countries," *Annual Review of Energy and the Environment* 18 (1993): 529–566.

Note: HDI = Human Development Index from United Nations Development Programme based on socioeconomic indicators from the late 1980s or early 1990s. High HDI includes Malaysia, Portugal, and Yugoslavia (pre-breakup); mid-HDI includes Brazil, China, Philippines, Thailand; low HDI includes India, Pakistan, Ghana, and Indonesia.

able, these estimates are derived by dividing Environmental Protection Agency (EPA) estimates of residential fuel combustion emissions by the corresponding number of occupied housing units.[43] (Fuel combustion is the major source of residential emissions for the traditional pollutants.) Moreover, residential emissions contribute a relatively high fraction of that portion of total human exposure acquired outdoors because these emissions are exhausted at low heights where people live and spend most of their time.

These improvements could be underestimates because they ignore several factors that may have further reduced indoor concentrations over the past fifty years: increases in the average house size; improvements in capturing, exhausting, and filtering combustion gases; and increased vacuuming of dust (as opposed

TABLE 10-3   Percentage of Total Exposure to Air Pollutants Attributable to Indoor Environment

Group of Nations	Percentage of Total Human Exposure Due to Indoor Air Quality
Developed	
Urban	93
Rural	90
Developing	
High HDI	
Urban	85
Rural	NA
Mid HDI	
Urban	80
Rural	92
Low HDI	
Urban	75
Rural	90

Source: Ibid.
Note: HDI = Human Development Index, from United Nations Development Programme. Total human exposure incorporates amount of time spent in a microenvironment multiplied by the average concentration in that microenvironment.

to sweeping). A complicating factor is that through the years, the air exchange rate between the indoors and outdoors has been reduced for the average home because of greater reliance on air-conditioning (particularly in summer) and the increased emphasis on energy conservation. Lowered air exchange rates may result in a cleaner in-home environment if the concentration outdoors is higher than indoors; otherwise, it would have the opposite effect.

By the same measure volatile organic compounds (VOC) concentrations in a "nonsmoking house" due to fuelwood combustion also declined 90 percent. However, this improvement could be offset by increases in indoor emissions of solvents and other volatile substances stored or used in the house.

Improvements in indoor air quality stem from the same factors that improved outdoor air quality in urban areas: fuel switching, technological change, and affluence. These improvements probably constitute the greatest reduction in the general population's exposure to traditional air pollutants and began decades before the promulgation of federal legislation. When the need is obvious and cause and effect determinable with confidence, people will voluntarily take measures to improve their personal environment without the government's intervention and at some expense to themselves. Because they were undertaken voluntarily, the costs of these improvements are not included in any accounting of air pollution control costs even though they are probably the most effective measures for reducing total human exposure to pollutants.

Outdoor Air Pollution

Air Pollutant Emissions in the United States, 1990–1992

Data on traditional air pollutants in the United States from 1900 to 1992 indicate the following:[44]

- Except for NO_x, these emissions grew at a much slower rate than population, and a much slower rate than GNP or GNP per capita. SO_2, VOC, and NO_x emissions grew by 128 percent, 193 percent, and 787 percent, respectively, while population increased by 235 percent and the GNP (in constant dollars) by 1,489 percent (figure 10-4).[45] The best news is that between 1940 and 1992, CO and PM-10 emissions declined 4 percent and 62 percent, respectively.

- Key emissions that have risen in recent years have all peaked and are now declining on a per capita basis, in conformance with the notion of an environmental transition. These transitions occurred at per capita GNPs (in 1987 dollars) of $6,250, $10,800, and $17,600 for per capita emissions of SO_2, VOC, and NO_x, respectively.[46] Per capita PM-10 and CO emissions have been declining at least since per capita GNP level exceeded $6,850; lack of longer-term data precludes a more precise determination of these transitions. This confirms that the pollutants controlled first were those deemed to be the biggest public health and nuisance problems, the cheapest to control, and for which individuals could easily see that the costs they incurred resulted in obvious benefits.

- For the most obvious pollutants, cleanup started before the 1970 Clean Air Act's provisions became effective, based on trends in emissions per capita. SO_2 emissions per capita peaked around 1920 (figure 10-5). There was a dip in the 1930s probably due to the depression, another peak in 1945 due to the war effort, and a smaller one around 1970. It is now down 60 percent from its peak. Both PM-10 and CO emissions per capita have been declining more or less continuously since World War II; per capita PM-10 is down 79 percent since

FIGURE 10-4   The Relationship of U.S. Emissions, GNP, and Population VOC, SO₂, and NOₓ, 1900–1992 (1900 level=1)

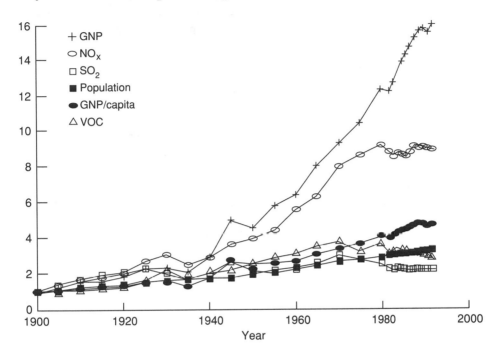

Population = increase in U.S. population

Sources: Based on data from: U.S. Environmental Protection Agency, *National Air Pollutant Emission Trends, 1900–1992*, EPA 454/R-93-032, Office of Air Quality Planning and Standards (Research Triangle Park, N.C.: EPA, 1993); U.S. Bureau of the Census, *Historical Statistics of the United States, Colonial Times to 1970* (Washington, D.C.: Department of Commerce, 1975); U.S. Bureau of the Census, *Statistical Abstract of the United States 1992* (Washington, D.C.: Department of Commerce, 1992).

1940 and CO 53 percent since 1945. VOC emissions per capita stayed roughly constant between 1925 and 1970, before they declined; in 1992 they were down 36 percent from the peak. NOₓ per capita peaked in 1980; it is now down about 10 percent from its peak.

• The timing of these reductions indicates that a substantial portion of the initial reductions was undertaken voluntarily. These voluntary actions included switching to cleaner energy sources and new technologies for combustion equipment in all sectors. These were supplemented by regulations imposed by local and state jurisdictions for sulfur-in-fuel, incinerators, and industrial sources in urban areas, which in many cases preceded the 1970 Federal Clean Air Act.

FIGURE 10-5   U.S. Emissions per Capita, 1900–1992

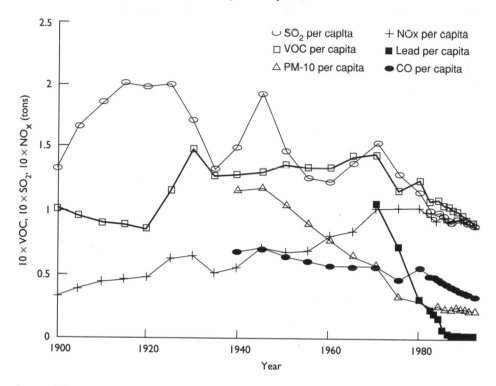

Sources: Ibid.

Trends in Ambient (Outdoor) Air Quality in the United States

Various U.S. government publications provide nationwide ambient data only for a few (ten to twelve) years at a time.[47] There are some problems associated with splicing these data together to construct longer-term trends.[48] For example, the number and precise locations of monitoring stations that form the basis of the reported national trends keep changing continually, as do measurement methods and instruments. Moreover, changes and fluctuations in meteorologic conditions can mask changes due to emissions.[49] Nevertheless, many of the series overlap in time, and for air quality in the United States, at least, averages based on different but large ensembles of monitors indicate that while the mean values may differ between ensembles, their trends are more or less parallel.[50] Thus, it is possible to piece together several of these series, each extending over a few years, to construct longer-term trends, but it must be recognized that these trends are qualitative rather than quantitative.

The results are broadly consistent with the emission trends discussed earlier. They indicate that nationwide air quality has improved for all the traditional pollutants. The largest improvements have been for SO_2, TSP, and CO and the smallest for NO_2. For reasons discussed earlier, TSP and SO_2 improvements came first, with a significant portion coming prior to federal requirements' becoming effective—about 1972 or earlier.[51] Because many of the emission reductions were at low-level sources while emission increases were at sources with taller stacks, improvements in ambient TSP, SO_2, and CO air quality levels are greater than would be indicated by trends in total emissions.

Particulate Matter   Based on the mean of the annual average concentrations for several monitors, TSP levels declined about 40 percent between 1960 and 1990.[52] The national "average," which used to be above the old primary (public health related) NAAQS of 75 $\mu g/m^3$, is now 37 percent below that. As the previous discussions regarding Pittsburgh and New York indicated, urban areas saw the most improvement.

In 1987, the TSP NAAQS were replaced by the PM-10 standards that include only PM less than 10 micrometers in diameter. This change recognizes that PM-10 is a better indicator of the health impact than is TSP, since smaller particles

Pollution control initially takes a back seat to industrial development in countries, like India, that are not yet affluent.

are more likely to be inhaled deeper into the lungs. The national composite annual average for PM-10 declined 17 percent, from 33.6 to 27.8 µg/m³, between 1988 and 1992.[53] For comparison, the primary annual NAAQS is 50 µg/m³.

Sulfur Dioxide   Some data for SO₂ are available from 1962 to 1969 and from 1974 onward.[54] SO₂ concentrations declined quite dramatically in the 1960s. Earlier I noted that annual levels declined about 85 percent in New York City between 1964 and 1972, from 570 to 81 µg/m³. By 1992, the level had dropped to about 50 µg/m³.[55] Other cities also showed substantial, though not as dramatic, improvements. Between 1962 and 1969, based upon twenty-one urban monitors, the mean annual average dropped about 40 percent, from 69.4 to 42.5 µg/m³. The corresponding primary NAAQS is 80 µg/m³. Between 1974 and 1992, the national "average" dropped 50 percent, with about two-thirds of that drop occurring in the first ten years. It currently is about 19 µg/m³.[56]

Carbon Monoxide   CO air quality began improving at least in urban areas in the 1960s due to reductions at stationary sources.[57] These improvements continued as the number of vehicles subject to the Federal Motor Vehicle Emission Control Program, first effective for the 1968 model year, kept increasing. Between 1970 and 1992, the national mean CO concentration (based on the mean of the second highest eight-hour concentration) decreased about 60 percent.

Ozone   Ozone, in contrast with other traditional pollutants, is not emitted directly. It is formed by a series of complex chemical reactions between VOCs and NO$_x$ in the presence of sunlight. The speed of these reactions and the movement and concentration of ozone depend on meteorological factors such as temperature, wind speed, height of inversion layer, cloudiness, and precipitation. Ozone is one of the major components of Los Angeles's smog. Because it is the product of a very complex set of reactions, which, moreover, are mediated by meteorology, it is difficult to predict changes in ozone levels due to reductions in their precursors (VOC and NO$_x$) with much accuracy. Another feature of ozone is that it can be formed and transported hundreds of kilometers away from the original sources of its precursors. To complicate matters still further, NO$_x$ reductions may either increase or decrease ozone levels depending on the ratio of NO$_x$ to nonmethane hydrocarbons. It is possible that, as a result, ozone levels would decline in one place but increase in another.[58] The fact that NO$_x$ reductions were being obtained even as VOC was being reduced to attain the ozone NAAQS may have aggravated ozone in some center city areas while reducing it in outlying suburban areas.[59] Finally, for a long time EPA and other regulatory agencies have tended to ignore the contribution of natural (biogenic) VOC emissions to the ozone problem.[60] In fact, biogenic VOC emissions during the peak ozone season exceed anthropogenic emissions (12.2 million tons versus about 6.0 million tons, respectively, in 1987).[61] For these reasons, attaining the NAAQS for ozone has proved to be difficult and expensive.

Most official data on ozone air quality are provided in terms of that statistic most readily compared with the NAAQS: the magnitude of the second highest one-hour reading in the year (after eliminating the day with the highest one-hour number). The national mean for this statistic shows[62] a roughly 30 percent improvement between 1975 and 1992. Between 1983 and 1992 this statistic declined about 21 percent. However, another statistic, the average number of days when the NAAQS was exceeded, which has greater significance for public health, improved by a greater amount. Based on two-year running averages, this number decreased about 56 percent between 1983–1984 and 1991–1992.[63]

Nitrogen Dioxide NO_2 peaked around 1978–1979 and has declined slightly since then.[64] Between 1983 and 1992, the mean annual average declined about 8 percent. In the United States, Los Angeles is the only designated NO_x nonattainment area. However, no monitor exceeded the NAAQS in 1992 or 1993.[65] Between 1988 and 1993, the highest reading in Los Angeles County declined 19 percent, from 0.0613 to 0.0499 parts per million (ppm). The standard is 0.053 ppm.

Acid Rain   Acid precipitation is caused by the transformation of SO_2 and NO_x (primary pollutants) in the atmosphere into more acidic substances such as sulfuric or nitric acids, which are then deposited (through rain or other processes) hundreds of kilometers away from the emission sources. Increased SO_2 and NO_x emissions, particularly in the quarter century since World War II, resulted in substantial increases in acid precipitation. These increases were aggravated by two factors.[66] First, the increased use of tall stacks, while beneficial from the public health standpoint, increased the amount of time the average SO_2 molecule spent in the atmosphere, thus increasing both the total amount of transformation products and the distance over which they were deposited from the emission source. Second, reductions in TSP emissions removed material that could scavenge and neutralize SO_2 and acidic sulfates. In addition, this increased the distance acid pollutants could travel before deposition. Thus, more acidic material was formed, less of it was neutralized, and it was spread farther, leading to widespread predictions of imminent ecological calamity.

A comprehensive $500 million, ten-year study of acidic deposition in the United States under the auspices of the National Acidic Precipitation Assessment Program (NAPAP) concluded that acid rain in the United States at the levels found in the late 1980s is more of an aesthetic than an ecological problem.[67] The study found that acidic deposition has not been responsible for damaging crops. It may have played a role in the dieback of high-elevation eastern spruce–fir forests (comprising 0.1 percent of total U.S. forest area), but not elsewhere. By contrast, ozone affects both forest and crop productivity. In fact, because sulfates and nitrates are plant nutrients, acidic deposition may well have benefited both forests and crops. Finnish researchers in a survey of European forests found that forest productivity over the past two decades has been enhanced 25 to 30 percent, a phenomenon they attribute in part to the transport of soil nutrients from acidic deposition.[68]

A National Surface Water Survey for acid-sensitive areas found that only 2 percent of the lake area (about 4 percent of the lakes) were acidic, as were 4 percent of the stream miles.[69] Deposition was the major source of acidification for 75 percent of the acidified lakes and 50 percent of the acidified streams. This study also indicated that in the Adirondacks, the area with the most acidification due to deposition, 32 percent of the lakes that historically had brook trout no longer support it: 11 percent due to acidic deposition and another 21 percent because of other causes. NAPAP found that in many situations, the lake acidification itself and its effect on fish populations could be effectively and cheaply reversed by liming and stocking water bodies with fish (or similar mitigation techniques). Where successful, such mitigation could be achieved in a matter of months—as opposed to several decades if emission reductions are used as the sole method of helping those water bodies recover.

The NAPAP study found that at current levels, the major negative effects of acidic deposition are probably reduced visibility and accelerated deterioration of outdoor cultural resources.

Experience in the Developed World and Europe

The U.S. experience is more or less being mirrored in virtually every other developed country for the traditional pollutants, especially PM and SO_2.[70] Both absolute and per capita SO_2 emissions declined from 1970 to the present for all areas, except for the area comprising the erstwhile East Germany—a powerful comment on the social and economic system that the state labored under for over forty years. Japan has the lowest emission rates per capita among the developed countries. Somewhat analogous to the U.S. situation, SO_2 emission reductions in Western Europe preceded the signing of the Convention on Long Range Transboundary Air Pollution in 1980 and the adoption of the SO_2 protocol in 1984 that called for a 30 percent decline. They have continued to decline since.

Urban Air Pollution in World Megacities

Based on Global Environment Monitoring System (GEMS) data from the late 1980s, a survey of air quality in twenty megacities of the world indicates that PM is the major pollutant problem, particularly where SO_2 is simultaneously very high.[71] That report next ranks, in order, ozone, lead, CO, and NO_2, at the bottom. Table 10-4 summarizes the gravity of the air pollution situation in each

TABLE 10-4   Severity of Urban Air Pollution in World Megacities

	SPM	SO_2	O_3	P_6	CO	NO_2
Bangkok	S	L	L	M	L	L
Beijing	S	S	M	L	I	L
Bombay	S	L	I	L	L	L
Buenos Aires	M	I	I	L	I	I
Cairo	S	I	I	S	M	I
Calcutta	S	L	I	L	I	L
Delhi	S	L	I	L	L	L
Jakarta	S	L	M	M	M	L
Karachi	S	L	I	S	I	I
London	L	L	L	L	L	L
Los Angeles	M	L	S	L	M	M
Manila	S	L	I	M	I	I
Mexico City	S	S	S	M	S	M
Moscow	M	I	I	L	M	M
New York	L	L	M	L	M	L
Rio de Janeiro	M	M	I	L	L	I
São Paulo	M	L	S	L	M	M
Seoul	S	S	L	L	L	L
Shanghai	S	M	I	I	I	I
Tokyo	L	L	S	I	L	L

Source: World Health Organization/United Nations Environment Programme (WHO/UNEP), *Urban Air Pollution in the Megacities of the World,* published for Global Environment Monitoring System (GEMS), WHO, and UNEP (Cambridge, Mass.: Blackwell Reference, 1992). Based on GEMS data for the late 1980s. Definitions used by that report are: S = serious problem (WHO guidelines exceeded by a factor of two), M = moderate to heavy pollution (WHO guidelines exceeded up to a factor of two), L = low pollution (WHO guideline normally met, though short-term guidelines may be occasionally exceeded), and I = insufficient information.

of twenty cities. However, based on the report, we can be optimistic that these levels in megacities can be reversed before they hit levels as high as those recorded in the infamous episodes of the Meuse Valley, Donora, and London.

Sulfur Dioxide Highest SO_2 concentrations were in Beijing, Mexico City, and Seoul with peak daily concentrations exceeding 700 $\mu g/m^3$, or more than four times World Health Organization (WHO) guidelines. Annual concentrations exceeded WHO guidelines by a factor of three. The data indicate, however, that increases in annual concentrations may have been arrested for Beijing and Mexico City and even reversed for Seoul. Two cities, Rio de Janeiro and Shanghai, had moderate to heavy levels. In the former, the increase in concentrations has slowed, but in Shanghai increases in SO_2 concentrations were continuing, at least in a residential location in the central city. Data were unavailable for Buenos Aires, Cairo, and Moscow. The remainder (twelve cities), including all the megacities in the sample from the developed countries (London, Los Angeles, New York, and Tokyo), had relatively low levels of pollution even though short-term WHO guidelines may occasionally have been exceeded.[72]

Particulate Matter Twelve of the megacities had serious PM problems, with annual averages between 200 to 600 $\mu g/m^3$ and peak daily concentrations in excess of 1,000 $\mu g/m^3$. Nevertheless, many of these cities seem to have reduced, if not reversed, the rate of increase of annual PM concentrations. These cities include Bangkok, Beijing, Bombay, Mexico City, Seoul, and Shanghai. For this pollutant too, all three megacities having low pollution levels were in developed countries (London, New York, and Tokyo). However, Los Angeles was indicated as having only moderate air quality.

Carbon Monoxide This too is dependent on traffic and traffic density. Only Mexico City was deemed to have "serious" levels. However, seven cities, including London, Los Angeles, and New York, had moderate to heavy levels. As indicated earlier, levels in those cities have dropped substantially since the 1970s.

Ozone Highest concentrations are in Los Angeles, Mexico City, São Paulo, and Tokyo. Beijing, Jakarta, and New York were deemed moderate to heavy. Bangkok, London, and Seoul had low levels, and in ten of the cities there were insufficient data. Los Angeles, despite heroic efforts, is still one of the worst polluted cities in terms of ozone. Only Mexico City has worse ozone air quality. While some of the reasons undoubtedly have to do with relative levels of affluence, there are other, natural reasons, especially geography and meteorology.

Nitrogen Dioxide This is a moderate problem, at its worst in Los Angeles, Mexico City, Moscow, and São Paulo.

Global Trends for Arctic Haze and Carbon Monoxide

Arctic Air Pollution

Measurements of Arctic air pollution (Arctic haze) in Barrow, Alaska, which are primarily affected by emissions in Europe and Asia, peaked in 1982 and by 1992 had declined by a factor of two, probably due to the SO_2 reductions in Western Europe as well as the decreases in economic activity in the former communist states of Eastern Europe and the former Soviet Union.[73]

Global Carbon Monoxide Levels

There seems to have been a significant turnaround in global CO levels from the 1981–1986 to the 1987–1992 periods.[74] In the earlier period, the global average concentration increased about 1.2 percent (±0.6 percent) per year and decreased about 2.6 percent (±0.8 percent) per year for the latter period. This turnaround, more pronounced for

Long-term air quality trends in individual countries confirm that richer is cleaner, middle income is dirtier, and, sometimes, though not always, poorest is the dirtiest.

the Southern Hemisphere, was also evident for the Northern Hemisphere. The downward trend in the Northern Hemisphere is consistent with the discussion regarding emission controls in the developed countries and the combination of an economic slowdown and more realistically priced energy in the former communist countries. The Southern Hemisphere trend could be due to reductions in biomass burning in the tropics.

Summary

Long-term air quality trends in individual countries confirm that richer is cleaner, middle income is dirtier, and, sometimes, though not always, poorest is the dirtiest. Certainly for indoor air quality, poorer does seem to be dirtier.

In recent decades, both indoor and outdoor air quality has improved dramatically in the developed world, except possibly for ozone in some areas. In developed countries, deadly pollution episodes such as those that struck Donora or London are now a thing of the past. With improvements in air quality, it is uncertain whether and, if so, to what extent current ambient levels contribute to any premature deaths. That remains an issue for further scientific research and debate.[75] In developed countries, the focus increasingly is on welfare effects (effects on visibility, materials, and ecosystems) and morbidity effects. There are also efforts to address so-called hazardous air pollutants, but their effects at current levels are also a matter of debate. Cost-benefit analyses for Los Angeles and the entire United States suggest that the major benefit of any further improvement in air quality may be that due to reducing particulate emissions further.[76] Surprisingly, even for the Los Angeles area, despite its high population and ozone levels, additional measures to improve ozone air quality did not meet the cost-benefit test.

In developing countries, air pollution remains a significant public health hazard. Compared to a quarter of a century ago, the situation has worsened in many areas because of increasing population densities, urbanization, and industrialization. Indoor air pollution due to heating and cooking with coal, wood, and dung or other biofuel probably always has been a part of life in developing countries, and resulting indoor PM levels often rival, if not exceed, magnitudes that caused the Donora and London episodes.[77] To this now must be added the risks due to higher outdoor levels, particularly PM and SO_2.

Clearly the world has come far with respect to particulate matter, yet we have far yet to go. The first priority must be to improve indoor air quality in developing countries. The second priority should be outdoor air quality in urban residential areas in developing countries. By comparison, the air pollution problems of developed countries seem trivial.

The problems of both indoor and outdoor PM could be relieved substantially by switching to cleaner fuels such as low sulfur oil, natural gas, and electricity in the home and in central city areas and to new, more efficient heaters and cookers. However, such switching is possible only if the general level of affluence increases. As the experience of developed nations has shown, such switches could come more or less voluntarily with affluence and technology (including knowledge). Sulfur-in-fuel regulations, such as were promulgated for New York in the 1960s, could accelerate such trends, yet because they increase the cost they could dissuade some poor people from switching to cleaner forms of energy.

As developing nations' reliance on electricity grows—and it will, with affluence—emissions from thermal power plants will increase. To avoid aggravating existing urban air pollution problems any further, here too developing nations could take a page from developed nations by encouraging such power plants to locate away from population centers and rely on taller stacks to reduce local ambient air concentrations, and therefore local public health-

Mexico City suffers some of the world's worst air pollution. However, the state of Mexico appears poised to go through the "environmental transition" that leads to decreasing levels of air pollutants.

related problems. New plants could be designed and constructed with allowances for later retrofits to allow for the fact that if all goes as hoped and incomes rise and technology changes, then environmental standards may not only become more demanding but also meeting them would become more affordable.

Conclusion

Affluence and the natural environment are intimately related but not in the way many believe. Although affluence may increase consumption of material goods, it does not inexorably increase environmental degradation, because at the same time it also increases the "consumption" of environmental quality. Moreover, affluence and technology are highly symbiotic: the institutional frameworks that foster one also foster the other. In fact, we see that generally richer is cleaner, middle income is dirtier, and sometimes, though not always, poorest is dirtiest. This is confirmed not only by cross-country analyses but also by long-term trends in numerous air quality indicators for individual advanced developed nations such as the United States, Japan, and other member nations of the Organization for Economic Cooperation and Development. These trends indicate that in many cases, environmental quality initially suffers as incomes climb; then the quality goes through an environmental transition and finally improves with increased affluence.

The level of affluence at which this transition occurs varies with the known or perceived benefits of pollution control relative to social and economic costs. And that, in turn, depends on available hardware and software technologies; the latter include knowledge of impacts, benefits, costs, management options, and operating rules. Since technology improves with time, the environmental transition for some indicators in poor countries occurs today at an earlier developmental stage than it did for developed countries. In fact, the transition is virtually nonexistent for some indicators; environmental quality associated with these indicators improves almost immediately with increased incomes.

From a global perspective, the environmental priorities for improving air quality, all focused on the developing world, should be, first, reducing indoor air pollutants in homes using biofuels and coal for cooking and heating, and, second, reducing urban outdoor particulates and SO_2. Finally, anything that unduly retards economic growth in developing countries—including inefficient policies, no matter how well intentioned—will ultimately retard net environmental progress and imperil human lives.

NOTES

1. P. Ehrlich and A. Ehrlich, *The Population Explosion* (New York: Touchstone, 1990).

2. G. Grossman and A. Krueger, *Environmental Impacts of a North American Free Trade Agreement*, Discussion Paper 158 (Princeton, N.J.: Woodrow Wilson School, Princeton University, November 1991); World Bank, *World Development Report 1992* (New York: Oxford University Press, 1992); N. Shafik and S. Bandyopadhyay, *Economic Growth and Environmental Quality: Time Series and Cross-Country Evidence*, Policy Research Working Papers (Washington, D.C.: World Bank, June 1992); N. Shafik, *Economic Development and Patterns of Change*, Oxford Economic Papers (forthcoming).

3. This analysis adjusted incomes for purchasing power parity (PPP), which accounts for the differing ability of consumers in different countries to purchase a variety of goods when local currencies are converted to a common currency using prevalent exchange rates. PPP is an inherently superior approach because there are many goods and services for which exchange rate–based costs are more in one place than another. For example, the cost of food in India is substantially less (in terms of the exchange rate) than in, say, Switzerland. Thus, exchange rate conversions are not appropriate for comparing relative well-being in these two countries.

4. Grossman and Krueger, *Environmental Impacts*, also noted an upswing for ambient SO_2 levels at very high GDP. However, they cautioned against reading too much into that. Once again, examination of trends for the advanced developed countries does not indicate any upswing. See the discussion in this chapter on air quality. However, an upswing, while not detected heretofore for any environmental indicator, is quite plausible for two reasons: if the limit on each process or activity has been reached, emissions would grow once again with any increase in that activity, or if a nation concludes, based on new information or changes in societal values and attitudes, that past control efforts, for whatever reasons, went too far or were unnecessary.

5. World Health Organization, *Our Planet, Our Health*, Report of the WHO Commission on Health and the Environment (Geneva, Switzerland: WHO, 1992); World Bank, *World Development Report 1992*.

6. I. M. Goklany, "Adaptation and Climate Change" (paper presented at the annual meeting of the American Association for the Advancement of Science, February 1992).

7. Organization for Economic Cooperation and Development (OECD), *The State of the Environment* (Paris: OECD, 1991).

8. D. O'Connor, *Managing the Environment with Rapid Industrialization: Lessons from the East Asian Experience*, Development Centre Studies (Paris: OECD, 1994), p. 118; Environmental Protection Agency (EPA), *National Air Quality and Emission Trends Report*, EPA 454/R-93-031, OAQPS (Research Triangle Park, N.C.: EPA, 1993).

9. J. H. Ausubel, "Does Climate Still Matter?" *Nature* 350 (1991): 649–652.

10. Shafik and Bandyopadhyay, *Economic Growth*.

11. World Bank, *World Development Report 1984* (New York: Oxford University Press, 1984); A. Sen, "The Economics of Life and Death," *Scientific American* 268 (1993): 40–47; M. Livi-

Bacci, *A Concise History of World Population*, English ed., trans. C. Ipsen (Cambridge, Mass.: Blackwell, 1992).

12. Goklany, *Adaptation and Climate Change*; I. M. Goklany and M. W. Sprague, *An Alternative Approach to Sustainable Development: Conserving Forests, Habitat and Biological Diversity by Increasing the Efficiency and Productivity of Land Utilization*, Office of Program Analysis, U.S. Department of the Interior (Washington, D.C.: Government Printing Office, 1992); I. M. Goklany, *Key Issues Related to Sustainable Development, Issue: Reconciling Human Demands on Land and Other Natural Resources with Those of Nature*, Background Paper, Department of the Interior, Office of Program Analysis (Washington, D.C.: Government Printing Office, October 1993).

13. Goklany and Sprague, *An Alternative Approach*.

14. World Commission on Environment and Development, *Our Common Future* (Oxford: Oxford University Press, 1987). See also Intergovernmental Panel on Climate Change (IPCC), *The IPCC Response Strategies* (Washington, D.C.: Island Press, 1991); I. M. Goklany, "Adaptation and Climate Change" (paper presented at the Annual Meeting of the American Association for the Advancement of Science, Chicago, February 1992).

15. P. Brimblecombe, "Attitudes and Responses towards Air Pollution in Medieval England," *Journal of the Air Pollution Control Association* 26 (1976): 941–945; "Smoke and Smoke Prevention," *Encyclopaedia Britannica* (1959), 20: 841–843.

16. F. W. Lipfert, "Sulfur Oxides, Particulate, and Human Mortality: Synopsis of Statistical Correlations," *Journal of the Air Pollution Control Association* 30 (1980): 366–371. See also F. W. Lipfert, *Air Pollution and Community Health* (New York: Van Nostrand Reinhold, 1994).

17. One method of specifying air quality is based on measuring the weight of pollutant (generally in micrograms—μg) in a fixed volume of air (1 cubic meter, or m^3), that is, in terms of $\mu g/m^3$. An alternative approach is to specify the number of molecules of the pollutant in a million molecules of air, or parts per million (ppm).

18. Standards for air quality around the world are established in terms of air pollutant concentrations at a fixed point in the outdoors over a specified period of time. For instance, in the United States, the annual public health–related (or primary) air quality standard for sulfur dioxide is a never-to-be-exceeded average of 80 micrograms per cubic meter or 0.03 part per million (ppm) at any point outdoors. Similarly, the short-term public health and welfare standards for SO_2 are 365 micrograms per cubic meter (0.14 ppm) for twenty-four hours and 1,300 micrograms per cubic meter (0.5 ppm) for three hours, respectively. Short-term standards are not to be exceeded more than once a year for any single outdoor point.

19. H. H. Schrenk et al., *Air Pollution in Donora, Pa.: Epidemiology of the Unusual Smog Episode of October 1948, Preliminary Report*, PHS Bulletin 306 (Washington, D.C.: Public Health Service, 1949).

20. R. E. Waller and B. T. Cummins, "Episodes of High Pollution in London 1952–1966, " in *Proceedings: Part I. International Clean Air Congress, London, 4–7 October, 1966*, International Union of Air Pollution Prevention Association (London: National Society for Clean Air, 1966), pp. 228–231.

21. F. Pearce, "Back to the Days of Deadly Smogs," *New Scientist*, December 5, 1992, pp. 25–28.

22. J. J. Scheuneman, "A Roll Call of the Communities—Where Do We Stand in Local or Regional Air Pollution Control," *Proceedings: The Third National Conference on Air Pollution*, PHS Publication 1649 (Washington, D.C.: Government Printing Office, 1966), pp. 386–399.

23. C. D. Yaffe, "A Roll Call of the States—Where Do We Stand in State and Interstate Air Pollution Control," in *Proceedings: The Third National Conference on Air Pollution*, pp. 359–363.

24. U.S. Bureau of the Census, *Historical Statistics of the United States, Colonial Times to 1970* (Washington, D.C.: Government Printing Office, 1975).

25. EPA, *National Air Pollutant Emission Trends, 1900–1992*, EPA/450/R-93-032, Office of Air Quality Planning and Standards (Research Triangle Park, N.C.: EPA , 1993).

26. The short time frames for developing, analyzing, and executing state implementation plans precluded any real analysis, particularly with the threat of sanctions. It led to widespread paper showings of attainment, and many states adopted technology-based standards for broad source categories as a substitute for credible analysis showing attainment. See Pacific Environmental Services, *An Overview of the SIP Review Process at the State Level and the SIP for Particulate Matter, Sulfur Dioxide and Ozone*, report to the National Commission on Air Quality, NTIS PB81 1127573 (1980).

27. C. I. Davidson, "Air Pollution in Pittsburgh: A Historical Perspective," *Journal of the Air Pollution Control Association* 29 (1979): 1035–1041.

28. Ibid.; EPA, AIRS database, available electronically, December 1994.

29. The 1972 level is based on the highest one-day reading reported in the city in EPA's data summary for 1972. See EPA, *Monitoring and Air Quality Trends Report 1972*, EPA-450/1-73-004 (Research Triangle Park, N.C.: EPA, 1973). The 1993 data point is the highest reading reported in the AIRS database.

30. H. Schimmel and T. J. Murawski, "SO_2—Harmful Pollutant or Air Quality Indicator," *Journal of the Air Pollution Control Association* 25 (1975): 739–740.

31. Ibid.; C. F. Simon and E. Ferrand, "The Impact of Low Sulfur Fuel on Air Quality in New York City," in *Proceedings of the Second International Clean Air Congress* (New York: Academic Press, 1971), pp. 41–50; EPA, *Monitoring and Air Quality Trends Report 1972*; EPA, AIRS database.

32. EPA, *Monitoring and Air Quality Trends Report 1972*; EPA, AIRS database.

33. S. R. Craxford, M. Clifton, and M-L. P. M. Weatherley, "Smoke and Sulphur Dioxide in Great Britain: Distribution and Changes," in *Proceedings: Part I. International Clean Air Congress*, pp. 213–216.

34. S. Eggleston et al., "Trends in Urban Air Pollution in the United Kingdom During Recent Decades," *Atmospheric Environment* 26B (1992): 227–239.

35. World Health Organization/United Nations Environment Programme (WHO/UNEP), *Urban Air Pollution in the Megacities of the World*, published for Global Environment Monitoring System (GEMS), WHO, and UNEP (Cambridge, Mass.: Blackwell Reference, 1992).

36. EPA, *Report to Congress on Indoor Air Quality*, vol. 2: *Assessment and Control of Air Pollution*, EPA/400/1-89/001C, Office of Air and Radiation, Environmental Protection Agency (Washington, D.C.: EPA, 1989).

37. GEOMET, *Comparison of Indoor and Outdoor Air Quality* (Palo Alto, Calif.: Electric Power Research Institute, 1981), p. ES-5.

38. EPA, *Air Quality Criteria for Carbon Monoxide*, EPA/600/890/045F (Washington, D.C.: EPA, 1991).

39. EPA, *Air Quality Criteria for Oxides of Nitrogen*, Vol. 1: EPA/600/8-91-049aF, Office of Research and Development (Washington, D.C.: EPA, 1993).

40. K. R. Smith, "Fuel Combustion, Air Pollution Exposure, and Health: The Situation in Developing Countries," *Annual Review of Energy and the Environment* 18 (1993): 529–566.

41. Ibid.; K. Smith and Y. Liu, "Indoor Air Pollution in Developing Countries," in *Epidemiology of Lung Cancer*, ed. J. M. Samet (New York: Marcel Dekker, 1994), pp. 151–184.

42. Smith, "Fuel Combustion."

43. EPA, *National Air Pollutant Emissions Trends, 1900–1992*; Bureau of the Census, *Historical Statistics of the United States: Colonial Times to 1970*, and *Statistical Abstract of the United States 1992*.

44. All U.S. emission estimates used in this chapter are obtained from EPA, *National Air Pollutant Emissions Trends, 1990–1992*. This report describes the methodologies used to derive these emissions. Note that the 1980 CO, NO_x, and VOC estimates were affected by methodological changes for estimating highway and off-highway vehicle emissions. This may have contributed to the noticeable bump in 1980 NO_x and CO emissions.

45. I use GNP rather than the theoretically more appropriate GDP because the difference is slight for the United States. The two are, on the average, within 0.5 percent of each other based on a comparison from 1935 to 1992.

46. Data on GNP and population are obtained from Bureau of the Census, *Historical Statistics of the United States, Colonial Times to 1970* and *Statistical Abstract of the United States 1992*.

47. Data sources used in this section include EPA's annual reports on air quality trends, the Council on Environmental Quality's annual reports, entitled *Environmental Quality*, and various *Statistical Abstracts of the United States*. The latter two rely exclusively on data from EPA.

48. Measurements made today may not be comparable with those made even a decade or so ago. Monitoring methods are constantly refined, if not altered, as better and more accurate instruments are developed. For example, measurements of NO_x in the 1960s used the Jacob-Hochheiser method, which was shown to overestimate NO_x concentrations at low levels. This invalidated many of the NO_x data collected prior to 1972 when a modified measurement method was deployed as part of the National Air Surveillance Network (NASN). See, e.g., National Academy of Sciences (NAS), *Air Quality and Stationary Emission Control*, prepared for the Committee on Public Works, U.S. Senate, Serial No. 94-4 (Washington, D.C.: Government Printing Office, 1974). Older methods may be replaced without undertaking detailed side-by-side comparisons with newer ones to allow trends to be developed in the future. Locations of monitoring stations change continually; about 10 percent of National Air Monitoring System stations are phased out or replaced each year (W. Freas, Office of Air Quality Standards and Planning, EPA, personal communication, 1994). Microenvironments—the environment in the immediate vicinity—may have changed for monitoring stations; the area adjacent to the stations may have been paved over, its land use altered, or an industry may have moved near,

or away from, it. Over time, individuals and laboratories gathering and analyzing the data are replaced, resulting in inconsistent handling and potentially trading one set of biases for another.

49. Year-to-year and longer-term variability in meteorological and hydrological (natural) factors at all geographic scales can affect both air and water pollutant concentrations. These factors include, for air pollution, temperature, precipitation, insolation, and wind speeds, and for water pollution, temperature, runoff, evaporation, and flow rates. Variability in these factors can partially mask long-term trends, particularly for "extreme" short-term concentrations (e.g., the highest, second highest, or highest 1 percent of the concentrations for three hours, eight hours, or twenty-four hours) because these extreme levels are just as likely to be determined by meteorological extremes as by emissions. But, in fact, many ambient environmental quality standards are specified in terms of extreme concentrations for very short time periods. This makes it difficult to confirm attainment or nonattainment of the ambient standard until several years of monitoring data have been collected so that the effects of meteorology and hydrology can be averaged or filtered out mathematically. However, since emissions (or pollutant loadings) can also fluctuate, though often not as dramatically as natural variability, averaging out several years of data may be insufficient to eliminate the effects of the latter, and mathematically filtering the latter's effect can be tricky. Filtering techniques include adjustment of water quality indicators to reflect typical, rather than actual, flow rates. Such adjustments may be more difficult for air pollutants, though for ozone, in particular, EPA and some researchers have used mathematical models for adjusting monitor data to reflect typical, rather than actual, meteorology. See, e.g., EPA, *National Air Quality and Emission Trends Report*; A. Davidson, "Update on Ozone Trends in California's South Coast Air Basin," *Air and Waste* 43 (1993): 226–227.

50. EPA, *National Air Quality and Emission Trends Report*.

51. EPA, *National Air Quality Levels and Trends in Total Suspended Particulate and Sulfur Dioxide Determined by Data in the National Air Sampling Network*, NTIS PB 227059, U.S. EPA, Office of Air Quality Planning and Standards (Research Triangle Park, N.C.: EPA, April 1973).

52. *Environmental Quality 1971, 1979, 1981, 1991*; EPA, *National Air Quality and Emissions Trends Report, 1990*; Freas, personal communication.

53. EPA, *National Air Quality and Emissions Trend Report 1992*; Freas, personal communication.

54. *Environmental Quality 1981*; *Statistical Abstract of the United States 1981, 1986*; Freas, personal communication.

55. EPA, AIRS database.

56. EPA, *National Air Quality and Emissions Trend Report 1992*.

57. *Environmental Quality 1971, 1981, 1984*; *Statistical Abstract of the United States 1981*; Freas, personal communication.

58. National Research Council, *Rethinking the Ozone Problem in Urban and Regional Air Pollution* (Washington, D.C.: National Academy Press, 1991).

59. I. M. Goklany et al., *Critique of Air Quality Analysis Methods Used to Establish Motor Vehicle Emission Standards*, prepared for Volkswagen of America, Warren, Mich. (Concord, Mass.: Environmental Research and Technology, 1981); E. L. Meyer, Jr., *Review of Control Strategies for Ozone and Their Effects on Other Environmental Issues*, EPA-450/4-85-001, OAQPS (Research

Triangle Park, N.C.: EPA, 1986); J. B. Milford, A. G. Russell, and G. J. McRae, "A New Approach to Photochemical Pollution Control: Implications of Spatial Patterns in Pollutant Responses to Reductions in Nitrogen Oxides and Reactive Organic Gas Emissions," *Environmental Science and Technology* 23 (1989): 1290–1301; National Research Council, *Rethinking the Ozone Problem*; J. S. Roselle, T. E. Pierce, and K. L. Schere, "The Sensitivity of Regional Ozone Modeling to Biogenic Hydrocarbons," *Journal of Geophysical Research* 96 (1991): 7371–7394.

60. Goklany et al., *Critique of Air Quality Analysis Methods*; National Research Council, *Rethinking the Ozone Problem*; Roselle, Pierce and Schere, "The Sensitivity of Regional Ozone Modeling."

61. Based on data from EPA, *National Air Pollutant Emission Trends, 1900–1992*, pp. 6-14, 7-1, 7-11.

62. *Environmental Quality 1981, 1984*; *Statistical Abstract of the United States 1986*; Freas, personal communication.

63. Freas, personal communication.

64. Environmental Quality 1981, 1984; *Statistical Abstract of the United States 1981, 1988*; Freas, personal communication.

65. EPA, AIRS database.

66. National Acid Precipitation Assessment Program (NAPAP), *1990 Integrated Assessment Report* (Washington, D.C.: NAPAP, 1991), I. M. Goklany and G. F. Hoffnagle, "Trends in Emissions of PM, SO_2, and NO_x, and $VOC:NO_x$ Ratios and Their Implications for Trends in pH near Industrialized Areas," *Journal of the Air Pollution Control Association* 34 (1984): 844–846; L. O. Hedin et al., "Steep Declines in Atmospheric Base Cations in Regions of Europe and North America," *Nature* 367 (1994): 351–354.

67. NAPAP, *1990 Integrated Assessment*.

68. P. E. Kauppi, K. Mielikainen, and K. Kuusela, "Biomass and Carbon Budget of European Forests, 1971 to 1990," *Science* 256 (1992): 70–74.

69. The NAPAP study defined an acidic water body as one that has no ability to neutralize any additional input of acidic material. These numbers show that smaller lakes are more easily acidified.

70. Data on OECD nations used in this chapter are from Commission of the European Communities (CEC), *The State of the Environment in the European Community*, catalog CB-CO-92-151-EN-C, ISBN91-77-42828-7 (Luxembourg: CEC, 1992), vol. 3; OECD, *OECD Environmental Data 1993* (Paris: OECD, 1993); OECD, *State of the Environment*; OECD, *Environmental Indicators* (Paris: OECD, 1991).

71. WHO/UNEP, *Urban Air Pollution*.

72. World Bank, *World Development Report 1992*; OECD, *OECD Environmental Data 1993*.

73. B. A. Bodhaine and E. A. Dutton, "A Long-term Decrease in Arctic Haze at Barrow, Alaska," *Geophysical Research Letters* 20 (1993): 947–950.

74. M. A. K. Khalil and R. A. Rasmussen, "Global Decrease in Atmospheric Carbon Monoxide Concentrations," *Nature* 370 (1994): 639–641.

75. L. A. Cifuentes and L. B. Lave, "Economic Valuation of Air Pollution Abatement: Benefits from Health Effects," *Annual Review of Energy and the Environment* 18 (1993): 319–342.

76. A. J. Krupnick and P. R. Portney, "Controlling Urban Air Pollution: A Benefit-Cost Assessment," *Science* 252 (1991): 522–528; Office of Technology Assessment (OTA), *Catching Our Breath: Next Steps for Reducing Urban Ozone* (Washington, D.C.: OTA, 1989). Krupnick and Portney's analysis for Los Angeles indicated additional health benefits of $2 billion due to 2,000 premature deaths per year, valued at $1 million each, and $0.7 billion and $0.3 billion due to PM- and ozone-related morbidities. Costs were of the order of $13 billion annually, in addition to what was already being expended.

77. Using a conversion factor of TSP = 3 × RPM levels; (see Cifuentes and Lave, "Economic Valuation of Air Pollution Abatement"), data from Smith ("Fuel Combustion, Air Pollution Exposure, and Health") indicate daily indoor PM levels due to biofuel and coal burning could be as high as 7,800 and 3,600 µg/m^3, respectively. By comparison, levels during the Donora and December 1952 London episodes, PM levels were 5,320 and 4,460 µg/m^3 respectively. See Table 10–1 in this chapter.

EPILOGUE
Reappraising Humanity's Challenges, Humanity's Opportunities

Fred Smith

"We now have in our hands—in our libraries, really—the technology to feed, clothe, and supply energy to an ever-growing population for the next 7 billion years," says Julian Simon. Not so, responds Vice President Al Gore: "Humankind has suddenly entered into a brand new relationship with our planet. Unless we quickly and profoundly change the course of our civilization, we face an immediate and grave danger of destroying the worldwide ecological system that sustains life as we know it."[1]

The doomsayers and the cornucopians who have dominated the debate over the plight of the earth to date differ on almost every dimension, most significantly in the extent to which they see the need to transform Western society radically. The doomsayers' viewpoint, shared by many conventional environmentalists, argues the human dilemma in Terrible Toos terms: There are *too* many of us! We consume *too* much! We rely *too* heavily on technology, which we understand *too* poorly! The human impact on the environment is directly related to our numbers, our affluence, and our reliance on technology.[2] Cornucopians, in contrast, argue that humanity faces no real problems; technological and institutional advances have and will continue to make it possible to address any shortages.

Doomsayers dominate the published literature, with noted figures such as Paul Ehrlich, Barry Commoner, Herman Daly, Carl Sagan, and Jeremy Rifkin all commanding major audiences throughout the world.[3] Still, the cornucopian literature has been significant, with individuals such as the late Herman Kahn and Julian Simon publishing upbeat volumes.[4] Public opinion polls, at least in the

United States, suggest that to date the doomsayers have been more persuasive. Polling data indicate that most Americans seem to believe that the earth's condition is worsening, for example, overwhelming percentages believe that pollution is increasing.[5]

There are major differences in emphasis and orientation between the two schools of thought, as well as in the beliefs or paradigms that undergird their disparate views. To the doomsayers, the earth is an extremely fragile system, akin to a marble balanced on a globe: any disturbance will lead to disaster.[6] All change is of concern, specifically economic and technical changes, which being manmade are novel and thus more worrisome since the earth will not have endured comparable stress in the past. Most doomsayers see the "creative destruction" of capitalism as incompatible with the survival of our planet. Doomsayers are quick to note any evidence of global disaster and to see this as evidence that the Malthusian downturn has begun.

Doomsayers, especially more recently, have focused less on the original Malthusian concerns over food and minerals and turned instead to resources falling outside the human system of private exchange arrangements: air and water pollution, the resources in the oceans, the noneconomic resources collected under the biodiversity banner. A typical doomsayer comment is that by Lester Brown, head of the Worldwatch Institute:

> The environmentally destructive activities of recent decades are now showing up in reduced productivity of croplands, forests, grasslands, and fisheries; in the mounting cleanup costs of toxic waste sites; in rising health care costs of cancer, birth defects, allergies, emphysema, asthma, and other respiratory diseases; and in the spread of hunger.[7]

This emphasis on the damage people have done to the earth and to long-term public health concerns is typical of the neo-Malthusian orientation, although Brown more than most others still seeks to find evidence for the original Malthusian prediction of declining agricultural productivity. Doomsayers tend also to view nature as inherently good; they see the economy as threatening the ecology, people as threatening nature.

In contrast, cornucopians view the earth as robust—more akin to a marble resting at the bottom of a bowl. Perturbations within even the broadest limits likely to stem from any human activity will result in mild disruptions, with a swift return to equilibrium. Change, say the cornucopians, provides great opportunities to innovate, to solve new and more complex problems. They point to history as evidence that we solve far more problems than we create. Cornucopians emphasize the misuse of science and economics by the doomsayers and have the better of the intellectual argument; however, they have rarely discussed pollution, biodiversity, or those resources left outside the system of human exchange save to note the lack of evidence that problems exist in such areas. They act as if technological advances will resolve any problem—any problem, that is, that becomes truly relevant to humanity. In part, their emphasis reflects their focus on resources already integrated into the market exchange system—or on resources that can readily be replaced by resources integrated into that system.

For example, if asked to discuss the problems of the ocean's fisheries, most cornucopians would probably note the rapid expansion of private aquaculture and suggest that if, in fact, oceanic resources do disappear, the world will have already created adequate substitutes. Cornucopians have spent little time addressing environmental quality or resources falling outside the system of private property and market exchange, either because they view such resources as unimportant or because they believe that if such elements become important, the same feedback arrangements that have resolved the Malthusian problem will come into play. Cornucopians neglect the institutional arrangements that facilitate or block creative responses to "common property" tragedies, taking an implicit attitude that "somehow it will all work out."[8]

Thus, to some degree, the two groups are talking past each other. Their disagreement reflects different perspectives, different ways of viewing the same situation. Is the cup half filled or half empty? What we expect is often what we see. Some parts of this book are clearly cornucopian, but not all. Taken as a whole, this book charts a middle course between the doomsayers and the cornucopians. It can best be described as following a line of thought explored by the Reverend Thomas Malthus himself in various revisions of his famous essay. His initial work seemed to argue that in a world where demand grew exponentially and supply grew linearly at best, disaster was inevitable. Yet Malthus himself soon moved away from that initial pessimism as he came to realize that impending scarcity sometimes stimulated creative responses—conservation efforts, active efforts to find and develop new supplies, and innovative efforts to find technological substitutes—that mitigated or eliminated the shortfall. Malthus was an honest researcher and realized that our ability to feed, clothe, and house ourselves had steadily improved. His conclusion? Supply might—under the right institutional arrangements—grow more rapidly than demand. This book supports that conclusion, finding that in many important ways, the world has indeed improved. Human lifespans have dramatically increased in this century; human wealth and nutritional standards have improved almost everywhere for many years. Yet the doomsayers are right to point out that resources falling outside the system of private management such as wildlife or the ocean's fisheries do face problems.

That some institutional arrangements better protect resources than others should not be surprising. The critical challenge is to develop creative institutional arrangements that will encourage people to consider the status of valued resources and the risks to them. Unfortunately, this more tempered perspective of the Reverend Malthus has yet to influence public policy. Neither doomsayers nor cornucopians have focused on the framework within which resources are used—or misused. We must examine these issues and determine what factors might explain the successes and failures discussed in the main body of the book. Our ultimate goal is to eliminate the distinction between economic and ecological values, to encourage institutions that ensure the widest possible scope for human valuation of environmental values.

A Brief Historical Overview and Some Organizing Concepts

Human prehistory is a story of exploitative resource management. In the beginning, people hunted and gathered in small, wandering bands, with little concern how such policies might affect future resource availability. Coming upon a laden fruit tree, the band might well pull down the branches to ease the task of gathering the bounty. Similarly, a hunting party coming upon a trapped herd of deer might well kill them all, thus savoring the tastier cuts, even though much of the meat would go to waste. In the original world, nothing was owned; everything was a common property resource. Ownership was conveyed only when one killed the species, harvested the fruit. In this world, the only rational policy was a use-it-or-lose-it strategy.

Such management approaches are feasible only for small populations visiting an area infrequently. As people settled in villages, abandoning the hunter-gatherer nomadic existence, the future became more relevant: a damaged fruit tree would reduce the future welfare of all villagers, as would excessive depletion of local game. "Open access" rules became increasingly inefficient. The resulting problem was explored by Garrett Hardin in a now-famous article that appeared in 1968, "The Tragedy of the Commons."[9] To Hardin, a wide range of environmental problems might best be viewed as arising from a system of open access to a commonly owned resource. The result of such policies is disaster: "Ruin is the destination toward which all men rush, each pursuing his own best interest in a society which believes in the freedom of the commons. Freedom in a commons brings ruin to all."[10]

Hardin used the example of a medieval commons—a pasture in the center of the village on which all townspeople grazed their cows. When the population is low and the number of cows few, there is little problem; the carrying capacity of the pasture is greater than the stress placed on it. But once the village becomes wealthy and the number of cows exceeds that capacity, each additional cow has a major negative consequence for all users. Yet self-interest still motivates each villager to add more cows; this person gains the benefits while the costs are shared among all the other villagers. Since all villagers face the same situation, overgrazing inevitably results. Under a commons management scheme, benefits are privatized, while costs are socialized. Self-interest leads to societal disaster.

Hardin dismissed the idea that the tragedy of the commons resulted from a lack of ecological concern—that if only people cared more about the environment, all would be well. He noted that ecological altruism in a common property situation would create "a selective incentive system that works toward the elimination of conscience from the race."[11] In such a world, the sacrifice of the altruist would almost certainly be futile, as others took advantage of his or her restraint by grazing more of their cattle. To Hardin, either private or political management of the commons was essential.[12]

Private ownership avoids the tragedy of the commons rather easily. In a communally owned forest, for instance, the incentive is to cut down as many trees as possible as quickly as possible. But a person who owns property can exclude others from using it. He can make rational decisions not just for his own but for others' benefit. He can consider the future. If trees are becoming scarce, he is motivated to protect his own trees because they will become more valuable.[13]

The lesson of the tragedy of the commons is clear: commons must be managed. Moreover, it is clear that management strategies have evolved as need emerged rather than being imposed. An example from Canada's history is illustrative: the protection of the beaver.[14] In precolonial times, beaver were plentiful throughout what would become the nation of Canada. Native American demands on the beaver population were low and therefore represented little threat to the species. But when French fur trappers arrived, the situation changed. There was a great demand for beaver pelts in Europe, and the French provided technologies, guns, and traps, which greatly reduced the costs of catching beaver. Beaver hunting escalated rapidly, and the result was a sharp decline in the beaver population.

The indigenous populations were cognizant of this problem and met to resolve the issue. Historically, beaver had been treated as a common property resource: any beaver could be taken by anyone. That system had worked in a period of minimal demand and vast supply, but with the entrance of the Europeans, it no longer ensured sustainability. To account for the increased demand, the Indians created a system of private property rights, allocating to each family grouping an area containing one or more beaver lodges. In effect, the beaver were privatized. Rules and procedures for adjudicating disputes and policing these property rights also evolved. A family that overexploited its resource base would feel that impact immediately and suffer the economic consequences, while those that used their beavers in a sustainable manner prospered in concert with the beaver population. This decentralization of resource management encouraged more careful stewardship of the resource: each family benefited directly by preventing poaching on its lands, by inventorying its beaver population to guard against overharvesting. The overall effect of this property rights–based management scheme was to restore the balance between people and nature, to ensure sustainable development. That balance survived until the English immigrants arrived. These newcomers neither understood nor recognized Indian property rights and began to trap beaver directly. As a result, the system collapsed (neither the British nor the Indians understood or respected the property rights rules governing the other), and the beaver were quickly hunted to the verge of extinction. This story illustrates well the value and the pressures that led to the evolution of property rights—and their vulnerability to political forces.[15] No property rights system can survive if the underlying legal system fails to protect and defend it.

Human prehistory has greatly complicated modern environmental policy. As Frederick Hayek once noted, centuries of existence as hunter-gatherers had left people with a romantic attachment to the idea that common property management regimes are somehow more moral, more natural[16] and that open access resources are somehow superior to fenced, private property. The result is a bias toward political solutions to natural resource and environmental protection issues. The challenge has been to make political management a viable option—and here the story as told throughout this book is not so positive.

Political management of the commons is difficult, troublesome, and rigid. Political managers avoid innovation, are often shortsighted, and reward the powerful few at the expense of the unorganized many. Management budgets and priorities reflect interest group pressures rather than sound environmental policies. Administrators must respond to political considerations. In such a system, good intentions lead to good policy only by serendipity, especially since there is little incentive to examine carefully the consequences of actions. In effect, politicizing the situation does not resolve the tragedy of the commons; rather, it institutionalizes it.

Evidence for this view is best provided by the experiment conducted over the last hundred years among the world economic powers. As Mikhail Bernstam notes, market economies advanced both economic and ecological objectives far more effectively than did the centrally planned economies of the world. He presents data showing that per dollar of gross national product (GNP), socialist economies had consumed nearly three times as much energy as market economies. This comparison persisted even when two nations were very similar. The former East Germany consumed 40 percent more energy per person and more than three and a half times as much energy per dollar of GNP as West Germany. North Korea uses 70 percent more energy per capita than South Korea. Market economies create more targeted efficiency incentives. Waste is always costly, but nonmarket economies diffuse the costs of such wastes over vast bureaucracies; in contrast, under capitalism the less efficient firm immediately loses in comparison with its more efficient competitor. The result is decentralized private property–based economies that have dramatically outperformed their centralized socialist counterparts. Communal institutions find it hard to create the incentives needed to mobilize the knowledge dispersed among their members, to organize their varied strengths and talents.[17]

Decades of such comparative advantages have convincingly demonstrated that it is in the decentralized market systems of the West where environmental problems are less serious. Bernstam notes a little observed fact:

> In the 1970s and the 1980s, an amazing divergence took place in the trends in resource use and pollution within the developed industrial world. This divergence between Western market economies and the socialist economies of the USSR and Eastern Europe went virtually unnoticed. Yet, it may have signified the most important reversal in economic and environmental history since the Industrial Revolution. . . . The amounts of throughput of major resources and the ensuing discharges of air, water, and soil pollution began to decline rapidly in those nations with competitive market economies. This is true notwithstanding further economic growth in Western market countries. During the same two decades, the throughput of resources and environmental disruption was rapidly increasing in the USSR and European socialist countries even though their economies slowed down and eventually stagnated.[18]

This dematerialization of Western economies was in part a response to gradually changing resource costs—encouraging ever greater efficiencies in material and energy utilization—and in part to the growing costs of environmental protection and waste disposal, these latter affected by the level of material throughput. Lacking any equivalent market feedback mechanism, the planned economies lagged in this effort. Their earlier achievements can best be viewed as an effective effort at replicating the established technologies of the past; they were less equipped to track the rapidly innovating market economies of the post–World War II era. Institutions matter, and in this area, the superiority of property-based market arrangements is massive.[19]

This history also suggests that successful institutional arrangements are rarely imposed from the top down but more commonly evolve as experience indicates the strengths and weaknesses of various approaches. Strict property systems are not always required. Sometimes, the simple availability of an exchange market may "save" a resource.

Consider the collapse of the commercial whaling industry in the nineteenth century. At that time (and, for that matter, still today), neither whales nor the oceans were owned at all. Both were common property resources. As the world became a richer place, the demand for home illumination increased, and the resulting demand for whale oil lamps exploded. Whales were "free" but increasingly expensive to locate and kill in the vast expanse of the world's oceans. As whales became scarce and search cost increased, whale oil prices soared—to over $137 a barrel, a price vastly higher than petroleum prices today.[20] The higher prices encouraged creative innovations in the whaling industry and outside the industry to develop alternative lighting technologies. This process produced coal gas, then kerosene, then natural gas, and eventually the electric light. The transformation was not smooth; whale oil was an extremely clean-burning fuel, while the early coal gas and kerosene lamps smoked badly. Nonetheless, the cost factor made them competitive, and soon technological improvements made it possible for such "artificial" illumination technologies to dominate the day.[21]

Note, however, that this happy situation need not prevail. Generally as the number of a common-property species or the quantity of a common-property resource declines, there will be some increased search and capture costs; however, there is no guarantee that such cost increases will offset increased demand. The salvation of the whale depended on the relationship of search costs, whale oil prices, and the cost and quality of competing lighting technologies. Had sonar and radar come available at an earlier date and reduced whale capture costs quickly enough, the whale might well have disappeared. Indeed, land-based species have suffered losses of this type both because search costs are less and habitat conversion is a more serious threat.

The fisheries of the world, as noted in this book, are suffering from overexploitation. Mariculture, unlike the petroleum industry, has not yet made fishing noneconomic. The whaling situation occurred when governments were less quick to subsidize declining industries; today, governments sometimes willfully subsidize noneconomic practices that harm the environment. Had government moved to subsidize the costs of whale oil, demand might well have remained high. Indeed, coal extraction in Germany and Spain remains high for exactly this reason. Had whale oil been subsidized, moreover, the incentives to develop alternative illumination technologies would have been far weaker. Finally, there were few restrictions on technological innovation in nineteenth-century America. Had the petroleum industry faced the barriers now encountered by the biotechnology industry, for example, change might have been delayed, and the plight of the whale might have become far greater. Technological change takes time even when government does not interfere, and such delays affect the stress placed on the environment. The deforestation of England, in part, resulted from the difficulty of replacing charcoal with coal (in the form of coke) in steelmaking.

For such reasons, the tragedy of the commons is most likely to proceed to its gloomy end in poorer societies, societies discouraging technological innovation, societies reluctant to allow private property ownership, and societies in which government subsidies are common. Unfortunately, it is precisely in such societies where much of the world's resources still reside.

At this point some readers may be objecting that *any* attempt to integrate people and nature is inherently suspect and denigrating to nature itself. Nature is wild and free—not fenced and owned, they say. This is a utopian vision left over from hundreds of thousands of years past. The brute fact is, as the late Kenneth Boulding once argued, that as humans become an ever more dominant presence on this earth, it becomes inevitable that "any species which cannot in some sense be domesticated is doomed."[22] Boulding's use of the *domestication* term was expansive, of course; his definition included pets, wildlife, and even wilderness. The point, however, is critical: humans are by far the most powerful creatures on the earth. To ignore that fact, to act as if people were not present or could be ignored, is to advance policies that have no prospect for success.

This brute fact, however, has a bright side. *Domestication* is a term that suggests ownership but not necessarily for economic purposes. Indeed, the value of ownership of resources that do not directly create wealth—cats and dogs, for example—may be the most creative area in which property rights might be expanded. Consider that no one is responsible for preserving pet biodiversity, for ensuring that Persian cats or Irish wolfhounds do not disappear from the earth. There is no endangered pet act. Were it not so silly, this might seem surprising, since, after all, far more people care about these species than those "protected" by the Endangered Species Act. The reason, of course, is that fears of this sort seem fanciful. No one worries about the extinction of pets because it is widely recognized that as a specific species declines, it becomes scarce and thus more eagerly sought after. Humans seek, in part, to differentiate themselves by their pets, and this creates a mechanism whereby scarcity generates its own redress.

The distinction between economic and noneconomic motives is largely arbitrary in any event. True, the primary motivation for pet ownership is noneconomic; few pet owners profit financially from ownership. Yet this noneconomic motivation has created a massive economic pet industry that provides services (feeding, medical care, housing, transportation, buying/selling). Also, this industry finds its worthwhile to invest in knowledge creation, to become more knowledgeable about pet diseases, superior grooming and handling techniques, and even animal tastes for housing and food. The stewardship relationship between owner and pet, between humans and nature as it were, creates linkages that motivate and mobilize human talent to make the world a better place for pets. The earlier Kenneth Boulding quotation suggests that domestication may be inevitable; the example of pets suggests that it may be an extremely effective way of linking human welfare to nature, of mobilizing the dispersed interest of humanity to advance an environmental objective. One may regret that fact or find it ennobling, but it remains a reality that cannot be ignored.

A policy option flowing from these comments is suggested by Stephen Edwards in chapter 7 in this book. He notes that extinctions most frequently occur among island populations. In part, this reflects the greater vulnerability of an isolated population to stress, in part the fact that such island populations are sometimes better cataloged. However, that exact isolation also suggests that human intervention might prevent such extinctions. People have a great experience in moving species from one habitat to another, and the applicability of this strategy to environmental resources has yet to be fully exploited. Those seeing species movements as "not what we mean by environmental protection" should consider the comparable movement of children from wartime London. Providing safe havens for endangered species may be critical, especially as the Third World undergoes demographic transitions. Some nations will remain stable during this era; others already are unstable. Neither nature nor humanity is secure in a war zone, and policy should recognize that fact.

This point should be reemphasized. A major theme that unites this book is the realization that while many of the stresses placed on this earth are the result of human activities (that is, that humans are part of the problem facing earth), human-centered solutions offer the only hope of resolving these problems. Stephen Edwards makes that point most eloquently in his comment: "Faith in fundamental human respect for the environment leads me to believe that far more biodiversity will be conserved if people are given rights to use wild resources than if we continue our present course."

Conclusion: Institutions Matter

In seeking to ensure the well-being of our planet, we should consider the statement by Winston Churchill about the English cathedral: "First we shape our institutions; and then they shape us!" That comment is even more applicable in considering the rich array of legal and political institutions that determine the treatment of natural and environmental resources. Our challenge is to design institutions suitable for humanity as it is—not as we would have it be. Altruistic behavior should be recognized, and even honored, but our greatest reliance in organizing society should be placed on that more common attribute, self-interest.

Resource management is a challenge of extending to natural and environmental resources the same protections now afforded more conventional resources. Far too many critical environmental resources—the oceans, the airsheds, much of the world's flora and fauna, almost all its groundwater, and, in much of the world, even forestry and mineral resources—still are managed as common property resources. Even partial private property and exchange arrangements can mitigate the disasters forecast by Hardin. The problems discussed in this book and the fact that environmental expectations have risen rapidly in the developed world suggest that improved institutional arrangements are overdue. Property rights are a fiction, of course, unless the legal system acts to protect and defend them. And, obviously, resource management of any sort is most feasible in a society largely at peace. War zones rarely encourage long-term thinking by anyone regarding anything.

The case for free market environmentalism is rather simple for those resources in which property rights are easily defined and exchanged. However, there are more difficult environmental areas where defining property rights appears to be impossible. As one critic, Robert Stavins, asked, "Does anyone really believe that acid rain can be efficiently controlled by assigning private property rights for the U.S. airshed and then effecting negotiations among all affected parties?"[23] Certainly it is one thing to put a fence around one's land or to patrol it to deter poachers, litterers, and other undesirables. It is quite another to keep unauthorized fishing boats out of one's stretch of ocean or to identify the source of pollution that is damaging one's orchard (or lungs).

How do we fence the airshed, groundwater, or the oceans? This feat appears as difficult to us now as did the fencing of the western grazing lands in the nineteenth century. In those windswept arid plains, substantial acreage was needed to sustain a family, and building wooden fences or stone walls to privatize land was prohibitively expensive. An 1850s Stavins would have argued that in such a situation, no property rights solution was feasible, just as the real Stavins does today with regard to air and water. Yet the problem of property rights in the West was resolved through voluntary actions. Institutions evolved that defined and protected property rights. Ultimately a technology—barbed wire—greatly reduced the costs of marking property boundaries.[24]

Technologies now exist that make it possible to determine, within limits, the quantity and types of air pollution entering a region. Lasimetrics, for example, could map atmospheric chemical concentrations from orbit. In time, that science might provide a sophisticated means of tracking cross-boundary pollution flows. Also, large installations such as power plants could add (or be required to add) chemical or isotopic "labels" to their emissions to facilitate tracking. Such "labeling" has long been routine in the manufacture of explosives, to help trace explosives used in crime or terrorism.[25]

The institutional arrangements now used to address technological approval questions, to ensure that the risks created and reduced by technology are handled in a balanced way, are far from ideal. A comparative study of these institutions suggests that although the risks of innovation are weighted heavily, the risks of technological stagnation receive almost no attention. As a result, the developed world is now spending large sums to address risks associated with change while neglecting the risks posed by technological stagnation.

Among the most important institutions needed to ensure sustainable development and an improved environment are those that generate and preserve knowledge. Indeed, among the earliest of human institutional innovations was the creation of a priestly class to ensure that hard-won knowledge was not lost. That capability was strengthened when writing was invented and knowledge became more readily stored and transmitted. Although we take for granted such institutions today, we should realize how important for both economic and ecological purposes was the ability to plant at appropriate times, to understand which crops did best on which soils, and so forth. Moreover, the ability to process foods made it possible to reduce human stress on the environment. Domestication in which people took over some of the functions provided by the wild species—in return for a greater yield—is another critical innovation. These institutions should be recognized and strengthened as we seek to bring in from the cold the various resources still neglected in our modern world.

A world of voluntary arrangements may not be perfect, but it is the best solution to our ecological problems. Only under a system where resources are privately held will people have the ability to express their environmental value accurately. Only through a price system will those values be conveyed to entrepreneurs, who can in turn satisfy those values.

It is time to launch the second wave of environmentalism. Environmental values will survive only with the same protections now afforded to economic resources. We cannot deny the existence of the house of humanity. Our goal instead must be to integrate the house of humanity with the house of nature.

NOTES

1. Norman Myers and Julian L. Simon, *Scarcity or Abundance? A Debate on the Environment* (New York: W. W. Norton, 1994), p. 65; Al Gore, "An Ecological Kristallnacht. Listen," *New York Times*, March 19, 1989.

2. The relationship was formalized by Paul Ehrlich in his *I = PAT* equation (impact equals population times affluence times technology). This equation and its limitations was discussed by Indur Goklany in chapter 10.

3. Paul R. Ehrlich, *The Population Bomb* (New York: Ballantine Books, 1968), and *The Population Explosion* (New York: Touchstone Books, 1990); Barry Commoner, *The Closing Circle: Nature, Man and Technology* (New York: Bantam Books, [1971] 1974); George J. Mitchell, *World on Fire: Saving an Endangered Earth* (New York: Charles Scribner's Sons, 1991); Anita Gordon and David Suzuki, *It's a Matter of Survival* (Cambridge: Harvard University Press, 1991); Al Gore, *Earth in the Balance: Ecology and the Human Spirit* (New York: Plume, [1992] 1993).

4. Julian L. Simon, *Population Matters: People, Resources, Environment, and Immigration* (New Brunswick, N.J.: Transaction Publishers, 1990), and *The Ultimate Resource* (Princeton, N.J.: Princeton University Press, 1981); Julian L. Simon and Herman Kahn, *The Resourceful Earth* (New York: Basil Blackwell, 1984).

5. A 1991 poll found that "74 percent of Americans think the greenhouse effect is a problem" and "41 percent believe it's a serious problem." See Kenneth T. Derr, "Oil and Its Critics: The Facts and the Future" (speech delivered to the Cambridge Energy Research Associates Executive Conference, Houston, Texas, February 8, 1994).

6. This linkage between views on earth's stability and cultural values is a major element in the significant work conducted by Mary Douglas and the late Aaron Wildavsky. For a discussion of how views of nature relate to values, see "A Credible Biosphere," in Mary Douglas, *Risk and Blame: Essays in Cultural Theory* (New York: Routledge Press, 1992).

7. Lester R. Brown, *State of the World* (New York: W. W. Norton, 1993), pp. 4–5.

8. The term *common property* refers to resources available for all to use, resources subject to open access. Such resources often are overused.

9. Garrett Hardin, "The Tragedy of the Commons," *Science*, December 13, 1968, pp. 1243–1248.

10. John Baden and Garrett Hardin, *Managing the Commons* (San Francisco: W. H. Freeman, 1977), p. 20.

11. Ibid., p. 25.

12. Randy Simmons and I have examined the respective merits of these two solutions in a paper, "The Tragedy of the Commons Revisited Again," Competitive Enterprise Institute Working Paper (April 1989).

13. See Harold Demsetz, "Toward a Theory of Property Rights," *American Economic Review* (May 1967): 347–360.

14. Ibid.

15. A major challenge is how one might integrate traditional property rights regimes into the legal property rights regimes of modern states. As the Canadian example illustrates, this may be difficult. The task is complicated by the fact that often traditional property rights can be exchanged only within the tribe or local region; outsiders are unable to engage in trades and thus have no peaceful way of expressing their values. Too often, this leads, as in the case discussed, to the native rights' being destroyed—and the resource depleted.

16. Frederick Hayek, "The Atavism of Social Justice," in *New Studies in Philosophy, Politics, Economics, and the History of Ideas* (Chicago: University of Chicago Press, 1978), pp. 57–68.

17. Mikhail Bernstam, *The Wealth of Nations and the Environment* (London: Institute for Economic Affairs, 1991), p. 14.

18. Ibid.

19. Failure to innovate is not, of course, restricted to fully planned economies. Key resource and industrial sectors in many developed nations are heavily regulated (pharmaceutical, public utilities), subsidized (agriculture, urban transportation), and in some cases even nationalized (the post office, Amtrak). These sectors rarely achieve the efficiencies common in the fully private sector.

20. The whale oil price of $2.50 per gallon is taken from Daniel Yergin, *The Prize* (New York: Simon and Schuster, 1991), p. 22.

21. Pressures on a host of other wild species have been reduced by similar cost-induced technological substitutions. Mink, for example, were once trapped in the wild; however, increased quality demands and higher trapping costs encouraged ranch mink, and today ranched mink dominate the fur market. Similarly, pressures on oceanic resources are already lessening as mariculture expands.

22. Kenneth Boulding, "Economics and Ecology," in *Future Environments of North America*, ed. F. Fraser Darling and John P. Milton (Garden City, N.Y.: Natural History Press, 1966).

23. Robert Stavins, letter to *Policy Review* (Summer 1989): 95–96.

24. Terry L. Anderson and P. J. Hill, "The Evolution of Property Rights: A Study of the American West," *Journal of Law and Economics* 12 (1975): 163–179.

25. These and other free market environmental approaches are discussed in Fred L. Smith, Jr., "A Free-Market Environmental Program," *Cato Journal* (Winter 1992): 457–475, and "The Market and Nature," *Freeman* (September 1993): 350–356.

BENCHMARKS

The Ecological and Economic Trends That Are Shaping the Natural Environment and Human Societies

Figures compiled by Jonathan Adler and Peter Cazamias

Analysis by Jonathan Adler, Peter Cazamias, and David Monack

I

WORLD DOMESTIC PRODUCT

Global economic output has increased dramatically since World War II. World gross domestic product (GDP) nearly doubled from 1970 to 1993, rising from just over $9.6 trillion to just under $18.8 trillion. During the 1970s, 1980s, and 1990s, global gross national product per capita has also increased significantly. Individual productivity rose considerably in the latter half of the 1980s, spurred by advances in technology and a wave of market liberalization around the globe. The world GDP chart on the facing page reflects the level of global prosperity by measuring the cumulative domestic product values for all national economies. Growth in GDP per capita slowed between 1979 and 1982 due to a worldwide global recession and the debt crisis that affected many developing countries, but it has since rebounded.

While North America and Western Europe have enjoyed steady and significant economic growth for the past two centuries, many of the world's less developed regions have begun to catch up at an accelerated rate, taking advantage of modern methods of production.

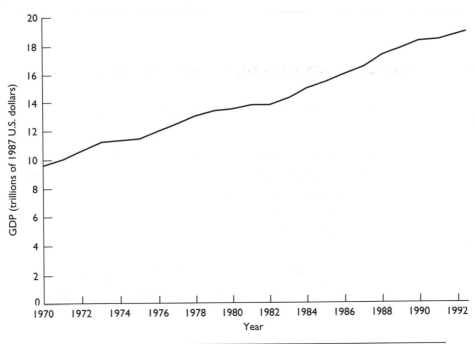

Year	World GDP (trillions of 1987 U.S. $)
1970	9.7
1971	10.0
1972	10.5
1973	11.2
1974	11.3
1975	11.4
1976	11.9
1977	12.4
1978	12.9
1979	13.4
1980	13.6
1981	13.8
1982	13.8
1983	14.2
1984	14.9
1985	15.4
1986	15.8
1987	16.4
1988	17.1
1989	17.7
1990	18.1
1991	18.3
1992	18.6
1993	18.8

Source: International Bank for Reconstruction and Development (IBRD)/World Bank, *World Tables 1994*, data on diskette (Washington, D.C.: IBRD/World Bank, 1994).

2

ANNUALIZED POLLUTION CONTROL EXPENDITURES, UNITED STATES

Increased concern for environmental quality and the potential impact of pollution on human and natural environments has fueled the rapid growth of environmental regulations, particularly in the industrialized world. Western nations have created large regulatory agencies to enforce compliance by businesses and individuals with proliferating environmental regulations.

An increasing amount of wealth is devoted to pollution control. In the United States, pollution control expenditures have increased nearly fivefold since 1972. Well over 2 percent of the gross domestic product of the United States is now spent on pollution control and compliance with environmental laws. Similar trends could be observed in other developed, Western nations.

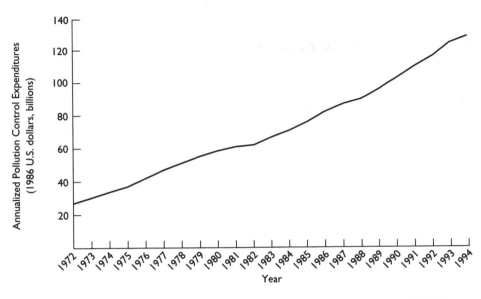

Year	Expenditures (1986 U.S. $)
1972	$ 26,481
1973	30,261
1974	33,614
1975	36,842
1976	41,572
1977	46,509
1978	50,482
1979	54,824
1980	57,969
1981	60,539
1982	61,273
1983	65,477
1984	69,925
1985	74,021
1986	80,046
1987	85,290
1988	88,490
1989	94,280
1990	100,167
1991	107,867
1992	114,181
1993	123,735
1994	127,039

Source: U.S. Environmental Protection Agency, *Environmental Investments: The Cost of a Clean Environment,* Office of Policy Planning and Evaluation, EPA 230-12-90-084 (Washington, D.C.: Government Printing Office, December 1990).

POPULATION PROJECTIONS, WORLD

World population has more than doubled since 1950, chiefly because of the large reduction in worldwide death rates, not a major increase in global birthrates. The decline in death rates should continue as economic growth and scientific advances improve human health.

The United Nations issues a series of population projections for the twenty-first century, but population projections vary widely depending on the assumptions used. Without further understanding of how the projections were calculated, one might assume that the "medium projection" is the most likely. In fact, the medium UN projection assumes that acceleration or deceleration of death and fertility rates will continue as they have in recent years. However, much evidence suggests that this assumption is wrong. Historically, in a given society, after a lag period, a rapid reduction in fertility rates follows the drop in mortality rates due to economic progress. When economic development modernizes an economy by improving women's education, per capita incomes, and the infant mortality rate, fertility rates quickly decline to the point at which the population's growth rate is at or below zero. This process has already been completed in many Western nations, and even in Asia and Latin America, and some parts of Africa, fertility rates are dropping rapidly. For example, India's fertility rate dropped 25 percent in the 1980s, and China's dropped by 60 percent from 1970 to 1990. Future reductions in the world fertility rate are likely to produce populations more in line with the UN's low projection. In any event, population will slow at some point; even the highest projected UN trend has the world's population stop growing in 2075.

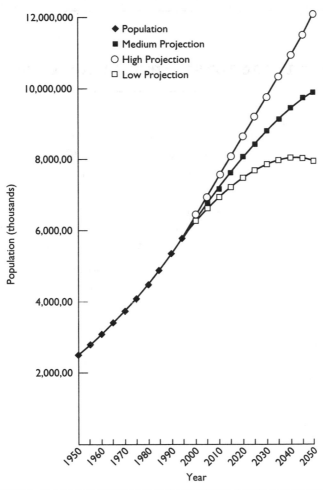

	Population	Medium Projection	High Projection	Low Projection
1950	2,519,742			
1955	2,754,189			
1960	3,021,478			
1965	3,337,812			
1970	3,697,131			
1975	4,096,974			
1980	4,444,340			
1985	4,846,320			
1990	5,284,816	5,284,816	5,284,816	5,284,816
1995		5,716,407	5,741,568	5,688,944
2000		6,158,030	6,234,813	6,081,292
2005		6,594,379	6,744,211	6,448,957
2010		7,032,268	7,273,968	6,790,801
2015		7,468,896	7,826,155	7,104,255
2020		7,887,826	8,392,384	7,372,140
2025		8,294,310	8,978,696	7,603,152
2030		8,670,581	9,566,714	7,780,625
2035		9,013,842	10,153,537	7,900,775
2040		9,318,190	10,734,294	7,958,748
2045		9,587,279	11,315,486	7,959,963
2050		9,833,167	11,912,306	7,917,503

Source: United Nations, *World Population Prospects: The 1994 Revision—Annex Tables* (New York: United Nations Population Division, Department for Economic and Social Information and Policy Analysis, 1994), tables A-1, A-2.

4

ESTIMATED AND PROJECTED TOTAL FERTILITY RATES, WORLD

Fertility rates worldwide have been declining over the past several decades. One of the clearest indicators of fertility, the total fertility rate (TFR), corresponds to the average number of births per woman over the course of childbearing ages. The world's TFR has dropped by nearly two-fifths since 1950/55, from roughly 5 children per woman to 3.26 children per woman today, and appears to be heading toward further decline. The TFR for the world's developed regions dropped by nearly a third, from 2.83 children per woman in 1950/55 to 1.91 children per woman in 1990/95. This is below the replacement level of 2.1 children per woman. The developing regions of the world have witnessed a steep 41 percent reduction in their TFRs.

These fertility declines are due, in part, to significant advancements in contraceptive technology since World War II. However, changes in desired fertility that have accompanied economic development appear to have been the dominant force behind lower fertility rates.

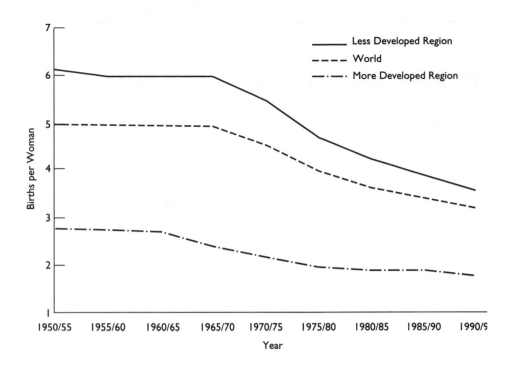

Year	World	More Developed Region	Less Developed Region
1950/55	5.00	2.83	6.19
1955/60	4.88	2.82	5.95
1960/65	4.98	2.69	6.09
1965/70	4.89	2.44	6.01
1970/75	4.46	2.21	5.42
1975/80	3.84	2.0	4.55
1980/85	3.64	1.93	4.23
1985/90	3.43	1.92	3.9
1990/95	3.26	1.91	3.64

Source: Ibid., pp. 122, 126.
Note: Medium variant projections used for 1990/95.

ESTIMATED AND PROJECTED LIFE EXPECTANCY AT BIRTH, WORLD

The twentieth century has witnessed an explosion in global health, as evidenced by the dramatic increase in human longevity. Life expectancy since 1950 has increased close to 40 percent, from 46.4 years to 64.7 years. Life spans for populations in the developing regions have increased over 50 percent, from 40.7 years in 1950 to 62.4 years currently. Populations from the world's more developed regions have extended their average life span by eight full years in the same period, from 66 years to 74 years.

Technological progress has been a driving force behind the extension of the human life span. Medical breakthroughs, infrastructure improvement, and innovations in communications and transport have improved the quality and capacity of medical relief to be administered to the world's populations. Improved medical care has factored into the nearly 60 percent decline of infant mortality rates, which in turn strongly affects life expectancy at birth. Agricultural innovations also have yielded an increase in the supply and availability of the world's food. Increased food availability has averted millions of deaths from starvation that would have otherwise occurred.

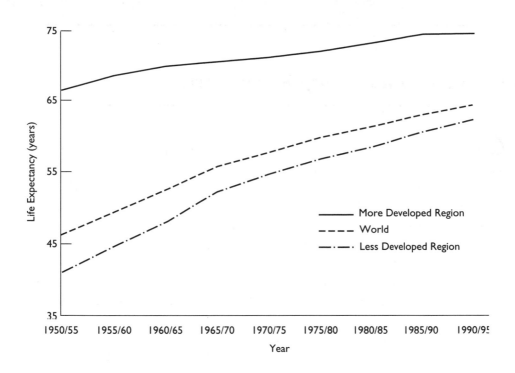

Year	World	More Developed Regions	Less Developed Regions
1950/55	46.4	66.0	40.7
1955/60	49.6	68.4	44.1
1960/65	52.4	69.8	47.4
1965/70	56.0	70.5	52.0
1970/75	57.9	71.1	54.5
1975/80	59.7	71.9	56.6
1980/85	61.4	72.7	58.6
1985/90	63.3	73.7	60.7
1990/95	64.7	74.0	62.4

Source: Ibid., pp. 166, 180.
Note: Medium variant projections used for 1990/95.

WETLAND LOSSES AND GAINS, UNITED STATES

One of the main concerns about economic development is the effect it has on land use. In particular, some fear that species habitat and environmentally sensitive areas will be converted to other uses. In the United States, much of this concern has focused on wetlands because they serve to filter water and can play an important role as coastal buffer zones. Many animal species, including whooping cranes, alligators, and two-thirds of North American species of ducks and geese, either live or breed on wetlands.

In the postwar period, wetlands have been converted to other uses at a dramatic rate in the United States: 300,000 to 450,000 acres per year. Most wetlands were lost by being drained and converted into farmland. As farm productivity has increased, however, the rate of wetland conversion has slowed dramatically, from an average of 458,000 acres per year between 1954 and 1974, to an average of 135,000 acres per year between 1982 and 1992. At the same time, wetland restoration has emerged as a significant trend in the United States. Beginning in 1987, various incentive and restoration programs have been responsible for the restoration and creation of several hundred thousand acres of wetlands per year. In 1994, wetland restoration occurred on 172,000 acres. The result is that there was no net loss of wetlands that year.

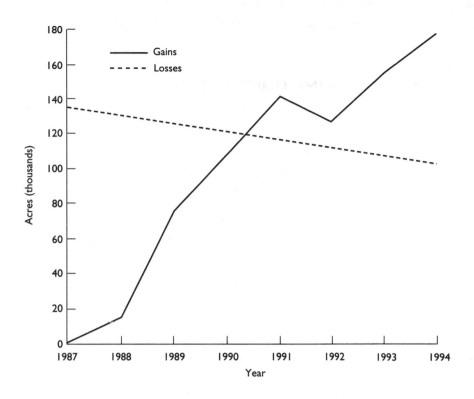

Year	Losses (thousand acres)	Gains (thousand acres)
1954–1974 (avg.)	458	
1974–1983 (avg.)	290	
1982–1992 (avg.)	135	
1987	135	2
1988	131	16
1989	126	75
1990	121	107
1991	116	139
1992	112	126
1993	107	151
1994	102	172

Source: U.S. Department of the Interior, *National Wetlands Inventory and Trends, Mid-1950s to Mid-1970.* (Washington, D.C.: Government Printing Office, 1982); U.S. Department of Agriculture, Soil Conservatior Service, *National Resources Inventory* (Washington, D.C.: Government Printing Office, various years) Jonathan Tolman, *Gaining More Ground: Analysis of Wetland Trends in the United States* (Washington, D.C. Competitive Enterprise Institute, October 1994).

ARABLE AND PERMANENT CROPLAND, WORLD

The amount of arable and permanent cropland worldwide has been increasing at a slow but relatively steady rate over the past two decades. Global cropland area expanded by less than 2 percent between 1980 and 1990. While the overall trend was toward more cropland, many regions saw a decrease. The amount of European land under crops declined by about 5 percent between 1970 and 1990, while North and Central America together registered a cropland decline of about 0.7 percent. The former Soviet Union's area under crops declined by 1 percent between 1980 and 1990. The greatest increases in cropland from 1980 to 1990 occurred in Oceania (11.6 percent), South America (10.9 percent), and Africa (4.4 percent). These increases account for the overall upward trend.

Before the twentieth century, the world increased its food supply chiefly by expanding the amount of land cleared and planted in crops. By dramatically increasing the amount of food grown on land already under cultivation, humanity has already managed to save up to 10 million square miles—the total area of North America—of rain forests, wetlands, and mountain terrain from being plowed down. Higher agricultural yields were achieved by substituting more productive crop varieties, pesticides, and fertilizers for extra acreage.

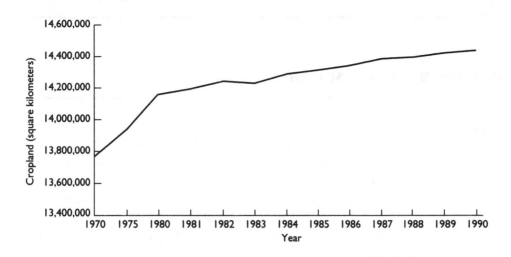

Year	Cropland (square kilometers)
1970	13,772,210
1975	13,941,490
1980	14,170,980
1981	14,204,020
1982	14,246,470
1983	14,227,590
1984	14,290,310
1985	14,313,720
1986	14,340,450
1987	14,387,430
1988	14,403,470
1989	14,423,130
1990	14,442,170

Source: Organization for Economic Cooperation and Development, *OECD Environmental Data, Compendium 1993* (Paris: OECD, 1993).

PRODUCTION OF CEREALS, WORLD

World cereal output increased from close to 885 billion metric tons in 1961 to over 1.95 trillion metric tons in 1992. Japan and the former Soviet Union are the globe's greatest net importers of cereal, while the United States, Canada, and France remain the world's leading net exporters of cereals. The world's developed countries accounted for most of this period's cereal production. However, the percentage of produced cereal from developing regions climbed from 46 percent in 1966 to 54 percent in 1990.

Cereal production expanded rapidly from 1974 to 1989 as Green Revolution technologies were adopted by farmers throughout developing regions, particularly in Asia. Since 1989, a slowdown in the growth rate of aggregate cereal production is discernible. This slowdown is primarily the result of grain surpluses that have driven down commodity prices and diminished the incentive to invest in cereal production, irrigation development projects, and agricultural infrastructure.

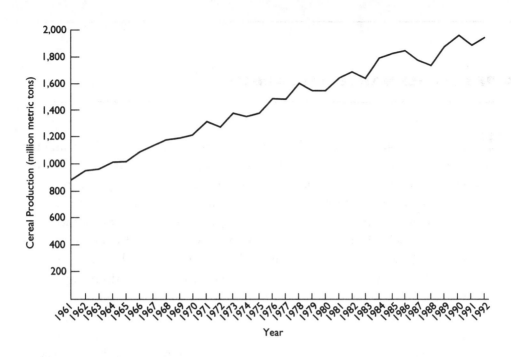

Year	Cereal Production (thousand metric tons)	Year	Cereal Production (thousand metric tons)
1961	884,773	1977	1,471,437
1962	941,782	1978	1,601,305
1963	955,564	1979	1,541,505
1964	1,010,102	1980	1,552,972
1965	1,006,154	1981	1,636,348
1966	1,089,463	1982	1,695,397
1967	1,133,650	1983	1,627,167
1968	1,172,133	1984	1,789,214
1969	1,181,000	1985	1,826,947
1970	1,204,890	1986	1,838,675
1971	1,311,658	1987	1,771,845
1972	1,270,333	1988	1,730,030
1973	1,373,577	1989	1,872,984
1974	1,340,497	1990	1,950,906
1975	1,372,602	1991	1,879,351
1976	1,480,201	1992	1,953,874

Source: World Resources Database, World Resources Institute, DSC Data Services, Washington, D.C., 1994.

AREA OF CEREALS HARVESTED, WORLD

The acreage devoted to cereal production increased from 1961 to 1981 and has since plateaued and even decreased slightly. The dispersion of Green Revolution farming technologies into Asia and other developing regions boosted cereal production through massive irrigation expansion, new high-yielding crop varieties, enhanced fertilizers and pesticides, and newer, more efficient farming methods. India was able to double its wheat yields in only a few years, and China now supports 22 percent of the world's population on just 7 percent of its arable land. Since the 1970s, global rice yields have risen 38 percent, wheat yields have risen 36 percent, and corn and sorghum have risen 30 percent.

The coming years promise even more productive cereal varieties. For example, the Veery wheats, particularly suited to Africa's subtropical climate, could boost the region's yields by up to 15 percent. Similarly, new Chinese hybrid rice and acid- and salt-tolerant plant varieties that can thrive in hot, arid, and previously uncultivatable areas are under development.

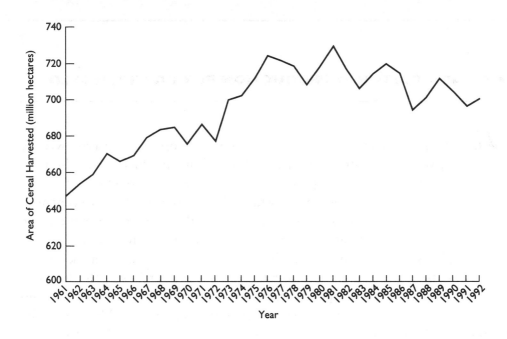

Year	Area Harvested (thousand hectares)	Year	Area Harvested (thousand hectares)
1961	647,192	1977	720,688
1962	652,837	1978	717,179
1963	658,100	1979	708,006
1964	669,703	1980	717,828
1965	665,994	1981	728,172
1966	668,602	1982	715,165
1967	679,117	1983	704,669
1968	683,251	1984	714,348
1969	684,474	1985	718,605
1970	674,771	1986	713,601
1971	685,726	1987	693,778
1972	676,836	1988	700,463
1973	699,325	1989	710,650
1974	701,720	1990	702,779
1975	712,565	1991	695,657
1976	723,480	1992	699,736

Source: Ibid.

INDEX OF AGRICULTURAL PRODUCTION PER CAPITA, WORLD

Agricultural production has consistently outpaced population growth over the past thirty years. The pattern of per capita agricultural production has also been fairly consistent: a period of impressive growth, followed by a brief period of stagnation or decline, followed by more growth. Although there has been a recent measurable slowdown of per capita agricultural output, this is the result of economic factors and not due to limitations on potential agricultural production. Tremendous surpluses in world food supplies have prompted North America and Europe to restrain their production. Moreover, the chaos that has followed the demise of communism in Eastern Europe and the former Soviet Union has dramatically reduced the region's contribution to global agricultural production. Meanwhile, the Third World has managed to increase agricultural production (recently) at an annual rate of 5 percent. It is noteworthy that each of the low points or dips in the cycle (1965, 1972, 1980, 1983, 1987, and 1992) is higher than the previous dip.

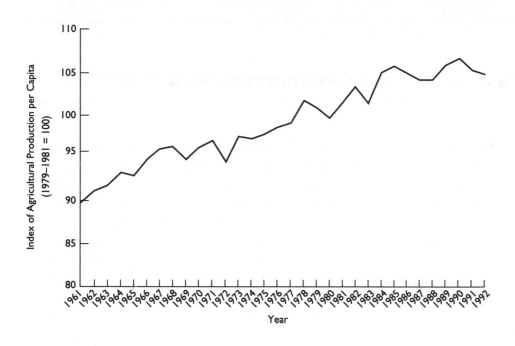

Year	Agricultural Production per Capita (1979–1981 = 100)	Year	Agricultural Production per Capita (1979–1981 = 100)
1961	89.56	1977	98.41
1962	90.94	1978	101.20
1963	91.57	1979	100.09
1964	93.23	1980	98.93
1965	92.89	1981	100.97
1966	94.38	1982	102.51
1967	95.73	1983	100.57
1968	96.15	1984	104.26
1969	94.43	1985	104.93
1970	95.81	1986	104.05
1971	96.60	1987	103.19
1972	93.97	1988	103.37
1973	97.08	1989	105.06
1974	96.77	1990	105.72
1975	97.20	1991	104.06
1976	98.04	1992	103.56

Source: Ibid.

INDEX OF AGRICULTURAL PRODUCTION, WORLD TOTAL

Almost without exception, agricultural productivity worldwide has risen year after year since 1961. Some of this increase can be attributed to cropland expansion, but most is the result of technological advances in farming, which have boosted yields per hectare of land exponentially over the past twenty-five years. The world's agricultural research system has proven international collaboration to be one of the most critical boons to global agricultural productivity. The International Rice Research Institute, the Centro Internacional de Mejoramiento de Maiz y Trigo, and the Consultative Group on International Agricultural Research work collectively to bring new scientific ideas to bear on the world's agricultural sectors.

The outlook for the future of yield performance is bright. The fact that the overall yield performance of most crops shows little sign of slowing down indicates that higher yields are still attainable. Developing countries showed considerable progress with certain crops in the 1980s and into the 1990s. For example, worldwide yields of maize grew at about 30 kilograms per hectare for each year in the 1980s, and fifteen developing countries enjoyed yield gains that were even higher than the world average. Chile experienced an astronomical gain of 300 kilograms per hectare per year in the 1980s, and there is every indication that plenty of room for growth remains.

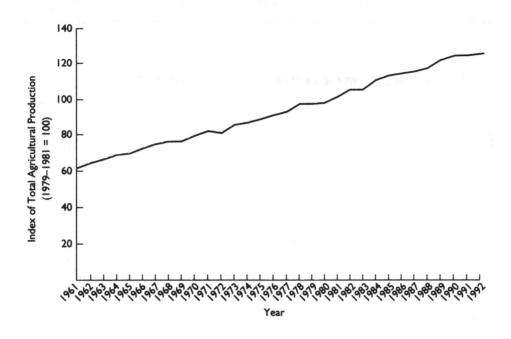

Year	Agricultural Production Index (1979–1981 = 100)	Year	Agricultural Production Index (1979–1981 = 100)
1961	61.95	1977	93.47
1962	64.14	1978	97.77
1963	65.91	1979	98.36
1964	68.50	1980	98.92
1965	69.68	1981	102.72
1966	72.28	1982	106.12
1967	74.84	1983	105.97
1968	76.74	1984	111.80
1969	76.93	1985	114.50
1970	79.67	1986	115.54
1971	81.99	1987	116.60
1972	81.38	1988	118.85
1973	85.75	1989	122.89
1974	87.13	1990	125.80
1975	89.14	1991	125.99
1976	91.53	1992	127.54

Source: Ibid.

12

INDEX OF FOOD PRODUCTION PER CAPITA, WORLD

The world's continuing growth in population has sparked some concern that food production might not increase fast enough to feed everyone. But if trends continue as they have over the past thirty years, people will be better nourished than ever in the future regardless of population gains. There is a fairly consistent upward trend in world per capita food production from 1961 to 1992, despite the fact that the world's population has more than doubled since 1950. Over this period, the amount of food produced per person increased 18 percent, mostly due to advances in farming technology.

The same kind of research that has brought about life-extending discoveries in vaccines, sanitation, and nutrition has also fostered advances in genetic engineering, irrigation, and pesticides that have kept food productivity well ahead of global increases in population. It is important to note that population growth is closely related to food abundance. More abundant food has helped reduce starvation and malnutrition and thus cut global death rates. This results in higher population growth rates. Economic growth and technological change have kept the food supply several steps ahead of the growing population.

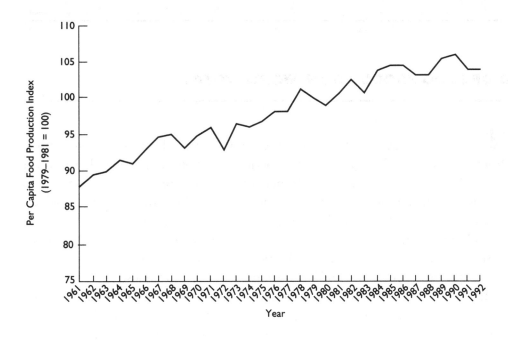

Year	Product Index per Capita (1979–1981 = 100)	Year	Product Index per Capita (1979–1981 = 100)
1961	88.30	1977	98.21
1962	89.56	1978	101.31
1963	90.11	1979	100.15
1964	91.88	1980	99.10
1965	91.23	1981	100.75
1966	93.31	1982	102.55
1967	94.83	1983	100.75
1968	95.35	1984	104.07
1969	93.55	1985	104.68
1970	95.19	1986	104.54
1971	95.99	1987	103.22
1972	93.08	1988	103.27
1973	96.55	1989	105.39
1974	96.20	1990	105.93
1975	97.04	1991	103.88
1976	98.26	1992	103.83

Source: Ibid.

13

INDEX OF FOOD PRODUCTION, WORLD TOTAL

There was a steady, nearly uninterrupted growth in total world food production from 1961 to 1992. The growth in production continued despite ever lower world food prices over the same period. The majority of this increase in production is due to the implementation of better agriculture technology resulting from new research conducted since the 1950s. This research, much of it conducted by the Consultative Group on International Agriculture Research, has fostered major advances in pesticides, genetic engineering, fertilization, prevention of soil erosion, crop rotation, irrigation techniques, and livestock production techniques.

In general, the rate of improvement seems to be increasing. After yields exceed 2,000 kilograms per hectare per year, it requires less time to achieve each next 1,000 kilograms per hectare per year in productivity. The reason is that the shift from subsistence agriculture to technological agriculture is an initially expensive procedure. After the shift is made, it is much easier to incorporate new scientific findings into farming practice. Most countries in the developing world have recently gone through this shift toward technology or will soon. The potential for increased implementation of the latest agricultural knowledge suggests that the growth in world food production will not slow in the near future.

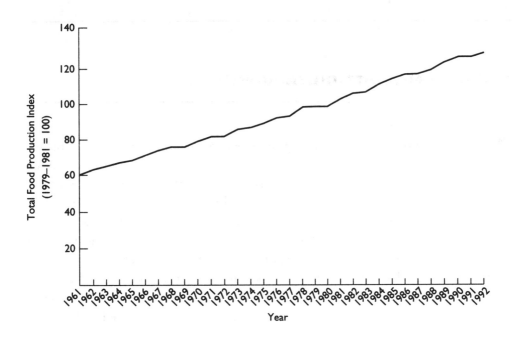

Year	Production Index (1979–1981 = 100)	Year	Production Index (1979–1981 = 100)
1961	61.08	1977	93.29
1962	63.17	1978	97.87
1963	64.85	1979	98.42
1964	67.51	1980	99.08
1965	68.43	1981	102.50
1966	71.46	1982	106.17
1967	74.14	1983	106.16
1968	76.10	1984	111.60
1969	76.22	1985	114.23
1970	79.16	1986	116.09
1971	81.48	1987	116.64
1972	80.62	1988	118.73
1973	85.29	1989	123.27
1974	86.62	1990	126.06
1975	89.00	1991	125.77
1976	91.74	1992	127.88

Source: Ibid.

14

TOTAL FOOD COMMODITY INDEX, WORLD

Global food prices have been dropping steeply for nearly two decades. This index of food commodity prices shows that throughout the 1960s, prices remained somewhat steady. Two peaks in price came in 1974 and 1977, followed by dramatic gains in food production that greatly outpaced population increases. The price surges in 1973/1974 and 1977 were related directly and indirectly to the oil crisis of those years, which increased the cost of some aspects of food production and gave the Soviet Union (a major oil exporter) the wealth to purchase grain on the world market for livestock production, radically increasing demand. These productivity improvements have led to a relatively steady increase in food abundance since 1977, when food prices were over three times what they were in 1992. This abundance has brought about a near end to mass famine. Those that have occurred in the past few decades have been caused by political strife, not a lack of resources.

The steady reduction in world food prices has been caused by several factors. One of the most significant has been the myriad of recent advances in agricultural technology. Another important factor in the drop in food prices was increased liberalization of global trade, which reduced the tariffs and price subsidies that previously had inflated food prices. Competition and increased crop specialization brought about a more efficient market in agricultural commodities.

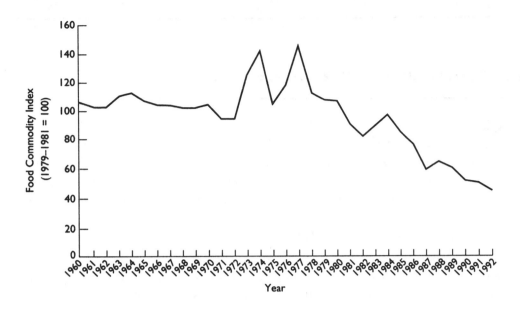

Year	Commodity Index (1979–1981 = 100)	Year	Commodity Index (1979–1981 = 100)
1960	106	1977	143
1961	102	1978	111
1962	101	1979	106
1963	110	1980	104
1964	112	1981	90
1965	107	1982	81
1966	104	1983	88
1967	103	1984	95
1968	102	1985	83
1969	101	1986	75
1970	103	1987	58
1971	93	1988	63
1972	93	1989	60
1973	124	1990	52
1974	140	1991	50
1975	103	1992	44
1976	117		

Source: World Resources Institute (WRI) and International Institute for Environmental Development, in collaboration with UN Environment Programme (UNEP), *World Resources 1988–89* (New York: Basic Books, 1988), table 14.3; WRI in collaboration with UNEP and UN Development Programme, *World Resources 1992–93* (New York: Oxford University Press, 1992); WRI, in collaboration with UNEP and UNDP, *World Resources 1994–95* (New York: Oxford University Press, 1994); for 1990–1992, data indexed by the Competitive Enterprise Institute.

15

FISH CATCH, WORLD

The growing demand for fish products has fueled the increase of the world's commercial fishing industry. Worldwide, the fishery harvest is well over 100 million metric tons per year, an increase of over 50 percent above the harvest of just two decades ago. Advanced technologies have enabled ocean harvesting on an unprecedented scale.

It appears that the current rate of harvest is depleting many marine populations. In 1994, the U.S. government shut down portions of George's Bank, historically one of the world's most fertile fishing grounds, due to fishery depletion. Many species of marine mammals, in particular, are facing serious population declines as well. Worldwide, the fish catch has declined modestly since 1989.

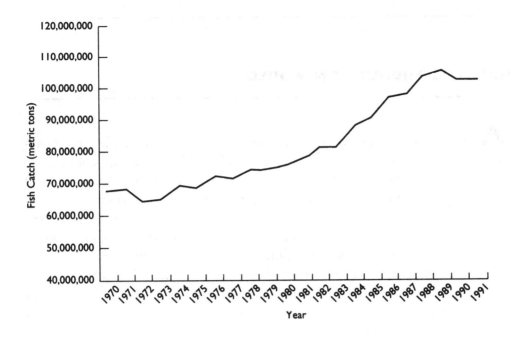

Year	Fish Catch (metric tons)
1970	66,969,420
1971	67,773,070
1972	63,855,330
1973	64,526,580
1974	68,185,310
1975	68,021,930
1976	71,509,720
1977	71,027,250
1978	73,422,560
1979	74,233,930
1980	75,587,120
1981	77,861,050
1982	80,050,870
1983	80,933,960
1984	87,685,160
1985	90,200,380
1986	96,666,970
1987	97,925,820
1988	103,149,600
1989	104,584,500
1990	101,755,000
1991	101,831,100

Source: World Resources Database, World Resources Institute, DSC Data Services, Washington, D.C., 1994.

16

AQUACULTURE PRODUCTION, WORLD

As demand for fish products has continued to climb, some people have developed methods to raise fish on the equivalent of farms. Known as aquaculture, this practice holds the potential to reduce pressures on marine fisheries; more fish would be raised for human consumption rather than caught on the open seas.

Aquaculture production increased by approximately 65 percent between 1984 and 1991, with the greatest increase occurring in Asia. Successful aquacultural techniques have been developed to farm salmon, tilapia, catfish, trout, abalone, oysters, crawfish, and shrimp, among others. Already in the United States, most trout and catfish served in restaurants are farm raised, as are significant portions of crawfish and oysters. Worldwide, shrimp farms produce approximately one-fifth of the shrimp sold on the market.

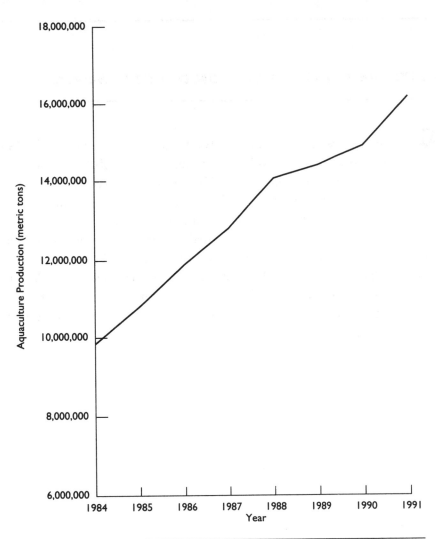

Region	Aquaculture (metric tons)						
	1984	1985	1986	1987	1988	1989	1990
Africa	24,500	40,200	40,800	44,600	58,800	79,300	57,400
South America	57,500	58,800	57,500	109,000	135,000	155,000	186,000
North America	334,000	332,000	383,000	396,000	381,000	401,000	350,000
Central America	66,800	69,400	68,800	81,700	89,600	92,800	96,400
Europe	916,000	1,010,000	1,040,000	1,080,000	1,140,000	1,190,000	1,190,000
Asia	8,420,000	9,340,000	10,300,000	11,100,000	12,300,000	12,400,000	13,000,000
Oceania	20,800	22,700	28,800	30,700	42,500	41,900	42,800
World	9,840,000	10,900,000	11,900,000	12,800,000	14,100,000	14,400,000	14,900,000

Source: Ibid.

17

PER CAPITA EMISSIONS OF CARBON DIOXIDE, WORLD

Global emissions of carbon dioxide (CO_2) have increased since the 1950s, primarily driven by the rise in fossil fuel consumption throughout the world. The dramatic increase in the 1960s and 1970s has since slowed, however, with CO_2 emissions actually declining from 1979 to 1983. In 1991, CO_2 emissions increased only 1.5 percent over the previous year, and much of that was due to the Kuwaiti oil fires. Indeed, had the fires not resulted in the emission of 130 million metric tons of CO_2, global CO_2 emissions would have declined in 1991.

The global emission of CO_2 has slowed, in large part due to the leveling off of per capita CO_2 emissions, which peaked at 1.23 metric tons of carbon per capita in 1979. Since then per capita CO_2 emissions have fluctuated mildly, with no significant increase or decrease.

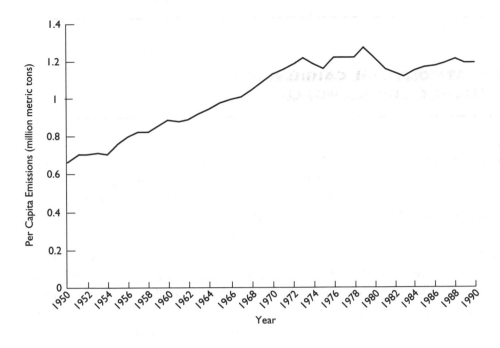

Year	Per Capita Emissions (million metric tons)	Year	Per Capita Emissions (million metric tons)
1950	0.65	1971	1.12
1951	0.69	1972	1.14
1952	0.69	1973	1.18
1953	0.7	1974	1.16
1954	0.69	1975	1.13
1955	0.74	1976	1.18
1956	0.78	1977	1.19
1957	0.8	1978	1.18
1958	0.8	1979	1.23
1959	0.83	1980	1.19
1960	0.86	1981	1.13
1961	0.85	1982	1.1
1962	0.86	1983	1.08
1963	0.89	1984	1.1
1964	0.92	1985	1.12
1965	0.95	1986	1.13
1966	0.97	1987	1.14
1967	0.98	1988	1.16
1968	1.01	1989	1.17
1969	1.05	1990	1.15
1970	1.1	1991	1.15

Source: G. Marland, R. J. Andres, and T. A. Boden, "Global, Regional, and National CO_2 Emissions," in T. A. Boden, D. P. Kaiser, R. J. Sepanski, and F. W. Stoss, eds., *Trends '93: A Compendium of Data on Global Change* (Oak Ridge, Tenn.: Carbon Dioxide Information Analysis Center, Oak Ridge National Laboratory, September 1995), pp. 505–508.

18

ESTIMATED ATMOSPHERIC RELEASES OF
CHLOROFLUOROCARBONS, WORLD

Chlorofluorocarbons (CFCs) were first used in the 1930s as refrigerants. After World War II, CFC-11 and CFC-12, the two most prominent CFCs, were also used as blowing agents for closed-cell foams (used for insulation) and as propellants for aerosol sprays. CFCs were popular because they are nonflammable and nontoxic, unlike the substances they replaced.

In the 1970s, concerns were raised about the potential impact of CFCs on the stratospheric ozone layer. Releases of CFCs into the atmosphere, it is believed, initiate a chain reaction that thins the ozone layer and potentially exposes the earth's surface to an increase in damaging ultraviolet solar radiation. These concerns led to the ban of CFCs for use in aerosols in the United States and an overall decline in the release of CFC-11 and CFC-12 into the atmosphere. However, this downward trend was quickly overtaken by a rapid increase in use of CFCs in the developing world. This trend was reversed with the ratification of the Montreal Protocol in 1987 (subsequently amended in 1990 and 1992), which calls for phasing out all CFC production. Due to the widespread use of CFCs, particularly in refrigeration units and air conditioners, the phaseout will come at considerable cost: an estimated $100 billion in the United States alone.

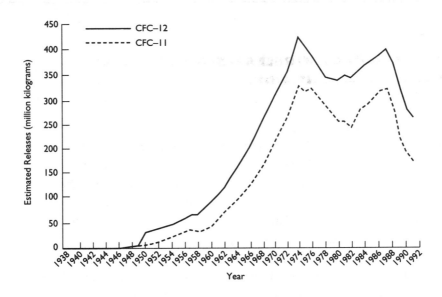

Year	Atmospheric Release (million kilograms)		Year	Atmospheric Release (million kilograms)	
	CFC-11	*CFC-12*		*CFC-11*	*CFC-12*
1938	0.1	0.1	1966	121.3	195.0
1939	0.1	0.1	1967	137.6	219.9
1940	0.1	0.1	1968	156.8	246.5
1941	0.1	0.2	1969	181.9	274.3
1942	0.1	0.4	1970	206.6	299.9
1943	0.2	0.5	1971	226.9	321.8
1944	0.2	0.8	1972	255.8	349.9
1945	0.3	1.2	1973	292.4	387.3
1946	0.6	1.7	1974	321.4	418.6
1947	1.3	2.3	1975	310.9	404.1
1948	2.3	3.0	1976	316.7	390.4
1949	3.8	3.7	1977	303.9	371.2
1950	5.5	29.5	1978	283.6	341.3
1951	7.6	32.4	1979	263.7	337.5
1952	11.0	33.7	1980	250.8	332.5
1953	15.0	37.9	1981	248.2	340.7
1954	18.6	42.9	1982	239.5	337.4
1955	23.0	48.2	1983	252.8	343.3
1956	28.7	56.1	1984	271.1	359.4
1957	32.2	63.8	1985	280.8	368.4
1958	30.2	66.9	1986	295.1	376.5
1959	30.9	74.8	1987	310.6	386.5
1960	40.5	89.1	1988	314.5	392.8
1961	52.1	99.7	1989	265.2	364.7
1962	65.4	114.5	1990	216.1	310.5
1963	80.0	133.9	1991	188.3	271.6
1964	95.0	155.5	1992	171.1	255.3
1965	108.3	175.4			

Source: Alternative Fluorocarbons Environmental Acceptability Study, 1994, cited in Boden et al., *Trends 93.*

19

METALS AND MINERALS, WORLD COMMODITY INDEX

There was a sustained, though erratic, decline in the world prices of metals and minerals from 1960 to 1992. Although metals and minerals exist in fixed quantities, prices have tended to decrease over time rather than increase. The chief reason for lower prices is that the supply of metals and minerals is increasing. New supplies of these resources have been discovered with improvements in technology and scientific knowledge that allow miners to locate sources for metals and minerals more precisely. Also, new technologies have made mining of less concentrated minerals, or minerals located in previously inaccessible areas, economically feasible. Improvements in mining technology were spurred by the volatility of the market, as depicted in the graph. Despite the long-term downward trend in prices, occasional increases in price caused by sudden scarcity made advanced technologies profitable. This encourages further research, and once demand is met, market forces fostered increases in efficiency, which in turn lower the price of new technology. This spiral of ever more productive technology increases supply faster than demand can grow.

On the demand side, the occasional price spikes encourage conservation efforts. More abundant substitutes for the commodity were found, technologies that use less of scarce resources were developed, and more efficient methods of waste recovery were discovered. Markets ceaselessly encourage the development of more efficient resource uses, making the limited supply of any physical commodity ultimately irrelevant.

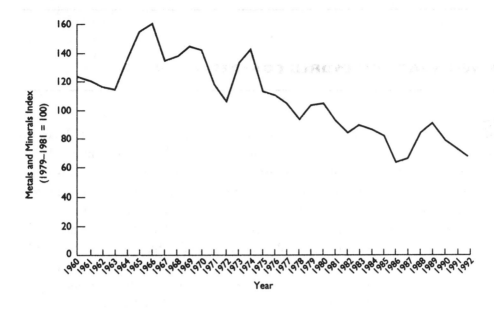

Year	Commodity Indexes (1979–1981 = 100)	Year	Commodity Indexes (1979–1981 = 100)
1960	124	1977	104
1961	120	1978	92
1962	115	1979	103
1963	114	1980	105
1964	137	1981	92
1965	155	1982	84
1966	160	1983	90
1967	134	1984	86
1968	138	1985	81
1969	144	1986	63
1970	142	1987	66
1971	117	1988	84
1972	106	1989	90
1973	133	1990	78
1974	144	1991	69
1975	113	1992	65
1976	110		

Source: WRI, International Institute for Environmental Development, and UNEP, *World Resources 1988–89,* table 14.3; WRI, UNEP, and UNDP, *World Resources 1992–93;* WRI, UNEP, and UNDP, *World Resources 1994–95;* Competitive Enterprise Institute.

20

CRUDE WELLHEAD PRICES, UNITED STATES

The price of crude oil declined steadily from 1949 to the early 1970s. Prices increased sharply in 1974 due to the Arab oil embargo and jumped again in 1979 in response to the Iranian revolution. After regional political crises and domestic energy regulations relaxed, the price of oil dropped steeply. In 1993, crude oil prices adjusted for inflation were the lowest they had been in nearly twenty years.

When the crude oil wellhead price exceeded $40 per barrel in 1981, new techniques for locating and drilling for oil were developed. This increased the supply, thus loosening the OPEC nations' ability to affect oil prices. New geological research boosted the exploration of oil in many countries that had previously produced little or none. Brazil, for example, eventually produced as much as a half a million barrels per day. Similar new production capabilities in countries around the globe greatly increased non-OPEC sources of oil.

Higher oil prices also encouraged switching to alternative sources of energy, chiefly coal and natural gas. New research in these competing energy sources yielded similar gains in production capacity, greatly increasing overall energy supplies. This diversification of energy sources and technologies has both cut oil prices to precrisis levels and made the oil market more resilient to other potential global crises.

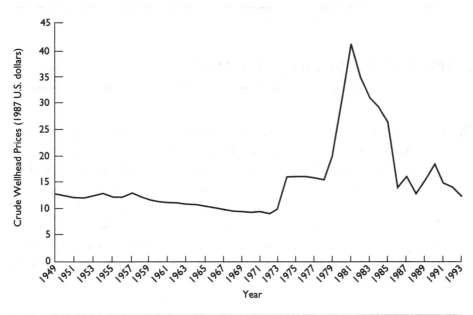

Year	Wellhead Prices (1987 U.S. $)	Year	Wellhead Prices (1987 U.S. $)
1949	12.76	1972	8.74
1950	12.43	1973	9.42
1951	11.88	1974	15.30
1952	11.77	1975	15.59
1953	12.18	1976	15.66
1954	12.52	1977	15.33
1955	12.10	1978	14.93
1956	11.82	1979	19.30
1957	12.66	1980	30.11
1958	12.09	1981	40.27
1959	11.33	1982	34.03
1960	11.08	1983	30.03
1961	10.99	1984	28.44
1962	10.78	1985	25.52
1963	10.63	1986	12.91
1964	10.40	1987	15.40
1965	10.07	1988	12.11
1966	9.80	1989	14.62
1967	9.64	1990	17.68
1968	9.25	1991	14.05
1969	9.25	1992	13.20
1970	9.03	1993	11.47
1971	9.14		

Source: Energy Information Administration, *Annual Energy Review 1993* (Washington, D.C.: U.S. Department of Energy, 1994), table 5.17.

21

TOTAL CONSUMPTION OF ENERGY, WORLD

As the world's economy has grown, so has its demand for energy. Demand has increased consistently, with only occasional lulls, such as the recession of the early 1980s that temporarily suppressed energy demand. Total final consumption of energy worldwide increased by nearly 50 percent from 1970 to 1990. This figure reflects the use of energy in all economic sectors—industrial, agricultural, residential, and commercial—as well as the nonenergy uses of fossil fuels. Over the same time period, the amount of electricity generated more than doubled, with the greatest increase occurring from the upswing in use of nuclear power to generate electricity.

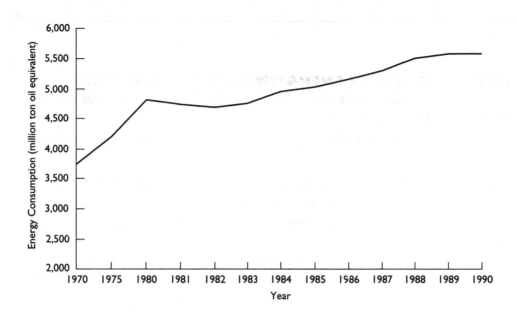

Year	Consumption of Energy (million ton oil equivalent)
1970	3,752.4
1975	4,186.3
1980	4,792.5
1981	4,746.1
1982	4,681.8
1983	4,731.0
1984	4,927.7
1985	5,027.2
1986	5,150.2
1987	5,304.3
1988	5,492.2
1989	5,572.2
1990	5,570.7

Source: Organization for Economic Cooperation and Development, *OECD Environmental Data Compendium 1993* (Paris: OECD, 1993), table 10.5A.

22

CONSUMPTION OF ENERGY PER UNIT OF GROSS DOMESTIC PRODUCT, ORGANIZATION FOR ECONOMIC COOPERATION AND DEVELOPMENT NATIONS

Energy use has increased substantially over the past two decades. However, as nations become more economically advanced, pursuing economic development through a market economy, energy efficiency increases significantly. Consider the trend in energy consumption per unit of gross domestic product (GDP) in developed countries. In Organization for Economic Cooperation and Development nations (Western Europe, the United States, Canada, Japan, Australia, and New Zealand), energy consumption per unit of GDP declined by nearly one-third from 1970 to 1989.

The same trend toward greater energy efficiency did not occur in the former communist nations. Centralized economies lack the market pressures that constantly encourage increased efficiency and innovation. As a result, the technological breakthroughs that allow industries to produce more using a constant resource or energy base fail to materialize, and potential efficiency gains are sacrificed.

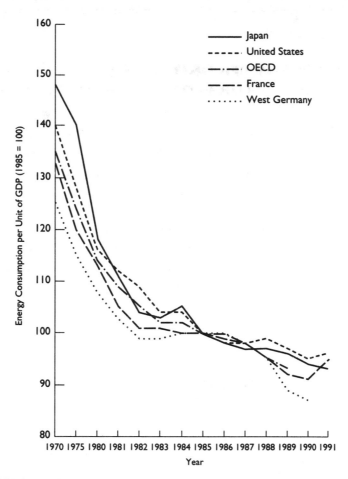

Year	Consumption of Energy per Unit of GDP (1985 = 100)				
	United States	*Japan*	*France*	*West Germany*	*OECD*
1970	140	148	133	125	135
1975	128	140	120	115	124
1980	116	118	113	108	114
1981	112	111	105	103	109
1982	109	104	101	99	105
1983	104	103	101	99	102
1984	104	105	100	100	102
1985	100	100	100	100	100
1986	98	98	99	100	98
1987	98	97	98	98	98
1988	99	97	95	95	95
1989	97	96	92	89	93
1990	95	94	91	87	
1991	96	93	95		

Source: Ibid., table 10.5D.

23

WATER QUALITY VIOLATIONS, UNITED STATES

The percentage of U.S. rivers and streams violating Environmental Protection Agency (EPA) standards for fecal coliform bacteria, dissolved oxygen, and phosphorus is declining. High concentrations of fecal coliform bacteria can cause a variety of infectious diseases, including cholera and typhoid. Common sources of this bacteria are insufficiently treated sewage and runoff from pastures, feedlots, and cities. The graph shows a fairly consistent decline in the rate of fecal coliform bacteria violations, especially during the period beginning in 1983, which showed a 41 percent decline over six years. The decline was largely due to increased use of effective wastewater treatment technology.

A violation of the dissolved oxygen standard means the tested water lacks oxygen concentrations high enough to support aquatic life fully. Low levels of dissolved oxygen can reduce the solubility of trace elements and affect the taste and odor of the water. The violation rate for this standard has dropped slightly even though larger population densities have increased oxygen demanding loads. Large technology investments in point source controls have helped keep this figure low.

Phosphorus in streams can add to oxygen depletion and increase the growth of aquatic vegetation, which can then clog water intake pipes. This figure dropped rapidly following limits put on the phosphate content in detergents in the late 1960s and early 1970s. Improvements in the 1980s can be attributed to a reduction in phosphorus fertilizer use and point source controls at sewage treatment, food processing, and other industrial plants.

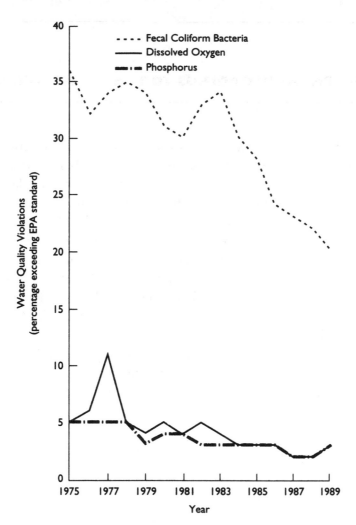

Year	Violation Rates (%)		
	Fecal Coliform Bacteria	*Dissolved Oxygen*	*Phosphorus*
1975	36	5	5
1976	32	6	5
1977	34	11	5
1978	35	5	5
1979	34	4	3
1980	31	5	4
1981	30	4	4
1982	33	5	3
1983	34	4	3
1984	30	3	3
1985	28	3	3
1986	24	3	3
1987	23	2	2
1988	22	2	2
1989	20	3	3

Source: Council on Environmental Quality, *Environmental Quality 1992* (Washington, D.C.: Council on Environmental Quality, 1992), table 32.

24

ESTIMATED PHOSPHORUS LOADINGS TO THE GREAT LAKES

The Great Lakes region is often viewed as a microcosm of the interaction between economic and environmental concerns. The Great Lakes Basin is the site of significant agricultural production, particularly in Canada, as well as a large portion of U.S. manufacturing. At the same time, there has been considerable concern over the ecological health of the Great Lakes in the past two decades.

Important environmental progress in this region has been made since the 1970s. Factory emissions—so-called point sources—now account for only one-tenth of the water pollution in the Great Lakes. Pollution of the Great Lakes, as measured by estimated phosphorus loadings, declined significantly in all five of the Great Lakes from 1976 to 1989. This reduction was greatest in Lake Erie, where phosphorus loadings declined by over 50 percent. Phosphorus is discharged from municipal sewage treatment plants and factories and can be found in agricultural runoff. Phosphorus is of concern because it stimulates excess blue-green algae growth, which can result in eutrophication, harming species and making water unfit for drinking or recreation.

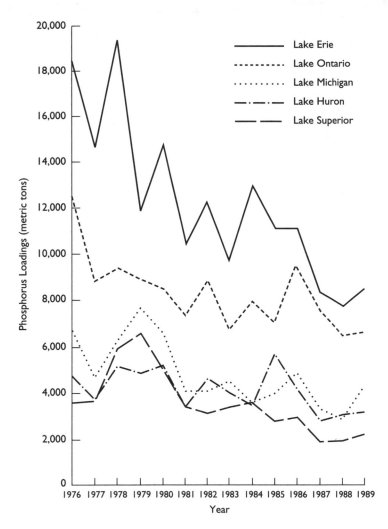

Year	Phosphorus Loading (metric tons)				
	Lake Superior	Lake Michigan	Lake Huron	Lake Erie	Lake Ontario
1976	3,550	6,656	4,802	18,480	12,695
1977	3,661	4,666	3,763	14,576	8,935
1978	5,990	6,245	5,255	19,431	9,547
1979	6,619	7,659	4,881	11,941	8,988
1980	6,412	6,574	5,307	14,855	8,579
1981	3,412	4,091	3,481	10,452	7,437
1982	3,160	4,084	4,689	12,349	8,891
1983	3,407	4,515	3,978	9,880	6,779
1984	3,642	3,611	3,452	12,874	7,948
1985	2,864	3,956	5,758	11,216	7,083
1986	3,059	4,981	4,210	11,118	9,561
1987	1,949	3,298	2,909	8,381	7,640
1988	2,067	2,907	3,165	7,841	6,521
1989	2,323	4,360	3,227	8,568	6,723

Source: Council on Environmental Quality, *Environmental Quality 1991* (Washington, D.C.: Council on Environmental Quality, 1991), table 33.

25

EMISSION AND ATMOSPHERIC CONCENTRATION TRENDS FOR SELECTED AIR POLLUTANTS, UNITED STATES

The United States has witnessed a tremendous improvement in air quality over the past two decades. Emissions of most regulated air pollutants are down, as are atmospheric concentrations. The vast majority of cities monitored by the Environmental Protection Agency meet the National Ambient Air Quality Standards (NAAQS) for carbon monoxide, nitrogen oxides, ozone, particulate matter, and sulfur oxides, which were established by the federal government to ensure that all cities in the United States surpass a set level of air quality. (All of the following graphs show this standard.) Indeed, at 90 percent of the measuring sites, mean concentrations meet or exceed the NAAQS for all the previously mentioned pollutants, except for ozone, which is produced by an interaction of volatile organic compounds, such as hydrocarbons, and nitrogen oxides. (The following graphs also specify the number of monitors measuring atmospheric concentrations.)

During the past two decades, increases in industrial efficiency, particularly in the combustion of fuels, have reduced the emissions per unit of production of many industries. In addition, the federal government established the NAAQS and imposed emission standards on new vehicles. Newer auto emission standards require the use of cleaner-burning fuels. A new car produced in 1994 emits over 96 percent less pollution than cars produced twenty-five years earlier.

Key to sources for tables and graphs on pages 444–453:

EQ: Council on Environmental Quality, *Environmental Quality, 1971, 1979, 1981, 1984, 1991* (Washington, D.C.: CEQ, 1971, 1979, 1981, 1984, 1991).

EPA: U.S. Environmental Protection Agency (EPA), *National Air Quality and Emission Estimates*, EPA 454/R-93-032, Office of Air Quality Planning and Standards (Research Triangle Park, N.C.: EPA, 1993).

WF 1994: W. Freas, Office of Air Quality Standards and Planning, EPA, personal communication, 1994.

SA: U.S. Bureau of the Census, *Statistical Abstract of the United States, 1981, 1986, 1988* (Washington, D.C.: Department of Commerce, 1981, 1986, 1988).

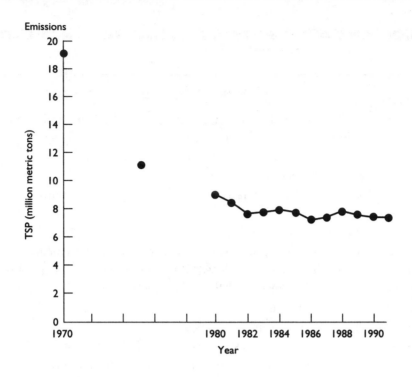

Year	TSP (million metric tons)
1970	18.99
1975	10.96
1980	9.06
1981	8.58
1982	7.67
1983	7.77
1984	8.08
1985	7.85
1986	7.31
1987	7.42
1988	7.94
1989	7.57
1990	7.4
1991	7.41

Source: EQ 1992.

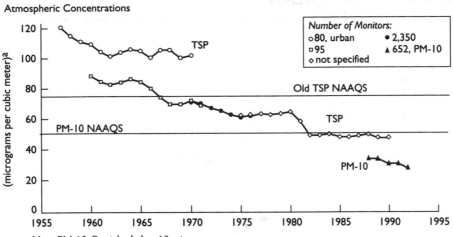

Atmospheric Concentrations

Number of Monitors:
○ 80, urban ● 2,350
□ 95 ▲ 652, PM-10
◇ not specified

TSP

Old TSP NAAQS

PM-10 NAAQS

TSP

PM-10

(micrograms per cubic meter)[a]

Note: PM-10: Particles below 10 microns.
[a] Mean annual average.

Year	TSP Concentrations (micrograms/cubic meter)[a]				
	Not specified	(80 urban monitors)	(95 monitors)	(2,350 monitors)	(652 monitor PM-10)
1957		120.5			
1958		114.4			
1959		110.7			
1960		109.1	88.0		
1961		103.8	84.0		
1962		101.0	82.0		
1963		103.6	84.0		
1964		105.6	86.0		
1965		104.7	84.0		
1966		100.7	80.0		
1967		105.2	74.0		
1968		104.9	69.0		
1969		100.6	69.0		
1970		101.7	72.0	70.4	
1971			68.0	69.6	
1972				67.1	
1973				65.4	
1974				62.5	
1975	61.9			60.8	
1976	62.8			61.8	
1977	62.9				
1978	62.4				
1979	63.1				
1980	64.2				
1981	57.4				
1982	48.7				
1983	48.4				
1984	49.9				
1985	47.7				
1986	47.6				
1987	48.6				
1988	49.7				33.6
1989	48.0				33.6
1990	47.3				30.7
1991					30.4
1992					27.8
1993					

Sources: EQ 1971, 1979, 1981, 1991; WF 1994; EPA 1990.
Note: Levels for 1979–1981 may be too high because of artifacts in the measurement procedure.
[a] Composite annual averages.

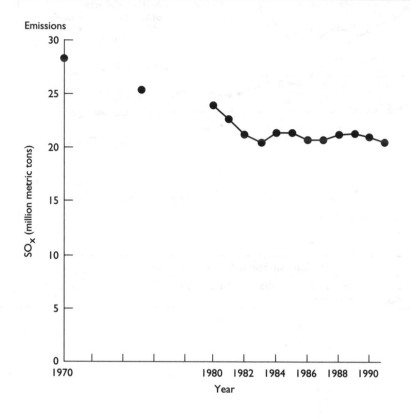

Year	SO$_x$ (million metric tons)
1970	28.42
1975	25.51
1980	23.78
1981	22.51
1982	21.21
1983	20.62
1984	21.47
1985	21.67
1986	21.15
1987	20.97
1988	21.3
1989	21.51
1990	21.05
1991	20.73

Source: EQ 1992.

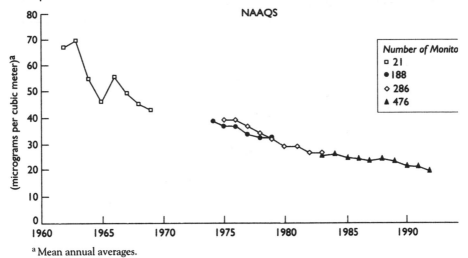

Atmospheric Concentrations of Sulfur Dioxide

^a Mean annual averages.

Year	SO$_2$ Concentrations (micrograms per cubic meter)[a]			
	(21 monitors)	(188 monitors)	(286 monitors)	(476 monit
1962	66.40			
1963	69.40			
1964	55.10			
1965	45.40			
1966	55.80			
1967	49.20			
1968	45.10			
1969	42.50			
1970				
1971				
1972				
1973				
1974		38.40		
1975		36.70	39.00	
1976		37.10	39.00	
1977		33.20	36.40	
1978		31.50	33.80	
1979		32.60	31.20	
1980			28.60	
1981			28.60	
1982			26.00	
1983			26.00	24.96
1984				25.74
1985				24.44
1986				23.66
1987				22.88
1988				23.40
1989				22.62
1990				21.06
1991				20.54
1992				19.24

Sources: EQ 1981; SA 1981, 1986; WF 1994.

^a Annual averages based on 24-hour averages.

CARBON MONOXIDE (CO)

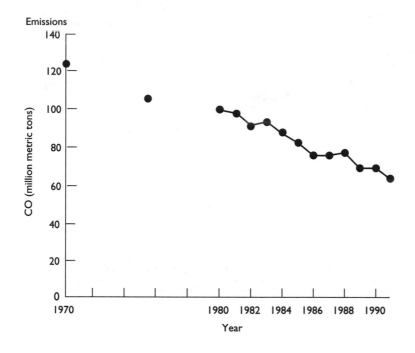

Year	CO (million metric tons)
1970	123.61
1975	104.76
1980	99.97
1981	98.04
1982	90.53
1983	92.31
1984	87.6
1985	83.12
1986	76.03
1987	75.05
1988	75.53
1989	68.32
1990	67.74
1991	62.1

Source: EQ 1992.

Atmospheric Concentrations

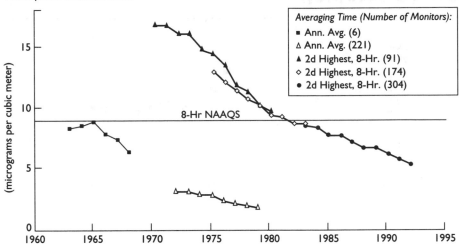

Year	CO Concentrations (milligrams per cubic meter)				
	(6 monitors)[a]	(221 monitors)[a]	(91 monitors)[b]	(174 monitors)[b]	(304 monitors)[b]
1963	8.36				
1964	8.59				
1965	8.93				
1966	7.90				
1967	7.45				
1968	6.53				
1969					
1970			17.0		
1971			16.9		
1972		3.4	16.3		
1973		3.4	16.2		
1974		3.1	15.1		
1975		3.0	14.7	13.20	
1976		2.7	13.8	12.47	
1977		2.5	12.2	11.79	
1978		2.4	11.7	11.11	
1979		2.2	10.6	10.63	
1980			10.1	9.86	
1981				9.76	
1982				8.89	
1983				8.79	8.68
1984					8.62
1985					7.76
1986					7.92
1987					7.43
1988					7.06
1989					7.02
1990					6.50
1991					6.16
1992					5.77

Sources: EQ 1971, 1984; SA 1981; WF 1994.

[a] Composite annual averages.

[b] Composite averages based on the second-highest, nonoverlapping, 8-hour average.

NITROGEN OXIDES (NOₓ)

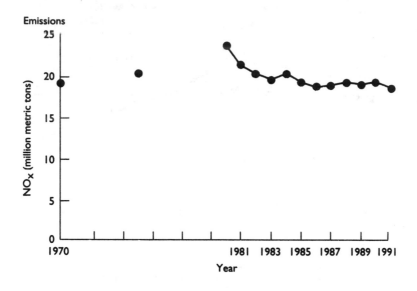

Year	NOₓ (million metric tons)
1970	18.96
1975	20.33
1980	23.56
1981	21.35
1982	20.37
1983	19.8
1984	20.11
1985	19.39
1986	18.83
1987	19.03
1988	19.65
1989	19.29
1990	19.38
1991	18.76

Source: EQ 1992.

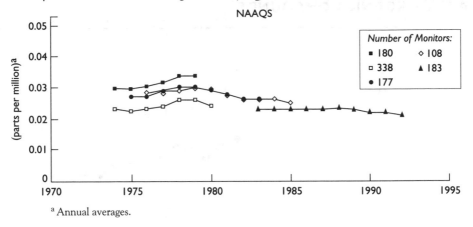

Atmospheric Concentrations of Nitrogen Dioxide (NO$_2$)

^a Annual averages.

Year	NO$_2$ Concentrations (parts per million)[a]				
	(180 monitors)	(338 monitors)	(177 monitors)	(108 monitors)	(183 monitors)
1974	0.030	0.023			
1975	0.029	0.023	0.027		
1976	0.031	0.023	0.027	0.028	
1977	0.032	0.024	0.028	0.029	
1978	0.034	0.026	0.030	0.029	
1979	0.034	0.026	0.030	0.030	
1980		0.024	0.029	0.029	
1981			0.027	0.028	
1982			0.026	0.026	
1983			0.026	0.026	0.023
1984				0.026	0.023
1985				0.025	0.023
1986					0.023
1987					0.023
1988					0.023
1989					0.023
1990					0.022
1991					0.022
1992					0.021

Sources: SA 1981, 1988; EQ 1981, 1984; WF 1994.
^a Composite annual averages.

VOLATILE ORGANIC COMPOUNDS

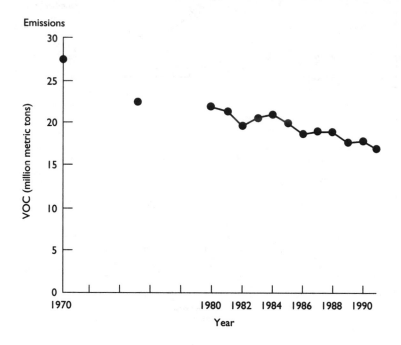

Year	Reactive VOC (million metric tons)
1970	27.4
1975	22.53
1980	21.75
1981	21.22
1982	19.5
1983	20.26
1984	20.99
1985	19.8
1986	18.45
1987	18.64
1988	18.61
1989	17.35
1990	17.58
1991	16.88

Source: EQ 1992.

Atmospheric Concentrations of Ozone

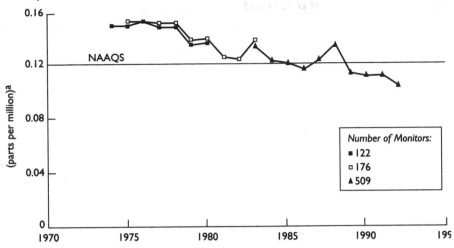

[a] Average of second highest daily maximum value.

Year	Ozone Concentrations (parts per million)[a]		
	(122 monitors)	(176 monitors)	(509 monitors)
1974	0.151		
1975	0.151	0.154	
1976	0.154	0.154	
1977	0.149	0.153	
1978	0.149	0.153	
1979	0.136	0.139	
1980	0.138	0.142	
1981		0.128	
1982		0.126	
1983		0.141	0.1362
1984			0.1254
1985			0.1236
1986			0.1193
1987			0.1264
1988			0.1369
1989			0.1164
1990			0.1143
1991			0.1157
1992			0.1071

Sources: EQ 1981, 1984; WF 1994; SA 1986.

[a] Means of maximum quarterly averages based on the second highest daily maximum 1-hour values.

APPENDIX

Limits of Statistical Certainty: The Case of Population, Food, and Income

Nicholas Eberstadt

Although we are accustomed to thinking of ours as an information age, considerably less reliable information is available on many of the topics of this book than is widely supposed. The limits to our knowledge of human prospects are very real and easily demonstrable; lengthy technical disquisitions on statistical detail are not needed to make the point. Even a brief review of the sources and quality of available data on global trends for population, food, and income will highlight the extent to which contemporary discussions of these phenomena betray a spurious precision, or rely on unsubstantiated and indeed unprovable assertions, or make a case through a dubious use of supposedly "hard figures."

Two very different sorts of problems constrain our ability to view global trends for population, food, and income. The first may be described as problems of poverty; the second may be thought of as problems of development. The former are self-evident: it is obviously difficult to compile accurate, detailed, and continuous information on living conditions in locales where illiteracy is prevalent, government is distant, and statistical organization is weak. The latter are perhaps less intuitively apparent, but they are even more intractable. Due to the tremendous changes in patterns of production and consumption in the modern era—and to the dramatic differences in consumption levels in the world today, both among countries and within them—it turns out to be extraordinarily hard to devise a single, consistent standard by which to measure unambiguously material progress or poverty. (Economists refer to this dilemma as the "index number problem.") Unlike the information bottlenecks attendant upon poverty, those attendant upon development per se are ultimately irresolvable—implying that our ability to measure and thus understand the material circumstances of humanity will be subject to irreducible uncertainties under even the most optimistic assumptions about the future.

But what of today's global data? Let us begin with population figures. By contemporary convention, international statistical authorities (the United Nations, the World Bank, and many others) offer projections for world population detailing totals down to the last hundred thousand—sometimes even to the last thousand. This is a misleading conceit.

Although the capabilities for enumerating the world's population have improved markedly during our century, especially over the past generation, it would be unwise to exaggerate them. By now almost every recognized country has conducted at least one national census. But some have held only one census to date, and governments that routinely conduct censuses commonly do so at ten-year intervals. Birth and death registration systems are even more problematic. In the estimate of the United Nations Statistical Office, fewer than a dozen developing countries with populations of a million or more field statistical systems with near-complete coverage of local births and deaths.[1] These states account for only a tiny fraction of the "Third World's" population—no more than 3 or 4 percent of its total.[2]

With such imposing gaps in coverage remaining to be filled, it should be clear that the figures proposed for population totals, birthrates, death rates, and growth rates are provisional or conjectural for much of the world. The tenuous nature of these estimates is underscored by some of the revisions occasionally required. In 1992, for example, Nigeria's census counted 88.5 million persons—20 million to 30 million fewer persons than international organizations were estimating at the time to live in that country.[3] Even after taking the possible scope of undercount into consideration, it would appear that international statistical authorities had been overestimating Nigeria's population by 15 percent or more around the time of the census.

In this instance, such errors were easy enough to understand: demographers had been doing their best to guess at trends for a populous country with essentially uncharted birthrates and death rates that had not held a full census for nearly thirty years. Some other current practices, however, are rather more difficult to justify. Although most of the countries in Africa, Asia, and Latin America still lack reliable vital registration systems, for example, a number of international organizations regularly publish figures for them on their year-to-year changes in birthrates, death rates, and infant mortality rates! One can only conclude that these organizations (UNICEF and the World Bank among them) are inventing their own numbers for their own reasons—for there is at present no possible way to measure what they claim to be measuring.

Estimates for world food output are even more problematic than world population figures. Among the easiest components to measure in the overall global food situation should be the grain trade between member countries of the Organization for Economic Cooperation and Development (OECD), yet as recently as the 1970s, serious discrepancies were evident between U.N. Food and Agricultural Organization (FAO) and U.S. Department of Agriculture (USDA) estimates for these totals.[4] Although those particular discrepancies have been subsequently resolved, fundamental questions about the accuracy of other estimates persist. In many low-income countries, for example, food produced for home or family use may constitute a considerable fraction of overall national agricultural output, but how this is to be reliably tracked is far from obvious.[5] In India, one Third World country with a highly regarded statistical system, significant differences are evident in the estimates of total grain output issued by the National Sample Survey, the state governments, and the central government.[6] Neither the FAO nor the USDA offers estimates of the margins of error on their world food figures (an oversight that is revealing in itself). Nevertheless, it would appear quite likely that the actual margins of error on food production and food availability figures are greater than the annual changes such numbers presume to measure in quite a number of countries. The problem is generally most pronounced precisely in those countries whose food situation generates the greatest humanitarian concern.

The troubles with estimates of global production and income are more daunting still. Some of these are apparent even in the most cursory of examinations. The World Bank, for example, recently placed the 1992 gross national product per capita of Mozambique at $60 per year.[7] The figure is preposterous on its very face: attempting to subsist solely on the goods and services that might be purchased in the United States in 1992 for $60, a person could not last out the year. The World Bank's demographic estimates for Mozambique, however, simultaneously indicate that this country with subsurvival income levels is undergoing a population explosion![8]

Given the public and official demand for facts and figures about the world economy, it may not be surprising to find that a global cottage industry has arisen for manufacturing nonsense numbers to satisfy anticipated inquiries. But the more scrupulous efforts at estimating the trend and composition of global output indicate the magnitude of the problems faced. Economists at the University of Pennsylvania, in collaboration with the United Nations's ongoing International Comparison Project (ICP), have been attempting to produce meaningful and detailed series on output and income ("national accounts") for all the countries of the world. In their most recent round of estimates, they graded the quality of the data they were using. Only 19 of the 139 countries and territories in their series were deemed to rate an A—all of them OECD members (and not even all of those; after all, even excluding Turkey and the former Yugoslavia, 22 states currently belong to the OECD). On the other hand, 65 of those 139 were given a D—their poorest ranking. This category included 37 of the 47 African countries listed and China, among others.[9]

In view of the apparent weakness of the data with which we must work, a measure of caution and humility would surely seem appropriate. Available statistics on world population, food, and income trends are not suitable for exacting calculations. And they are far from ideal when it comes to depicting the sorts of year-to-year changes for which they are so commonly used. Despite diverse technical advances, our information on these and other basic material trends for humanity is still distinctly limited.

NOTES

1. United Nations, *Demographic Yearbook 1991* (New York: United Nations, 1993), pp. 20–21.
2. Ibid., pp. 106–109.
3. See, for example, Kenneth B. Noble, "After Nigeria's Census, Skeptic Count Is High," *New York Times*, June 4, 1992, p. A13; "The Missing 20M," *Economist*, March 28, 1992, p. 37; Gina Porter, "The Nigerian Census Surprise," *Geography* 77 (October 1992): 371–374.
4. Leonardo Paulino and S. S. Tseng, *A Comparative Study of FAO and USDA Data on Production, Area, and Trade of Major Food Staples* (Washington, D.C.: International Food Policy Research Institute, 1980), pp. 67–72.
5. For more detail, see Nicholas Eberstadt, "Another Look at the World Food Problem," in *The Tyranny of Numbers* (Washington, D.C.: AEI Press, forthcoming).
6. For example, in 1972/73, the first year of the so-called "world food crisis" of 1972–1974, National Sample Survey–based estimates of gross grain production were 8 percent higher than the central government's estimates. The difference amounted to 7.7 million metric tons of grain—enough to feed (or starve) roughly 40 million people. P. C. Bansil, *Agricultural Statistics in India: A Guide* (New Delhi: Oxford and IBH Publishing Co., 1984), p. 163.
7. World Bank, *World Development Report 1994* (New York: Oxford University Press, 1994), p. 162.

8. Ibid., p. 210, where the annual rate of population growth in Mozambique is projected at 2.6 percent for 1980–1992, and 2.6 percent for 1992–2000.

9. Robert Summers and Alan Heston, "The Penn World Table (Mark 5): An Expanded Set of International Comparisons, 1950–1988," *Quarterly Journal of Economics* 106, no. 2 (May 1990): 327–368.

INDEX

PHOTOGRAPH CREDITS

Acknowledgments

Many staffers at the Competitive Enterprise Institute helped make this book possible: Brian Seasholes, Michael De Alessi, David Monack, and Paul Georgia for their research assistance, as well as Jennifer J. Green, Ike Sugg, Helen Hewitt and Jason Taylor for their assistance with other aspects of the Progress and the Planet Project, out of which this book developed. A special thank you is due to former CEI Director of Environmental Studies, Kent Jeffreys, for getting the Institute behind the project, its most ambitious to date, and to Peter Cazamias for handling the myriad administrative tasks that were thrown his way. The insights and unstinting support of CEI's Associate Director of Environmental Studies Jonathan Adler were invaluable in bringing this project to fruition.

The judicious advice and sharp editor's eye of Bruce Nichols at the Free Press helped refine the prose and clarify complicated issues and scientific data. Also, without the heroic efforts and unfailing good humor of Beverly Miller and Loretta Denner we would never have met the arduous and accelerated production schedule.

Thanks to literary agent Henry Dunow the book found the perfect publisher. And although he had every reason not to, Andrew Walworth graciously and patiently permitted me to take time from my job at New River Media to work on the book.

Finally, CEI would like to thank all those who have provided financial support for this project, especially the John M. Olin Foundation. Special thanks are due to Howard and Roberta Ahmanson and to Victor Porlier for their help in launching this book.